To David Huntley
with compliments.
Vasiliy, Nataliya

BAROCLINIC TIDES

Theoretical Modeling and Observational Evidence

Internal waves are water motions originating from density variations within the water body in which they propagate. When an oceanic tidal wave that is primarily active on the water surface impinges on the ocean shelf or passes a region with a seamount, this wave is modified by the topographic disruption. It is split into a less energetic surface wave and other internal modes with different wavelengths and propagation speeds. This cascading process, from the barotropic tides to the baroclinic components, eventually leads to the transformation of tidal energy into turbulence and heat. This is an important process for the dynamics of the lower ocean.

BAROCLINIC TIDES demonstrates how analytical and numerical methods can be used to study the generation and evolution of baroclinic tides and, by comparison with experiments and observational data, shows how internal waves can be distinguished and interpreted. Particular attention is paid to the investigation of strongly nonlinear solitary internal waves, which are generated by internal tidal waves at the final stage of their evolution. This book is intended for researchers and graduate students of physical oceanography, geophysical fluid dynamics, and hydroacoustics.

VASILIY VLASENKO received a Ph.D. from the the Marine Hydrophysical Institute of the National Academy of Sciences of Ukraine (Sevastopol) in 1987 and a Doctor of Sciences degree from the same Institute in 2000. Dr. Vlasenko was at the Institute of Mechanics of the Darmstadt University of Technology, Germany, from 1999 to 2003. He currently lectures in environmental modeling in the School of Earth, Ocean and Environmental Sciences at the University of Plymouth. Dr. Vlasenko is the co-author of three books and more than 70 papers on various aspects of baroclinic wave dynamics in the ocean. His research interests are in physical oceanography: internal gravity waves, coastal ocean dynamics, baroclinic tides, flow–topography interaction, and large-amplitude solitary internal waves.

NATALIYA STASHCHUK received a Ph.D. from the Marine Hydrophysical Institute of the National Academy of Sciences of Ukraine (Sevastopol) in 1990. From 1999 to 2004 she worked at the Institute of Mechanics of the Darmstadt University of Technology, Germany, and she came to the University of Plymouth in 2005.

Dr. Stashchuk has taken part in six cruises to the Mediterranean and the Atlantic Ocean and is the co-author of three books and more than 50 papers on oceanic internal waves. Her research interests are in mathematical modeling of oceanic processes: strait dynamics, internal waves, coastal ocean dynamics, baroclinic tides, and flow–topography interaction.

KOLUMBAN HUTTER received a Ph.D. from Cornell University, Ithaca, New York, in 1973 and has held the position of Professor of Mechanics at Darmstadt University of Technology, Germany, since 1987. His research interests are in geophysical fluid mechanics with applications in the dynamics of glaciers and ice sheets, the mechanics of granular materials, avalanching flows of snow, debris, and mud, and physical limnology and the foundations of continuum mechanics and thermodynamics. Professor Hutter is author or co-author of more than 340 papers and has written or edited 16 books; he also serves on the editorial boards of two journals. He was awarded the Max-Planck Prize of the Max-Planck Society and the Alexander von Humboldt Foundation (Germany) in 1994, the Alexander von Humboldt Prize of the Foundation of Polish Science in 1998, and the Seligman Crystal of the International Glaciological Society in 2003.

BAROCLINIC TIDES

Theoretical Modeling and Observational Evidence

VASILIY VLASENKO
SEOES, University of Plymouth

NATALIYA STASHCHUK
SEOES, University of Plymouth

KOLUMBAN HUTTER
Technische Universität Darmstadt,
Institut für Mechanik

CAMBRIDGE
UNIVERSITY PRESS

CAMBRIDGE UNIVERSITY PRESS
Cambridge, New York, Melbourne, Madrid, Cape Town, Singapore, São Paulo

Cambridge University Press
The Edinburgh Building, Cambridge CB2 2RU, UK

Published in the United States of America by Cambridge University Press, New York

www.cambridge.org
Information on this title: www.cambridge.org/9780521843959

Printed in the United Kingdom at the University Press, Cambridge

A record for this book is available from the British Library

Library of Congress in Publication data

ISBN-13 978-0-521-84395-9
ISBN-10 0-521-84395-2

To our parents

Contents

Tables

Preface

The scientific literature on baroclinic tides in the ocean is as abundant as is the general mathematical theory of internal gravity waves. The majority of research papers on this topic written since the late 1970s deal mainly with the attempts to provide a theoretical description of the vast number of experimental data obtained on the dynamics of baroclinic tides. Various analytical and semianalytical models were developed for the case of infinitesimal waves when the external tidal forcing is negligible. However, observations show that strong nonlinear baroclinic tides are the more common phenomena; they demand not only theoretical but also numerical methods to describe the wave processes adequately.

The idea of the book is to present the theoretical basis of baroclinic tides at all stages of their genesis and evolution (generation, transformation, dissipation), and to give, by comparison with experiments, advice and help regarding the theoretical interpretation and prognosis of baroclinic tides. Such an exhaustive and complete theoretical review does not exist at present, nor is there a scientific book which describes this phenomenon multilaterally. So we are intending to fill this gap.

A second reason for writing a book on this topic is to gather in one place our experience, which extends over more than 20 years, in investigating internal waves in general and baroclinic tides in particular. This includes the development of different linear and nonlinear mathematical models of internal tidal motions and the application of these to concrete oceanic situations. In so doing, we will report not only on our own research, but also on many other publications on the studied topic and subject them to a careful analysis. It is impossible to mention thousands of papers published on baroclinic tides, so our list of references is not exhaustive. We apologize to those colleagues whose references are not quoted.

This book is the outgrowth of our common research activity on the internal dynamics of the ocean and of lakes, in particular as exhibited by waves; this subject has been studied separately by Kolumban Hutter since the mid 1970s in Lakes Zurich, Lugano, and Constance, and by Vasiliy Vlasenko and Nataliya Stashchuk in the World Ocean since 1980. A lucky break brought us together in 1999 to work on research projects funded by the Deutsche Forschungsgemeinschaft on the propagation of waves in stratified rotating basins. Some topics in the present book are the direct result of our joint investigations. The research cooperation was so constructive and our achievements were so fruitful that after three years of collaboration we decided to summarize our work in a book. Vasiliy Vlasenko and Nataliya Stashchuk, who were the active researchers and less occupied by administrative duties, took on the burden of designing the first skeleton of the book's outline; they also were the sole writers of the first drafts of all the chapters. The role of Kolumban Hutter was to act as *advocatus diaboli* and to read, criticize and revise the various versions of the individual chapters. This process went back and forth at least three, more likely five, times. During approximately two years of writing, we learned to appreciate one another, and we now present a book that is a joint effort in which the first two authors carried the heavier burden than the third, but which all three of us agree with. We hope the product is fairly free from misconceptions and professional errors. However, we must ask the reader to bear with our English wording, a mixture of what two Ukrainians and a Swiss with their school English have tried to do. If the English is good, it is due to the copy-editorial staff of Cambridge University Press.

Acknowledgements

Work on this book was started at the Darmstadt University of Technology (Germany) with financial support from the Deutsche Forschungsgemeinschaft, and was continued at a later stage at Plymouth University (UK), although many of the scientific results were obtained in collaboration with colleagues from the Marine Hydrophysical Institute of the National Academy of Science of the Ukraine (Sevastopol). The authors express their gratitude to the co-authors of joint publications, whose valuable contributions made this book possible. Vlasenko and Stashchuk feel deeply indebted to Professor Leonid Cherkesov, who was their first teacher in wave theory. They profited not only from the generous sharing of his ideas and wisdom, but also from his friendship, support, and encouragement.

In the course of the preparation for the book we visited each other, and the hospitality of Plymouth University and Darmstadt University of Technology is gratefully acknowleged. Special thanks are due to the Head of the School of Earth, Ocean and Environmental Sciences, Professor David Huntley, who generously helped and encouraged our efforts.

Most figures included in this book have previously been published by the authors in various scientific journals. We are grateful to copyright holders for granting their permission to reproduce their figures in the text: the American Meteorological Society (Figures 1.1, 1.7, 5.9, 5.10, 5.12–5.19, and 5.27–5.34); Brill Academic Publishers (Figures 3.10, 4.10, 4.11, and 6.18–6.22); Elsevier (Figures 2.22, 4.9, 4.14, 4.15, 4.21, 6.1–6.10, and 7.1–7.14); the European Geophysical Union (Figures 5.20–5.26 and 5.35–5.46); MAIK "Nauka/Interperiodica" (Figures 4.1–4.5, 4.7, 4.8, 4.16–4.18, 5.2, and 6.23–6.26); Nauka (Figures 2.6, 2.7, 3.1–3.3, 4.12, 4.15, 5.5, 5.6, and 6.14–6.17); and Transworld Research Network (Figures 2.3, 2.4, 2.6, 2.7, 2.21, 4.19, and 4.20).

We also acknowledge the work of three anonymous reviewers who took the time to read the manuscript and who have made many fruitful comments on the improvement of the book's contents. Finally, we thank the staff of Cambridge University Press for their fruitful cooperation and attention to detail through the various stages of the publishing process.

Symbols

$A^{\mathrm{H}}, A^{\mathrm{V}}$	coefficients of horizontal and vertical turbulent viscosity
$A(x, t)$	profile of internal wave used in a weakly nonlinear theory
a_j	amplitude of stream function of jth mode generated in zone III
a_ξ	amplitude of vertical isopycnal displacement of solitary internal wave
$a_\xi^{(j)} = \dfrac{a_j k_j}{\sigma}$	amplitude of isopycnal displacement of jth mode generated in zone III
a_m^{inc}	amplitude of stream function of incoming mth mode in zone I
b_j	amplitude of stream function of jth mode generated in zone I
$b_\xi^{(j)} = \dfrac{b_j k_j}{\sigma}$	amplitude of isopycnal displacement of jth mode generated in zone I
c_1, c_2, c_3	coefficients of the model pycnocline
c_{p}	phase speed
c_{g}	group speed
$f = 2\Omega_{\mathrm{E}} \sin\varphi$	Coriolis parameter
g	acceleration due to gravity
$\mathscr{G}$	Green function

H	depth of ocean
H_0	average depth of ocean
$H_{\max}$	height of bottom obstacle
H_1, $H_2(x)$, H_3	water depth in zones I, II, and III, respectively
H_p, ΔH_p	depth and width of the model pycnocline
h_+, h_-	thickness of upper and lower layers in two-layer model
$h = z_1(-H)$	water depth in (x, z_1) variables
$h_0 = z_1(-H_0)$	average depth of a basin in (x, z_1) variables
$h_1 = z_1(-H_1)$, $h_2(x) = z_1(-H_2(x))$, $h_3 = z_1(-H_3)$	water depth in zones I, II, III in (x, z_1) variables
$h_{\max}$	height of bottom obstacle in (x, z_1) variables
$\mathscr{X} = x + z_1$	characteristic line
$J(a, b) = a_x b_z - a_z b_x$	Jacobian operator
K^{H}, K^{V}	coefficients of horizontal and vertical turbulent diffusivity
k_{1j}, k_{3j}	wave number of jth baroclinic mode in areas I and III, respectively
l	half-width of bottom obstacle
$\mathscr{L} = x - z_1$	characteristic line
$M = \dfrac{g}{\rho_0}\dfrac{\partial\rho}{\partial x}$	horizontal analog of buoyancy frequency
m	number of incident baroclinic mode
$N = \left(-\dfrac{g}{\rho_0}\dfrac{\partial\rho_0}{\partial z}\right)^{1/2}$	buoyancy frequency
N_0	value of buoyancy frequency for monotonic stratification
N_p	maximum value of buoyancy frequency
P	pressure
P_a	atmospheric pressure
$\tilde{P}$	wave disturbance of pressure
$q_j(z_1)$	eigenfunction of jth baroclinic mode

$\mathrm{Ri} = N^2/u_z^2$	Richardson number
$\mathcal{R}$	Riemann function
S	salinity
T	temperature
T	tidal period
$T_0 = \lambda/c_\mathrm{p}$	time scale for solitary internal waves
t_0	reference time
t_b	instant of wave breaking
U_0	amplitude of barotropic tidal velocity
u	component of velocity vector $\mathbf{v}$ in x-direction
u_+, u_-	velocities of upper and lower layer, respectively, in x-direction in two-layer model
v	component of velocity vector $\mathbf{v}$ in y-direction
v_+, v_-	velocities of upper and lower layer, respectively, in y-direction in two-layer model
w	component of velocity vector $\mathbf{v}$ in z-direction
$z_1 = \int_0^z \frac{ds}{\alpha(s)}$	z_1-transformation
$z_2 = \frac{\int_0^z N(s)\,ds}{\int_0^{-H(x)} N(s)\,ds}$	z_2-transformation
$\alpha, \alpha_+, \alpha_-$	slopes of characteristic lines
β	coefficient of dispersion
$\gamma = dH/dx$	bottom inclination
$\gamma_0 = dh/dx$	bottom inclination in $(x_1 z_1)$ variables
$\varepsilon = a_\xi/H$	nondimensional wave amplitude
$\varepsilon_0 = h_\mathrm{max}/h_0$	nondimensional height of bottom obstacle
ε_1	parameter of nonlinearity
$\theta = x - c_\mathrm{p}t$	variable in coordinate system moving with speed c_p

λ	wavelength
$\mu = (H/\lambda)^2$	parameter of dispersion
ξ	vertical isopycnal displacement
ξ_+, ξ_-	vertical displacements of free surface and interface, respectively (two-layer model)
ρ	density
$\rho_0(z)$	density in hydrostatic equilibrium
$\tilde{\rho}(x, z, t)$	wave disturbance of density
$\bar{\rho}_0 = \text{const.}$	reference density
ρ_+, ρ_-	densities of upper and lower layers, respectively (two-layer model)
σ	wave frequency
$\sigma_t = \rho - 1000$	conventional density
φ	latitude
φ_{c}	critical latitude
ψ, Ψ	stream function
ω	vorticity
Ω_{E}	angular rotation rate of Earth

Abbreviations

ADCP	acoustic Doppler current profiler
ADIM	alternative direction implicit method
BBL	bottom boundary layer
BSPF	Barents Sea Polar Front
BVP	boundary value problem
CFL	Courant Friedrich Levi
CTD	conductivity, temperature, depth (profiler)
ERS-1, ERS-2	European Remote-Sensing Satellite
GW	gigawatt
JASIN experiment	Joint Air–Sea Interaction Experiment
JUSREX-92	Joint United States–Russian Experiment
K–dV	Korteweg–de Vries
eK–dV	extended Korteweg–de Vries
ODE	ordinary differential equation
POM	Princeton Oceanographic Model
R/V	research vessel
SAR	synthetic aperture radar
SIW	solitary internal wave
TOPEX	Topography Experiment
TOPEX/Poseidon	Joint US–French orbital mission, launched in 1992 to track changes in sea-level height with radar altimeters
TW	terawatt
WKB	Wentzel, Kramers, Brillouin

Preamble

The World Ocean, considered as an active dynamical system, is in permanent motion. Most of its manifestations can be related to wave phenomena. Besides the well known surface waves, there are also waves of other nature; among these, internal gravity waves are particularly important. They exist due to the presence of vertical fluid stratification, they are permanently generated, and they evolve and are destroyed again in the deep ocean. The amplitudes of internal waves are usually much larger than those of surface waves, due to the weak returning force, and their amplitudes can sometimes reach values of 100 m and more (see refs. [5], [23], [120], [128], [180], [192], and [193]).

Numerous *in situ* measurements, carried out in all regions of the World Ocean (see, for instance, refs. [63], [119], [123], and [157]), have shown that internal gravity waves exist wherever a stable vertical stratification of a fluid is observed. They were discovered more than 100 years ago, and were understood by the scientists of the day to be a disappointing obstacle disturbing the "correct" structure and dynamics of the oceanic water masses. More than half a century passed before the importance of internal waves to the global dynamics of the ocean was realized. Because they penetrate water thicknesses from the free surface to the bottom, internal waves play an important, and sometimes a decisive, role in many hydrophysical processes occurring in the ocean.

Recent specific interest in internal waves has emerged because they influence the horizontal and vertical exchange processes in the ocean, their generation of small-scale turbulence, and their formation of both fine structures and oceanic stratification [67], [68]. According to the idea of global meridional oceanic circulation first formulated by Sandström almost 100 years ago [209], oceanic water masses are cooled in the polar areas and are downwelled into the abyss; they then propagate further with abyssal fluxes to the equator. In the equatorial regions they upwell to the surface, are heated, and are carried poleward with the surface currents. Munk and Wunsch [167] have concluded that it is not yet clear where and how the abyssal waters return to the surface. However, quite powerful sources of water mixing must exist in the World Ocean to maintain 2000 TW (1 TW = 10^{12} W) of the pole–equator heat flux associated with the global circulation.

In this mechanism of meridional overturning circulation, internal waves play a crucial role in the larger oceanographic context since they are thought to be associated closely with diapycnal mixing processes in the deep ocean, so providing a comparatively smooth observed vertical fluid stratification. In the absence of effective abyssal mixing (which is basically due to internal waves), the vertical thermohaline structure of the ocean would be represented by a thin (a few tens of meters), warm surface layer of a fluid, below which the main part of the stagnant, cold waters would be located (this follows from Sandström's theorem, which applies to an ocean that is heated and cooled at the free surface). Under such conditions, the vertical fluid stratification would be very close to that of a two-layer fluid with a very thin upper layer. The ecological consequences of such a situation would be catastrophic. Understanding where and how oceanic waters are mixed may therefore be crucial for realistic modeling of global long-term processes such as climate change.

All the above-mentioned facts point to the important role played by internal gravity waves in ocean dynamics, but they do not form a complete list. Internal waves arise within a wide range of frequencies – from short-periodic waves with periods $T \sim 10\ s^{-1}$ to inertial waves with periods of days. Those waves with astronomical tidal periods – the so-called baroclinic tides – occupy a central place because they are more intensive and more pronounced in the ocean than others. Calculations of power spectra for various sites of the World Ocean have shown that at practically all locations a peak can be found in the spectral density which is in the vicinity of the tidal frequency: the level of the spectrum in this frequency band usually exceeds the values of other frequencies by one to two orders of magnitude. Gregg and Briscoe [81] estimated that one-third of all vertical displacements related to internal waves can be assigned to baroclinic tides. An analysis by Müller and Briscoe [168] has confirmed the observation of spectral peaks at the inertial and semidiurnal tidal frequencies. Being more energetic, the baroclinic tides spread out their energy through the spectrum from large to small scales due to wave–wave interaction and thus develop universal spectral distribution, as was first recognized by Garrett and Munk [69], and specified more recently by Hibiya *et al.* [96].

Since baroclinic tides are one of the principal sources of oceanic internal waves, they assume a very important role in deep ocean mixing and ocean dynamics. In fact, there are indications that internal tides may be a dominant source of deep ocean mixing and a significant component in the thermohaline maintenance. On a global scale, the energy of the Earth–Moon and the Earth–Sun orbital systems is dissipated by lunar and solar tides, slowly increasing the length of the day (about

2 ms per century), while the orbital radius of the Moon increases by about 3.7 cm every year [166].

The total amount of tidal energy being dissipated in the Earth–Moon–Sun system comprises 3.7 TW, with 2.5 TW for the principal lunar tide M_2 [29], [113]. This value has been well determined using methods of space geodesy (e.g. altimetric measurements, satellite laser ranging, lunar laser ranging).

Within the oceans, the principal sink of energy occurs in shallow marginal seas due to the friction in the bottom boundary layer (2.6 TW); the traditional explanation is due to Jeffreys [110]. There is much evidence in favor of this statement [165]. Another possible energy sink occurs due to the conversion of energy into internal tides and other baroclinic waves. Based on satellite altimetry, Egbert and Ray [55] showed that as much as one-third of the global tidal dissipation (0.6–0.8 TW) occurs in the deep ocean. Models due to Kantha and Tierney [116] and Sjöberg and Stigebrandt [219] predict similar losses. As is summarized in ref. [167], up to one-quarter of all tidal energy in the open sea is transferred into baroclinic tides.

The main problem in Munk and Wunsch's analysis of the global oceanic circulation [167] was the definition of the basic sources of background turbulence, ensuring vertical mixing and formation of the observed vertical structure of the thermohaline fields. Theoretical estimations [164] show that, for the maintenance of the existing oceanic stratification, the coefficient of vertical turbulent exchange should be at the level of 10^{-4} m^2 s^{-1}, while the majority of direct *in situ* measurements performed far from the large-scale bottom roughness reveal a magnitude that is one order smaller.

The assumption that can be made from such considerations is that the basic sink of the tidal energy to turbulence forming the necessary degree of water mixing does not occur uniformly in all areas of the World Ocean but is concentrated in "hot spots" or "storm areas," where the main transfer of energy from the barotropic tidal motion to the internal waves and to turbulence takes place.

Dissipation in the open ocean is significantly enhanced around major bathymetric features. Although some of the shelf breaks such as the Bay of Biscay [192], [193] or the Queen Charlotte Islands [47] are sites of intense internal tide generation, surface tides tend to propagate along rather than across shelves, inducing only weak across-shelf currents to interact with shelf break topography. As was concluded in refs. [11] and [104], the generation of internal tides at the continental slopes is an insignificant sink, approximately 12 GW (1 GW = 10^9 W), for M_2 over the entire globe. But scattering of internal tides by deep-ocean topography may be more important than their generation at continental slopes [219].

Thus, an accurate identification of the regions of intense tidal dissipation in the oceans and its quantification is a very important, but still quite elusive, problem. The satellite altimeter data accumulated during the ten years of TOPEX/Poseidon indicated a possible way of detecting the areas of tidal dissipation in the World Ocean. Modern satellite altimetry can detect the sea-surface level to an accuracy of 1 cm. This is due to the fact that the tidal signal is time-coherent, which allows repeated measurements, so reducing noise. Taking into account that sea-surface elevations caused by baroclinic tidal waves with amplitudes of 5 m and more are of the same order makes it possible to map the "hot spots" in the World Ocean.

In doing this, Ray and Mitchum [204] report a coherent baroclinic tide propagating to the northeast from the Hawaiian Ridge. They found that both first and second baroclinic modes of the semidiurnal internal tides were present in the data, and that the signal of internal tide propagation could be tracked up to 1000 km from the Hawaiian Ridge. In a further study [205], they estimate that 15 GW of the semidiurnal internal tidal energy radiates away from the Hawaiian Ridge in the first baroclinic mode.

A further step was taken by Kantha and Tierney [116], who quantified the energy and energy dissipation rate in global baroclinic tides using a combination of precision altimetry and a numerical tidal model. They discovered more than 15 of the highest ridges of the World Ocean and found that approximately 15% of the barotropic tidal energy is transformed into the baroclinic component. A similar investigation was performed by Niwa and Hibiya [177] for the dissipation of the M_2 tide in the Pacific Ocean. It was found that the conversion rate from the semidiurnal tide to internal waves integrated over the whole Pacific Ocean amounts to 338 GW, 84% of which is located at the prominent topographic features. Thus, the global surveys of Kantha and Tiereny [116] and Niwa and Hibiya [177] show that the baroclinic tidal signals occur through the World Ocean and originate principally from large submerged ridge topographies. Recent *in situ* measurements have confirmed the existence of many of these "hot spots" (see refs. [3], [38], [48], [54], [64], [94], [126], and [127]). Through such "windows" the barotropic tidal energy is supplied to the ocean (it is transformed from the internal waves to turbulence), and is further distributed to all areas of the World Ocean.

Thus, a significantly difficult problem to be solved by oceanographers is to define "where," "how much," and "how" the tides dissipate their energy. Many works (e.g. the references listed above) address the questions of "where" and "how much." We try to answer the question "how" barotropic tidal energy transforms into the baroclinic component. For this purpose we shall highlight the most interesting aspects of the dynamics of baroclinic tides in a horizontally nonuniform ocean: generation and propagation of tidal internal waves over variable bottom topography and horizontal gradients of density (which are usually accompanied by bottom elevations).

A number of books published since the 1970s have considered different aspects of internal gravity waves. Some fragmentary data on internal waves and qualitative pictures of some particular problems are summarized in the corresponding chapters of the books by Monin *et al.* [159], Phillips [190], and Turner [236]. Despite the appearance of the fundamental work by Whitham [264], internal waves – due to their anisotropy and nonuniformity – still present a nontrivial object for investigations and for applying the methods outlined in these books. Specialists in wave dynamics must rely on what has been written in the more general textbooks by Le Blond and Mysak [132] and Tolstoy [232]. Other treatises are the books by Lighthill [139], which focuses on the general theory of the generation and propagation of internal waves, Gill [76], which discusses internal waves in atmospheric and oceanic processes, and Craik [46], which discusses wave interaction phenomena in fluids at rest and shear flows.

Only a few fundamental publications are known which deal entirely with internal waves: *Interne Wellen* by Krauss [123]; *Surface and Internal Waves* [37] and *Hydrodynamics of Surface and Internal Waves* [36] by Cherkesov; *Waves Inside the Ocean* by Konyaev and Sabinin [119]; Munk's *Internal Waves* [163] (an extensive literature review); and a more recent issue of Miropol'sky's *Dynamics of Internal Gravity Waves in the Ocean* [157], which is the English translation of the original version published in Russian in 1981. Another book, written by Baines, *Topographic Effects in Stratified Flows* [13], considers internal waves and hydraulic jumps that are generated by stationary flows interacting with bottom topography.

In contrast with the books given above, where the general wave theory is applied to study a wide range of baroclinic wave motions, this text focuses on the investigation of tidally generated internal waves. We will pay attention to linear as well as weakly and strongly nonlinear problems. The book presents results that have been obtained by the authors since the early 1980s.

1

General background

1.1 Introduction

Internal waves are present in the ocean because of the existence of water stratification. Properties of internal waves are different from the characteristics of waves which are observed at the ocean surface. Indeed, whereas ocean surface waves possess amplitudes of at most a few meters, internal wave amplitudes can be as large as 100 m or more. The following simple explanation can clarify the difference between surface and internal waves.

We consider wave motions in the Cartesian system of coordinates $Oxyz$, with Oxy within the undisturbed free surface of the fluid and the Oz-axis directed vertically upward. We assume that the water density consists of two parts: a stationary density $\rho_0(x, y, z)$ and density disturbances $\tilde{\rho}(x, y, z, t)$ introduced by the wave motions. In the real ocean, usually $\rho_{0z} \gg \{\rho_{0x}, \rho_{0y}\}$, which is why with good accuracy one can take $\rho_0 = \rho_0(z)$.[1] Some examples of such density distributions are presented in Figure 1.1(a). Imagine a unit volume of water is displaced vertically upward from its equilibrium position z_0 to $z_0 + \xi$ (Figure 1.1(b)). Newton's second law for this liquid particle reads

$$\rho_0(z_0)\,\frac{d^2\xi}{dt^2} = g\rho_0(z_0 + \xi) - g\rho_0(z_0), \tag{1.1}$$

where g is the acceleration due to gravity. The right-hand side of this equation represents the restoring force, which, in fact, is the difference between the Archimedean buoyancy force and the gravitational force.

Because the difference of the oceanic water density in (1.1) between $z = z_0$ and $z = z_0 + \xi$ is small (≤ 10 kg m^{-3}) in comparison with the density difference between water ($\rho_0 \sim 10^3$ kg m^{-3}) and air ($\rho_a = 1.210$ kg m^{-3}), the restoring force for internal waves is more than 100 times smaller than for surface waves. This

[1] Subscripts denote partial derivatives with respect to the subscripted variable.

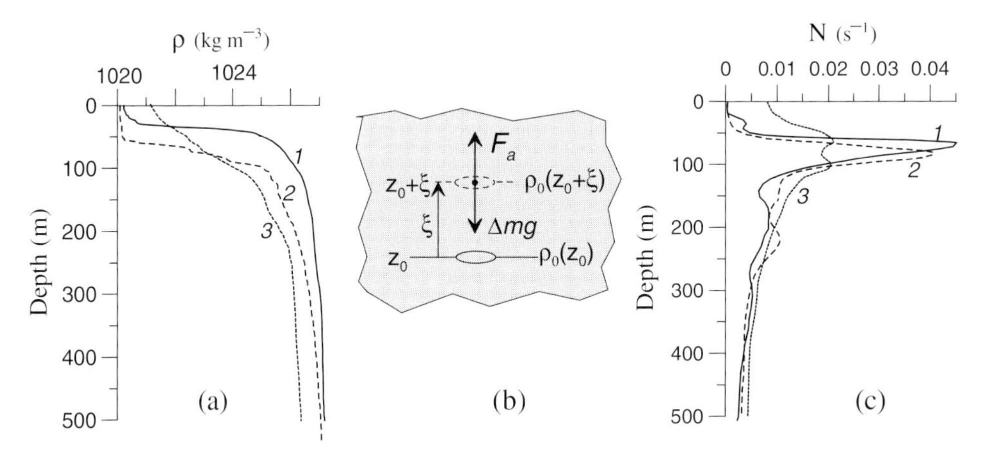

Figure 1.1. (a) Typical distributions of density and (c) buoyancy frequency measured in the Andaman Sea (1,2) and the Sulu Sea (3). (b) Schematic diagram of forces acting on a fluid particle as its position deviates from its state of equilibrium level, z_0. F_a represents the buoyancy force, the resultant of the pressure exerted by the ambient fluid on the particle's surface; Δmg is the weight of the fluid fragment.

is why the amplitudes of internal waves can be substantially larger than those of surface waves. Using Taylor series expansion one can express this as

$$\rho_0(z_0 + \xi) = \rho_0(z_0) + \xi \frac{d\rho_0}{dz} + O(\xi^2). \tag{1.2}$$

Thus, (1.1) can be rewritten as

$$\frac{d^2\xi}{dt^2} - \frac{g}{\rho_0(z)} \frac{d\rho_0}{dz}\xi = 0. \tag{1.3}$$

Depending upon the sign of $d\rho_0/dz$, solutions of this equation are bounded ($d\rho_0/dz < 0$) or unbounded ($d\rho_0/dz > 0$) as time approaches infinity. In fact, for stable stratification ($d\rho_0/dz < 0$), this equation describes vertical oscillations of fluid particles relative to their equilibrium positions with angular frequency $N(z) = [(-g/\rho_0(z_0))d\rho_0/dz]^{1/2}$, where N is known as the *buoyancy frequency*. This concept was introduced by Rayleigh [201] in his investigations of free convection processes. Alternatively, it is called the Brunt–Väisälä frequency because it was equally introduced by the Norwegian meteorologist Brunt [27] and the Finnish oceanographer Väisälä [240].

If the compressibility of the water is taken into account, the buoyancy frequency takes the form given by

$$N^2(z) = -\frac{g}{\rho_0(z_0)} \frac{d\rho_0}{dz} - \frac{g^2}{c_s^2}, \tag{1.4}$$

Table 1.1. *Important tides and their characteristics.*

Darwinian symbol	Period (hours)
K_1	23.9345
O_1	25.8193
M_2	12.4206
S_2	12.0000

where c_s is the speed of sound in water. Vertical profiles of $N(z)$ for the Andaman and Sulu Seas for the density profiles of Figure 1.1(a) are shown in Figure 1.1(c). They are typical for most oceanic situations. Of course, the above simplified example does not reflect the wide variety of internal gravity waves in the real ocean. It shows only the physical mechanism which is responsible for their existence. In fact, real oceanic internal waves consist of a continuous spectrum of oscillations, from the buoyancy frequency up to the local inertial frequency.[2]

It is interesting that scrutiny of a vast number of field observations at various sites of the World Ocean points to the existence of internal waves with tidal periods.[3] The energy of such waves very often exceeds the wave energy within other frequency bands [15], [167], [168].

An analogous situation occurs for surface waves: the barotropic[4] tide is one of the most remarkable and most pronounced wave phenomena. This similarity in manifestation does not mean, however, that barotropic and baroclinic tides are excited by identical mechanisms. Barotropic oceanic tides are mostly the result of the perpetual variation of the gravitational attraction between the Earth and the Moon. The latter is the astronomical body closest to the Earth, and the Sun is the most massive heavenly body in our planetary system. The periods of the most pronounced tidal harmonics which are mentioned in the present book are shown in Table 1.1 (a detailed tidal theory explaining these is presented in ref. [115]).

The direct generation of internal waves by tidal forcing in an ocean of constant depth is insignificant [123]. Therefore, the question of how internal waves of tidal periods are generated is very important for an adequate understanding of ocean dynamics. In oceans of variable depth, the generation of baroclinic tides is possible because of the interaction of the barotropic tidal waves with the bottom topography.

[2] This statement ignores all acoustic waves, which arise due to the compressibility of the water and propagate much faster than barotropic or baroclinic waves. We will not discuss acoustic waves in this book.

[3] See refs. [15], [35], [51], [77], [81], [89], [97]–[100], [119], [133], [167], [192], [193], [194], [212], [217], and [234].

[4] Surface and internal waves are oscillating phenomena belonging to two classes of water motion, defined as barotropic and baroclinic processes, respectively.

The possibility of this mechanism of baroclinic wave generation was first indicated almost 100 years ago for an ocean with a vertical density jump (two-layer model) by Zeilon [271]. The efficiency of this mechanism (for a continuously stratified fluid) was further theoretically corroborated by Cox and Sandström [43]. Numerous laboratory and field experiments subsequently confirmed this conclusion. Laboratory investigations, performed in wave tanks ([12], [16], [28], [149], [269]), as well as theoretical studies ([7]–[11], [147], [202], [210]) led to the isolation of the most important factors influencing the topographic generation of the wave fields. It was found that two basic parameters that control the dynamics of internal wave fields near bottom features are the inclination angle α of the characteristic lines of the hyperbolic wave equation

$$w_{zz} - \alpha^{-2} w_{xx} = 0 \tag{1.5}$$

(which will be derived in Section 1.4) and the inclination of the bottom relief $\gamma = dH/dx$. Here, w is the vertical velocity component, and the bottom topography is given by $z = -H(x)$. The relation between α and γ is very important from the point of view of the generation of internal waves. Baines [12] found by performing laboratory experiments on the generation of internal waves by barotropic fluxes over an underwater canyon that, for *subcritical* inclinations of the bottom, when $\alpha \gg \gamma$, the influence of the stratification on the movement of a fluid was very weak, and in the vicinity of a canyon the motion was basically barotropic. If the inclination of the bottom was close to *critical*, $\alpha \sim \gamma$, the movement was essentially baroclinic through the entire water column. With $\alpha < \gamma$ (*supercritical* case) the baroclinic energy propagated downwards along the canyon as radiated waves, thereby creating a wave "beam."

Similar features were observed during *in situ* measurements in the ocean, for example near the coast of Africa [77], [81], over the continental slope off Oregon [234], in the coastal zones of Australia [44], [45], [97]–[100], on a shelf off Vancouver Island [15], near the coast of California [228], in the Bays of Massachusetts [35], [87] and Biscay [192], [193], and in many other regions of the World Ocean. In these observations, subcritical as well as supercritical situations of internal wave generation were encountered. Using the theory of empirical orthogonal functions, analysis of the observational data allowed the conclusion [77] that over the subcritical continental slopes a significant part of the energy of the generated internal tides was concentrated in the first baroclinic mode,[5] which is propagating shoreward from the shelf break. The density of the wave energy in that direction increased, and in the shallow part of the shelf it reached saturation.

[5] For a given distribution of the buoyancy frequency, a countably infinite number of internal wave modes exists, each of which has its own speed of propagation. The mode with the fastest wave speed is called the *first* mode. This fact will shortly be proven.

It was also found that many internal waves completely dissipated prior to reaching a coastal line where they would reflect. A similar conclusion on the predominance of the lowest baroclinic mode in the fields of the K_1 and O_1 internal tides was also obtained for the near-critical bottom features off the eastern part of Florida [133].

In those areas, where the inclination of the bottom is supercritical, the characteristic feature of internal tides is the formation of a wave beam, i.e. the confinement of the wave field to a long and rather narrow region, where the basic part of the baroclinic tidal energy is concentrated. Usually, this wave beam arises at the shelf break and extends into the deep part of the ocean downwards along the continental slope (to be more precise, along the characteristic lines of the wave equation (1.5)). This structure of the internal tide was found in the Pacific Ocean near the coast of Oregon [234], in the Atlantic Ocean in the Bay of Biscay [192], [193], in the eastern North Pacific near the Mendocino Escarpment [3], and also in the Indian Ocean in the area of the JASIN experiment [51].

Nonlinear wave effects were also of specific interest during observational investigations of the baroclinic tides. Probably the most conspicuous manifestation of the nonlinear behavior that internal tides take is the formation of an internal hydraulic jump (or a baroclinic bore) and its subsequent evolution into a packet of short nonlinear waves. This phenomenon usually takes place under conditions of sufficient intensity of the tidal forcing, as will be discussed below. Such baroclinic bores were often observed at different sites of the World Ocean, for instance in the Bays of Massachusetts [35], [87] and New York [261], close to the Mascarene Ridge in the Indian Ocean [120], [206], in the Strait of Gibraltar [6], [263], on a shelf off Australia [98], and in many other regions. A vast amount of observational material was collected on the structure and spatial–temporal characteristics of solitary internal waves (SIWs) and wave packets generated during the nonlinear stage of the evolution of the baroclinic tides (see, for instance, the review [100]). They were sufficiently reliably registered by contact[6] as well as remote sensing[7] methods.

The many forms in which tidal waves appear increase the urgency of providing a theoretical description of their dynamics. Various linear and nonlinear theoretical models (both analytical and numerical) have been constructed since the 1960s in order to study the generation mechanisms of baroclinic tides. We will give an overview of many published theoretical models in the introductions to all chapters where an appropriate topic of baroclinic tidal dynamics is considered. The list of studied phenomena includes all kinds of wave motions that are generated by the oscillating tidal flow over topographic features, such as underwater ridges and

[6] See refs. [1], [78], [88], [102], [107], [119], [120], [128], [140]–[142], [176], [180], [191], [195], [206]–[208], [212].

[7] See refs. [2], [23], [66], [70], [71].

shelf-slope areas. We consider a wide range of the intensity of the periodic tidal forcing: from infinitesimal baroclinic internal waves at small values of the Froude number to the strong, nonlinear, unsteady lee waves, generated at large values of the Froude number. The effects connected with the influence of horizontal density gradients and three-dimensional effects are also considered. Our principal intention in writing this book is to clarify the many complicated issues concerning baroclinic tidal dynamics.

The present chapter is devoted to the description of the fundamental information used throughout the book: the basic equations and assumptions, the linear wave equation, the standard boundary value problem, and some preliminary information on the nonlinear approach.

1.2 Governing equations: basic assumptions and hypotheses

We consider wave motions of a continuously stratified fluid in a basin of variable depth, which rotates with the Earth (Figure 1.2(a)). First, we will formulate the governing equations describing the internal wave dynamics for the common case of nonlinear and nonadiabatic motions, which incorporate the full convective nonlinearities, and the dissipative and diffusive processes. To this end, the laws of balance of mass, momentum, and energy, the thermal equation of state, and some thermodynamic relations are required. Depending on the processes considered, this system will be simplified, or it will be used without changes (in its initial complete form). Wave motions are considered in a right-handed rectangular Cartesian system of coordinates, $Oxyz$, with Oxy at the undisturbed sea surface and the Oz-axis directed vertically upwards (Figure 1.2(b)). Newton's second law for a continuous

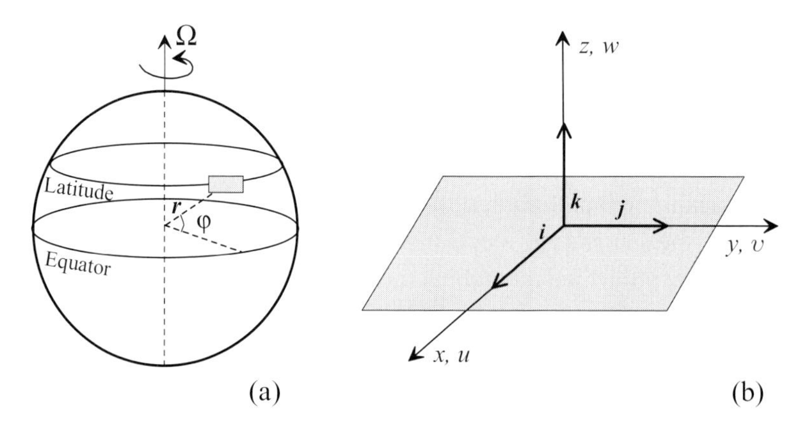

Figure 1.2. (a) The Earth's coordinate system. (b) Cartesian coordinate system used throughout this book. The direction of Ox (and consequently Oy) within the tangential plane is arbitrary.

medium in the noninertial system rotating with angular velocity $\mathbf{\Omega}$ has the form

$$\rho \frac{d\mathbf{v}}{dt} = -\nabla P - \rho g\mathbf{k} - \rho 2\mathbf{\Omega} \times \mathbf{v} + \mathbf{F}(\mathbf{v}). \tag{1.6}$$

It states that the product of the density ρ and the acceleration $d\mathbf{v}/dt$ ($\mathbf{v}$ is the velocity vector with components u, v, and w in the x, y, and z directions, respectively) is equal to the sum of the negative pressure gradient ∇P, the gravity force $\rho g\mathbf{k}$ (where $\mathbf{k}$ is the vertical unity vector), the Coriolis force $-\rho 2\mathbf{\Omega} \times \mathbf{v}$, and all other conservative and nonconservative forces $\mathbf{F}(\mathbf{v})$. In our case, $\mathbf{F}(\mathbf{v})$ includes the tidal forcing due to the Moon and the Sun and the dissipative force due to friction. In the latter case, for Newtonian fluids, $\mathbf{F}(\mathbf{v})$ takes the form

$$\mathbf{F}(\mathbf{v}) = \nu\rho[\nabla^2\mathbf{v} + \nabla(\nabla \cdot \mathbf{v})/3], \tag{1.7}$$

where ν is the coefficient of kinematic viscosity. As shown in ref. [123], the direct generation of internal waves by the tidal forcing due to the Moon and the Sun is negligible; consequently, for (1.6) we can use $\mathbf{F}(\mathbf{v})$ only in the form (1.7). In (1.6) we neglected the centrifugal force $\rho\mathbf{\Omega} \times (\mathbf{\Omega} \times \mathbf{r})$ [132] because it is only about 10^{-3} of the gravity force (here $\mathbf{r}$ is the vector from the center of the Earth to the considered point; see Figure 1.2).[8]

For oceanic internal gravity waves the balance law of mass has the standard form

$$\frac{\partial\rho}{\partial t} + \nabla \cdot \rho\mathbf{v} = 0, \tag{1.8}$$

or alternatively

$$\frac{1}{\rho}\frac{d\rho}{dt} + \nabla \cdot \mathbf{v} = 0. \tag{1.9}$$

The above equations (1.6), (1.7), and (1.9) establish a well posed set of four partial differential equations for the density and velocity fields. They must be complemented by a thermal equation of state and a thermodynamic equation.

Marine water is a solution of water in which the following main components are dissolved: $NaCl$, $MgCl_2$, $MgSO_4$, $CaSO_4$, K_2SO_4, $CaCO_3$, KBr, $SrSO_4$, H_2BO_3. Usually, i.e. far away from river inflows and at some isolated places of the shelf regions, the composition of these salts is stable, but the whole amount of dissolved salts can vary. The measure for this amount is the *salinity* S, which is defined as the ratio of the mass of dissolved solid substance to the mass of water in which this substance is dissolved. Conventionally, the salinity of marine water is measured in ppt, *parts per thousand*, or psu, *practical salinity units*, which is nearly equivalent to the total mass of dissolved solids in parts per thousand ($^\circ/_{\circ\circ}$).

[8] Alternatively, one may incorporate the centrifugal force in the gravity force, but in that case the gravity force has a component in the Oxy-plane that is ignored in (1.6).

The density of oceanic water thus depends on three variables: temperature T, salinity S, and pressure P, and the equation of state

$$\rho = \rho(P, \mathsf{T}, \mathsf{S}) \tag{1.10}$$

defines such a relation; generally it is nonlinear with respect to T, S, and P, and was found empirically. For the ocean, the ranges of these variables are as follows: for temperature, -2 to 40 °C; for salinity, 0 to 42 ppt; for pressure, 0–1000 bar, and for density changes, 1000–1040 kg m^{-3}. The equation of state, now commonly used, is expressed as a polynomial with empirical coefficients, and it defines the water density in a restricted range of input parameters. It is due to Millero and Poisson [156], and is given in the UNESCO Technical Paper in Marine Science [238] as follows:

$$\rho(\mathsf{T}, \mathsf{S}, P) = \rho(\mathsf{T}, \mathsf{S}, 0)[1 - P/K(\mathsf{T}, \mathsf{S}, P)]^{-1}, \tag{1.11}$$

where $\rho(\mathsf{T}, \mathsf{S}, 0)$ is the density of sea water at unit standard atmospheric pressure $(P = 0)$ and $K(\mathsf{T}, \mathsf{S}, P)$ is the secant bulk modulus; both are defined in tables in ref. [238]. In (1.11), T must be provided in degrees Celsius, S in ppt or psu, P in bars, and ρ is obtained in kg m^{-3}.

In some cases, the problem formulation allows a simpler (linear) representation of the equation of state [227]. For any preselected standard set of variables ρ_*, T_*, S_*, it then takes the form

$$\rho/\rho_* = 1 - \alpha_{\mathrm{T}}(\mathsf{T} - \mathsf{T}_*) + \alpha_{\mathrm{S}}(\mathsf{S} - \mathsf{S}_*), \tag{1.12}$$

where values of the *coefficient of thermal expansion*, α_{T}, and the *saline contraction coefficient*, α_{S}, are available from tables given in ref. [31].

The written system (1.6) (incorporating (1.7), (1.8) or (1.9), and (1.10)) is not complete: it contains five equations and seven unknown functions, namely the components of the velocity vector $\mathbf{v}$ (u, v, w), pressure P, water density ρ, temperature T, and salinity S. It must be completed by additional relations – the first law of thermodynamics, and an equation for the salinity balance.

According to thermodynamics, the change of the internal energy, dE, of the system can be written in the form

$$dE = \delta Q - \delta A, \tag{1.13}$$

where δQ represents the change of the internal parameters and the temperature of the system, and δA is the change of the external parameters accomplished by the work A. Unlike the energy E, the work A and the amount of heat Q depend on the trajectory of transformation.[9] For the problem of the generation of internal

[9] Expressed differently, dE is a perfect differential but δQ and are δA are not and are therefore path dependent.

tides, the term δA in (1.13) can be neglected because of the incompressibility of the fluid.

Incompressible fluid

The compressibility of a fluid is the reason that acoustic waves propagate in marine water with the velocity of sound:

$$c_s(P, \mathsf{T}, \mathsf{S}) = \left[\left(\frac{\partial P}{\partial \rho} \right)_{\epsilon} \right]^{1/2}, \tag{1.14}$$

where ϵ is the entropy. The phase speed of internal gravity waves, c_p, is defined by the dispersion relation, which derives from the governing equations. In the simplest case of long infinitesimal internal waves in a two-layer fluid, it reads

$$c_p = \left(g' \frac{h_- h_+}{h_- + h_+} \right)^{1/2}, \tag{1.15}$$

where h_+ and h_- are the thicknesses of the upper and lower layers, respectively, and $g' = g\Delta\rho/\rho$ ($\Delta\rho$ is the density difference between the upper and lower layers). The speed of sound, c_s, for oceanic conditions is about $1500\,\mathrm{m\,s^{-1}}$, whereas the phase speed of internal tidal waves, c_p, is less than $2\,\mathrm{m\,s^{-1}}$. Thus, the value of c_s is about three orders of magnitude larger than c_p. For an incompressible fluid, c_s is infinitely large. The difference between the two speeds is the reason why we can consider marine water to be a medium that is practically incompressible. Consequently, these two phenomena hardly interact. Expressed more directly in physical terms: there is hardly any energy loss from the internal wave motion to sound waves. Conversely, there is hardly any energy transfer from sound waves to internal waves. So, for the description of the oceanic baroclinic wave motions, assuming water to be an incompressible fluid is a very good approximation. In this case, the equation of balance of internal energy (1.13) can be presented in the form

$$\rho c_v \frac{d\mathsf{T}}{dt} = \nabla \cdot (k_{\mathsf{T}} \nabla \mathsf{T}) + Q_{\mathsf{T}}, \tag{1.16}$$

where c_v is the *specific heat* at constant volume, treated here as a constant, k_{T} is the *coefficient of thermal conductivity*, and Q_{T} denotes the sources and sinks of heat due to radiation and dissipation. Equation (1.16) ignores the power of working due to the pressure on the volume changes.

Next, a balance of mass for the salts allows deduction of the equation

$$\rho \frac{d\mathsf{S}}{dt} = -\nabla \cdot \mathbf{J}_{\mathsf{S}} + \rho Q_{\mathsf{S}}, \tag{1.17}$$

where $\mathbf{J}_\mathsf{S}$ is the diffusive flux of salt, generally given by a Fick-type relationship

$$\mathbf{J}_\mathsf{S} = -\rho k_\mathsf{S} \nabla \mathsf{S}, \tag{1.18}$$

in which k_S is the coefficient of molecular diffusivity, and Q_S represents the sources and sinks of salt. Combining (1.17) and (1.18) yields the balance equation for the salts,

$$\frac{d\mathsf{S}}{dt} = \nabla \cdot (k_\mathsf{S} \nabla \mathsf{S}) + Q_\mathsf{S}. \tag{1.19}$$

Thus, the system of equations (1.6), (1.8) or (1.9), (1.10), (1.16), and (1.19) is now complete and can be considered as a theoretical basis for the investigation of internal gravity waves. Nevertheless, in this complete form the governing equations are still rather complicated, and in this form they have never been used for theoretical analysis. Even the construction of numerical solutions is hardly ever attempted. One of the possible ways of achieving a physical understanding is to impose further reasonable assumptions and to postulate hypotheses which simplify the system. Obviously, when doing so, the errors, which inevitably are introduced by such simplifications, should be negligible for the description of the considered processes. The simplifications, which can be applied, depend strongly on the considered processes, and in each instance one must be very careful with them. With these provisos, let us consider possible additional assumptions and simplifications, thereby keeping in mind the purpose of the present book, namely the investigation of the dynamics of internal gravity waves in the horizontally nonuniform ocean.

Adiabaticity

One of the essential simplifications is to neglect the processes of heat conductivity and salt diffusivity.[10] For oceanic gravity waves with time scales from minutes to the inertial period, imposition of these assumptions is not a bad approximation in many cases in view of the time scales of the wave motions, which are substantially shorter than the characteristic time scales of the thermal and substantive diffusion processes. In such circumstances, we may put $k_\mathsf{T} = k_\mathsf{S} = 0$.[11]

[10] In thermodynamics, adiabaticity simply means that thermal diffusion is ignored. Oceanographers use the definition of adiabaticity to mean that the thermal diffusion *and* the salt diffusion are negligible.

[11] This argument can be made more rigorous. Let L_d be a typical length scale and c_d a typical speed of propagation for diffusive processes, and let c_p be the phase speed of internal waves (see (1.15)). Then

$$\frac{c_\mathrm{d}}{c_\mathrm{p}} := \frac{k_\mathsf{S}/L_\mathrm{d}}{c_\mathrm{p}} \cong \frac{k_\mathsf{S}}{L_\mathrm{d} c_\mathrm{p}}$$

measures, in dimensional form, the significance of the diffusive propagation to the wave propagation. With $k_\mathsf{S} \sim 10^{-4}$ m^2 s^{-1}, $L_\mathrm{d} \simeq 100$–1000 m, and $c_\mathrm{p} \sim 1$ m s^{-1}, this yields $c_\mathrm{d}/c_\mathrm{p} \simeq 10^{-6}$, implying that diffusive processes are not significant.

If, moreover, the sources Q_S and Q_T are absent, then (1.16) and (1.19) imply $d\mathsf{T}/dt = 0$ as well as $d\mathsf{S}/dt = 0$. Thus, from the thermal equation of state $\rho = \rho(\mathsf{T}, \mathsf{S})$ – the pressure does not influence this in an incompressible medium – we have

$$\frac{d\rho}{dt} = \frac{\partial\rho}{\partial\mathsf{T}}\underbrace{\frac{d\mathsf{T}}{dt}}_{0} + \frac{\partial\rho}{\partial\mathsf{S}}\underbrace{\frac{d\mathsf{S}}{dt}}_{0}.$$

For an incompressible fluid, this means that, in the absence of diffusive processes and when sources are negligibly small, the density of the fluid must be constant along particle trajectories,

$$d\rho/dt = 0, \tag{1.20}$$

and along with (1.9) this equation provides the following simple form of the continuity equation:

$$\nabla \cdot \mathbf{v} = 0. \tag{1.21}$$

Nonadiabatic processes

The continuity equation (1.21) is not valid without further scrutiny in the more general case when the wave motion is considered for a longer period of time during which diffusion processes can impose a sizeable influence. This means that the applicability of (1.21) ought to be scrutinized in each concrete situation. This can be done, for instance, by analyzing the orders of magnitude of both terms of (1.8) or (1.9). For oceanic internal gravity waves, the range of values which the controlling parameters take can be indicated, and so an estimation of the validity of (1.21) can be provided.

In order to write the mass balance equation (1.9) in nondimensional form, we introduce scales for all variables: $\mathbb{L}$ for the horizontal scale which is a measure for a wavelength λ, the total ocean depth $\mathbb{H}$ for the vertical coordinate, $\mathbb{T}$ for the time which corresponds to λ/c_p, where c_p is the phase velocity of internal waves; $\mathbb{U}$ and $\mathbb{W}$ for the horizontal and vertical velocities, respectively, and $\mathbb{R}$ for the density. It is convenient to write the water density ρ as a sum of the background density $\rho_0(z)$ and a perturbation $\tilde{\rho}(x, y, z, t)$. In dimensionless form this relation reads $\rho = \mathbb{R}(\hat{\rho}_0 + \delta\hat{\rho})$. Here δ is a small parameter ($\delta \ll 1$), and $\hat{\rho}_0$ and $\hat{\rho}$ are dimensionless quantities of order unity. In reality, the water density in the ocean changes from the free surface to the bottom by no more than 3–4%. So the inequality $\delta < 10^{-3}$ is a good estimate for characterizing the density variation for most oceanic internal wave processes.

Thus, the dimensionless counterpart of the mass balance equation (1.9) takes the form

$$(\delta/\mathbb{T})\hat{\rho}_{\hat{t}} + (\delta\mathbb{U}/\mathbb{L})(\hat{u}\hat{\rho}_{\hat{x}} + \hat{v}\hat{\rho}_{\hat{y}}) + (\mathbb{W}/\mathbb{H})(\hat{w}\hat{\rho}_{0\hat{z}} + \underline{\delta\hat{w}\hat{\rho}_{\hat{z}}})$$
$$+ (\mathbb{U}/\mathbb{L})(\hat{\rho}_0 + \underline{\delta\hat{\rho}})(\hat{u}_{\hat{x}} + \hat{v}_{\hat{y}}) + (\mathbb{W}/\mathbb{H})(\hat{\rho}_0 + \underline{\delta\hat{\rho}})\hat{w}_{\hat{z}} = 0.$$

It is stipulated that all variables carrying a caret, $\hat{\bullet}$, are of order unity. Then, since δ is small, the underlined terms in the factors with parentheses can be neglected in comparison with the first terms in these parentheses, so that

$$(\delta/\mathbb{T})\hat{\rho}_{\hat{t}} + (\delta\mathbb{U}/\mathbb{L})(\hat{u}\hat{\rho}_{\hat{x}} + \hat{v}\hat{\rho}_{\hat{y}}) + (\mathbb{W}/\mathbb{H})\hat{w}\hat{\rho}_{0\hat{z}}$$
$$+ (\mathbb{U}/\mathbb{L})\hat{\rho}_0(\hat{u}_{\hat{x}} + \hat{v}_{\hat{y}}) + (\mathbb{W}/\mathbb{H})\hat{\rho}_0\hat{w}_{\hat{z}} + O(\delta) = 0.$$

Henceforth we shall omit the error term $O(\delta)$. The second term in this equation can also be omitted, because it contains products of small quantities, whereas all others are linear in such terms. Moreover, it is negligible in comparison with the fourth term because it contains the small quantity δ. Then the simplified mass balance equation takes the form

$$(\delta\mathbb{L}/\mathbb{T}\mathbb{U})\hat{\rho}_{\hat{t}} + \hat{\rho}_0[\hat{u}_{\hat{x}} + \hat{v}_{\hat{y}} + (\mathbb{W}\mathbb{L}/\mathbb{U}\mathbb{H})\hat{w}_{\hat{z}}] + (\mathbb{W}\mathbb{L}/\mathbb{U}\mathbb{H})\hat{w}\hat{\rho}_{0\hat{z}} = 0. \quad (1.22)$$

The first term in (1.22) can be neglected, because the phase speed of internal waves, $\mathbb{L}/\mathbb{T}$, is approximately 1 m s^{-1}, and typical particle velocities, $\mathbb{U} \sim 0.1$–1 m s^{-1}, are of similar orders such that $\mathbb{L}/\mathbb{T}\mathbb{U} \leq 1$; so the Strouhal number, $(\delta\mathbb{L}/\mathbb{T}\mathbb{U}) \sim 10^{-2} \div 10^{-3}$, is small, and the corresponding term involving the time derivative $\partial\hat{\rho}/\partial\hat{t}$ in (1.22) may be ignored. Thus, (1.22) can be written approximately as

$$\hat{u}_{\hat{x}} + \hat{v}_{\hat{y}} + (\mathbb{W}\mathbb{L}/\mathbb{U}\mathbb{H})\hat{w}_{\hat{z}} + (\mathbb{W}\mathbb{L}/\mathbb{U}\mathbb{H})\hat{w}\hat{\rho}_{0\hat{z}}/\hat{\rho}_0 = 0. \quad (1.23)$$

Comparison of the last two terms indicates that they differ in order of magnitude by the factor $\hat{\rho}_{0\hat{z}}/\hat{\rho}_0$. Measurements of the density from profiles of the temperature, electrical conductivity, and pressure indicate that the maximum $\widehat{\rho_{0z}}/\hat{\rho}_0 \sim 10^{-3}$ is observed in the pycnocline layer. Thus, with an accuracy of at least $\sim 10^{-3}$ for oceanic internal gravity waves, the last term in (1.23) can be ignored, and the continuity equation may be used in the form given in (1.21).

Equations for wave disturbances

The preceding remarks concern the simplifications implied by the specific characteristics of the density field in a real ocean and they affect the mass balance equation. This idea can be exploited further for the simplification of the momentum balance equation. In order to do so, we decompose the pressure and density fields into two parts: the basic states $P_0(x, y, z)$ and $\rho_0(x, y, z)$, which are the values in a state of hydrostatic equilibrium, and wave disturbances $\tilde{P}(x, y, z, t)$ and $\tilde{\rho}(x, y, z, t)$ such that

$$\left.\begin{aligned} \rho &= \rho_0(x, y, z) + \tilde{\rho}(x, y, z, t), \\ P &= P_\text{a} + g \int_z^0 \rho_0(x, y, z)\, dz + \tilde{P}(x, y, z, t), \end{aligned}\right\} \tag{1.24}$$

where P_a is the atmospheric pressure. In the common case, the stationary density ρ_0 depends on three spatial variables, x, y, and z. Vertical gradients ρ_{0z} exist owing to the gravity force, and horizontal stationary density gradients ρ_{0x} and ρ_{0y} are maintained by geostrophic currents (more details on this are given in Chapter 3).

Generally speaking, the hydrostatic pressure in (1.24) must be calculated not from the undisturbed level of the sea surface $z = 0$ but from its displaced position expressed by the function $\xi(x, y, t)$. However, if we are interested only in baroclinic wave motions, the "rigid lid" condition[12] at the sea surface is an acceptable approximation. More about this can be found in Section 1.3. What we wish to emphasize here is that the "rigid lid" condition is not used throughout this book. Chapter 3, which is devoted to the dynamics of waves near stationary density fronts, does not use this assumption.

With these remarks in mind, equation (1.6) for the perturbations $\tilde{P}(x, y, z, t)$ and $\tilde{\rho}_0(x, y, z, t)$ takes the form

$$(\rho_0 + \tilde{\rho}\,)\frac{d\mathbf{v}}{dt} = -\nabla\tilde{P} + \tilde{\rho} g\mathbf{k} - 2(\rho_0 + \tilde{\rho}\,)\boldsymbol{\Omega} \times \mathbf{v} + \mathbf{F}(\mathbf{v}), \tag{1.25}$$

or

$$\left(1 + \frac{\tilde{\rho}}{\rho_0}\right)\frac{d\mathbf{v}}{dt} = -\frac{1}{\rho_0}\nabla\tilde{P} + \frac{1}{\rho_0}\tilde{\rho} g\mathbf{k} - 2\left(1 + \frac{\tilde{\rho}}{\rho_0}\right)\boldsymbol{\Omega} \times \mathbf{v} + \frac{1}{\rho_0}\mathbf{F}(\mathbf{v}), \tag{1.26}$$

where use has been made of the fact that, in the absence of motion, the fluid is under hydrostatic equilibrium.

Boussinesq approximation

As mentioned above, the characteristic scales of the density deviations in the ocean are of the order $\tilde{\rho}/\rho_0 \sim 10^{-2}$. So, with an accuracy of 10^{-2}, the term $\tilde{\rho}/\rho_0$ in (1.26) can be neglected in comparison with unity. In this case, the momentum balance equation (1.25) (or (1.26)) takes the form

$$\rho_0(x, y, z)\frac{d\mathbf{v}}{dt} = -\nabla\tilde{P} + \tilde{\rho} g\mathbf{k} - 2\rho_0(x, y, z)\boldsymbol{\Omega} \times \mathbf{v} + \mathbf{F}(\mathbf{v}). \tag{1.27}$$

[12] The "rigid lid" assumption supposes that the free surface does not move in the vertical direction. It is evident that it eliminates the surface waves from being analyzed.

So, in this assumption, the perturbations of the density field are neglected in comparison with the total density in the inertial terms and in the expression for the Coriolis force. However, the dependence of the gravity force on the variations of the density remains, and so buoyancy forces are accounted for. Needless to say, the mass balance equation is still given by $\nabla \cdot \mathbf{v} = 0$.

Yet another simplification, often used in studies of internal wave dynamics, is based on the fact that the total water density in the ocean varies in the range of no more than 3–4%. Such accuracy is sufficient for many tasks, and the density $\rho_0(z)$ in (1.27) is replaced by a constant averaged value $\bar{\rho}_0$. With this assumption, (1.27) simplifies further to

$$\bar{\rho}_0 \frac{d\mathbf{v}}{dt} = -\nabla \tilde{P} + \tilde{\rho} g \mathbf{k} - 2\bar{\rho}_0 \mathbf{\Omega} \times \mathbf{v} + \mathbf{F}(\mathbf{v}), \quad \bar{\rho}_0 = \text{const.} \tag{1.28}$$

In this form, the balance law of momentum is particularly useful for the construction of analytical as well as numerical solutions. To distinguish these two cases from each other, (1.27) is sometimes called the *free convection approximation* [157], whereas (1.28) is properly called the *Boussinesq approximation* [144].

Approximation of the Coriolis acceleration

The complete expression for the Coriolis acceleration $2\mathbf{\Omega} \times \mathbf{v}$ is $(-fv + \tilde{f}w, fu, -\tilde{f}u)$, where $f = 2\Omega \sin(\varphi)$ and $\tilde{f} = 2\Omega \cos(\varphi)$ are called the first and second Coriolis parameters, and φ is the latitude angle (see Figure 1.2(a)). Note that for mesoscale wave motion with spatial scales from several hundreds of meters to several dozens of kilometers under study in the present book, the Coriolis acceleration can be considered in the so-called f-plane approximation, in which the exact representation of the vector $2\mathbf{\Omega} \times \mathbf{v}$ is replaced by the vector $(-fv, fu, 0)$. Here the vertical component of the Coriolis acceleration is omitted; its weak effect can only be found in near-equator regions. We may also neglect the latitudinal dependence of the Coriolis force (β-effect) because the considered waves are much shorter than the distances over which the β-effect can exercise an essential input. So, the use of the f-plane approximation is quite reasonable and warrantable.[13] This has been done in the above without further justification. However, in a scaling analysis of the long-wave approximation, it can be rigorously proven that $|fv| \gg |\tilde{f}w|$, so that omitting all terms involving $\tilde{f}$ in the Coriolis acceleration is indeed justified.

Reynolds equations

Let us return to the momentum equation, in particular the last term in equation (1.6). This term, $\mathbf{F}(\mathbf{v})$, is responsible for the dissipation due to internal friction (we do

[13] Note that the second Coriolis parameter has dropped out in this approximation.

not consider any external gravity forces due to the variability of the positions of the Moon and the Sun that disturb the system). In the laminar case (i.e. under conditions prevailing in laboratory experiments), (1.7), which describes Newtonian fluids, gives the exact expression for the friction force; it is adequate for the description of viscous effects in water and in air.

The magnitude of the coefficient of kinematic viscosity ν of water is rather small, namely of the order 10^{-6} m^2 s^{-1}. This is why a *direct* influence of dissipation on oceanic internal waves is negligible. The spatial scales of oceanic internal waves are so large that the molecular viscosity cannot affect the force balance.

The situation becomes more complicated in the turbulent case. Because the kinematic viscosity of water is so small, there exists a continuous spectrum of turbulent motions in the ocean, which is maintained by the energy of the large-scale motions [69], [96]. The turbulent fluctuations tend to absorb energy from large scales and transmit it to the smallest scales, for which the influence of the viscosity must be taken into account. This *cascade* of energy transfer from the large to the small scale (i.e. effectively the loss of energy by large-scale processes) can roughly be taken into account by introducing the *effective force of turbulent friction*.

The governing equations, which incorporate the energy loss by large-scale motions due to turbulent fluctuations and which describe the dynamics of the background currents, can be obtained by an averaging procedure which is described in detail by many authors and belongs today to the routine literature in turbulence theory; see, for instance, the book by Pedlosky [187]. The idea, going back to Reynolds, is to regard the turbulent motion as a composition of a mean motion plus fluctuations, and the automatic supposition is to regard the mean motion as a "descriptor" of the essential physics and the fluctuations as "correctors" of the mean processes, i.e. an averaging process is performed. By such an averaging, any unknown field variable $\mathcal{F}$ is decomposed into a sum $\mathcal{F} = \langle\mathcal{F}\rangle + \mathcal{F}'$, where $\langle\mathcal{F}\rangle$ describes the average low-frequency background process and $\mathcal{F}'$ is its turbulent pulsation. The smoothing operation is performed in the Reynolds equations with the supposition that the average of any fluctuating quantity $\mathcal{F}'$ vanishes, $\langle\mathcal{F}'\rangle = 0$. In view of $\mathcal{F} = \langle\mathcal{F}\rangle + \mathcal{F}'$, this implies that $\langle\langle\mathcal{F}\rangle\rangle = \langle\mathcal{F}\rangle$, or more generally that $\langle\mathcal{F}\rangle = \langle\langle\ldots\langle\mathcal{F}\rangle\ldots\rangle\rangle$. This is a restrictive assumption known as ergodicity and requires the fluctuations to obey conditions of time independent, homogeneous statistics. This is not always the case, but for many turbulent processes it is a reasonable assumption that will be made here.

In the ensuing developments we shall not derive the Reynolds equations, but merely list them, directing the reader's attention to the specialized literature [213]. So, after averaging equations (1.10), (1.16), (1.19), (1.21), and (1.27), the following equations emerge:

$$\left.\begin{aligned}
&u_t + uu_x + vu_y + wu_z - fv \\
&\quad = -\tilde{P}_x/\rho_0 + A^{\mathrm{H}}(u_{xx} + u_{yy}) + (A^{\mathrm{V}}u_z)_z, \\
&v_t + uv_x + vv_y + wv_z + fu \\
&\quad = -\tilde{P}_y/\rho_0 + A^{\mathrm{H}}(v_{xx} + v_{yy}) + (A^{\mathrm{V}}v_z)_z, \\
&w_t + uw_x + vw_y + ww_z \\
&\quad = -\tilde{P}_z/\rho_0 - g\tilde{\rho}/\rho_0 + A^{\mathrm{H}}(w_{xx} + w_{yy}) + (A^{\mathrm{V}}w_z)_z, \\
&u_x + v_y + w_z = 0, \\
&\mathsf{S}_t + u\mathsf{S}_x + v\mathsf{S}_y + w\mathsf{S}_z = K^{\mathrm{H}}(\mathsf{S}_{xx} + \mathsf{S}_{yy}) + (K^{\mathrm{V}}\mathsf{S}_z)_z, \\
&\mathsf{T}_t + u\mathsf{T}_x + v\mathsf{T}_y + w\mathsf{T}_z = K^{\mathrm{H}}(\mathsf{T}_{xx} + \mathsf{T}_{yy}) + (K^{\mathrm{V}}\mathsf{T}_z)_z, \\
&\rho = \rho(\mathsf{S}, \mathsf{T}, P),
\end{aligned}\right\} \tag{1.29}$$

in which angled brackets denoting averaged quantities have, for brevity, been systematically omitted. This system will henceforth be referred to as the *Reynolds equations.* It should be completed also by expressions (1.24) because the equation of state is usually written for the full density ρ and pressure P, whereas the first three equations here are expressed in terms of wave disturbances $\tilde{\rho}$ and $\tilde{P}$.

Equations (1.29) warrant further comment. The first three equations are the averaged momentum equations written in the horizontal and vertical directions of a Cartesian coordinate system, and the last three terms on the right-hand sides represent the turbulent closures. In these, horizontal and vertical turbulent viscosities are distinguished by turbulent viscosities in the horizontal directions, A^{H}, being assumed constant, whilst those in the vertical direction, A^{V}, may vary with depth. Parameterization of the turbulent stresses in this form is invariant under rotations of the coordinate system about the vertical axis, but not under rotations about any axis of the three-dimensional space. This is not strictly correct, but it can be shown that in the long-wave approximation (shallow water approximation) such a Reynolds stress parameterization is consistent. In equations for the salinity and temperature, the turbulent mass and heat flux vectors are equally parameterized to account for transversely anisotropic turbulence. The horizontal mass or heat diffusivity, K^{H}, are again assumed to be constant, whereas the vertical diffusivity, K^{V}, may vary with position. The fact that heat and mass diffusivities are the same is a consequence of turbulent mixing.

Structurally, the form of the Reynolds equations (1.29) is very similar to that of the Navier–Stokes equations (1.6)–(1.7). Nevertheless, the principal difference between the two sets is that the Reynolds equations describe only average motions, but do not provide any information about the detailed dynamics of the small-scale motions. The dissipative effect of the small-scale turbulence on the *averaged* currents is parameterized by introducing large values for the coefficients of turbulent mixing: horizontal turbulent viscosity A^{H} and diffusivity K^{H} and their vertical counterparts A^{V} and K^{V}. These coefficients vary over a wide range, and their exact magnitudes strongly depend on the considered scales of the motions.

The turbulent parameterization introduced above lacks detail. Nevertheless, it permits us to describe correctly (on a qualitative and quantitative level) many phenomena related to internal gravity waves. The essential problem that one encounters when using such parameterizations is the correct choice of the values of the turbulent exchange coefficients. Results will depend on this choice, and this makes the correct prediction of wave phenomena sometimes rather difficult.

In this book we will call on the experience gained by oceanographers in the choice of the turbulent exchange coefficients. For instance, it is important to realize that the vertical turbulent exchange is much weaker than the corresponding horizontal exchange. Moreover, it is known that the vertical exchange is not uniform throughout the whole depth but depends on the vertical coordinate, z. This conclusion is in agreement with the stable vertical fluid stratification of the oceanic waters. Several parameterizations of the coefficients of vertical turbulent exchange have been proposed depending on the stratification and shear stresses; we will consider these below.

A last remark, worth mentioning here, concerns the equation of state. This equation, in general, is nonlinear, but it can also be presented by the linear law (1.12). In this case, one can combine the last three equations in (1.29) and obtain instead a single equation for the perturbation of the density, supposing that, in (1.24), $\rho_0 = \rho_0(z)$:

$$\tilde{\rho}_t + u\tilde{\rho}_x + v\tilde{\rho}_y + w\tilde{\rho}_z + w\rho_{0z} = K^{\mathrm{H}}(\tilde{\rho}_{xx} + \tilde{\rho}_{yy}) + (K^{\mathrm{V}}\tilde{\rho}_z)_z + (K^{\mathrm{V}}\rho_{0z})_z. \tag{1.30}$$

1.3 Problem formulation: boundary and initial conditions

Even with the simplifications described in Section 1.2, the governing system of equations is very complicated and defies any attempt to solve it analytically. In the above presented complete form (1.29), which is valid for time dependent unsteady motions in three dimensions, it can only be solved numerically; such calculations in general demand a considerable amount of time and effort. The usual way out of this dilemma is to impose additional simplifications. These must be such that the mathematical answer to the question of how a stratified fluid reacts dynamically can nevertheless be given. For instance, in numerical models of oceanic circulation a popular simplification is the *hydrostatic pressure hypothesis*. This assumes that the pressure at any depth is exclusively produced by the weight of the water column above the point considered (the terms involving the wave disturbances in (1.24) are neglected), and, instead of the third equation in (1.29) for the vertical velocity w, the simple relation $dP/dz = -g\rho_0(x, y, z)$ is considered.

Unfortunately, this hypothesis is inappropriate in general and cannot be implemented in the study of strongly nonlinear internal waves. When the nonhydrostatic effects are accounted for, an additional dispersion arises. This is very important in internal wave problems whenever nonlinearities are accounted for (such conditions prevail, e.g., in strong internal gravity waves, which are one of the topics of our study). In some cases, the nonhydrostaticity of the pressure plays a decisive role for stationary nonlinear internal waves such as solitons or cnoidal waves; they cannot exist without the operation of the nonhydrostatic dispersion.

Another simplification, often used in the theory of oceanic internal waves, was suggested by the analyses of observational data (field measurements and sea-surface images). As mentioned in the Preamble and in Section 1.1, one of the basic sources of internal waves in the World Ocean is the barotropic tides. Holloway and Merrifield [101], making numerical investigations with Gaussian-shaped seamounts, found that changing the horizontal aspect ratio of the topography from 1:1 (a seamount) to 3:1 (a ridge) increases the resulting baroclinic energy flux by nearly the order of an amplitude. Thus, the most conspicuous baroclinic tides are generated near the continental margins and oceanic ridges, where the average changes of the bottom relief predominate in one direction: roughly perpendicular to the shore, i.e. $H_x \gg H_y$, if x is directed off shore, and y along shore. As a consequence, the fronts of the internal tidal waves are elongated and are oriented basically parallel to the bottom topography. Therefore, in situations when the three-dimensional effects are weak, the wave fields can be considered to be two-dimensional. Examples are baroclinic tides in areas of continental slopes and oceanic ridges, intense oceanic solitary internal waves, and wave trains in the shelf areas, etc. Thus, we may assume and exploit conditions of invariability of all functions along one direction (for instance, along the Oy-axis), and we may impose the condition $\partial/\partial y \equiv 0$ for all physical variables.

Of course, this does not mean that the three-dimensional wave motion always reduces to two-dimensionality for internal gravity waves and that the third dimension can always be neglected. In many cases, for instance in refraction and diffraction problems, three-dimensional effects play a key role. Such phenomena are not the topic of the present book.

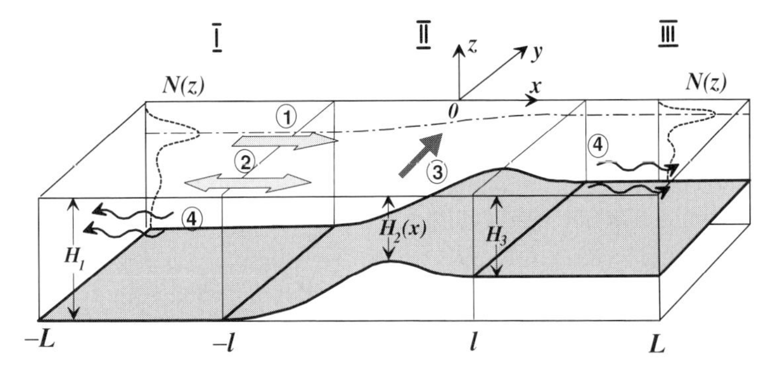

Figure 1.3. Rectangular region appropriate for the analysis of internal waves in this book. Regions I (the deep ocean) and III (the shelf) have constant depths H_1 and H_3, respectively; region II is characterized by topographic variations in only one direction normal to the shelf. Two profiles of the buoyancy frequency are shown in regions I and III. Arrows indicate: (1) the incident internal wave; (2) the oscillating barotropic tidal flux; (3) the geostrophic current maintaining the horizontal gradient of the density; (4) radiated internal waves.

Problem formulation

The physical processes and the physical model which are considered in this book are shown schematically in Figure 1.3, which reproduces a rectangular region of a basin of variable depth, which, conventionally, can be divided into three zones. In zone I ($x < -l$, $-\infty < y < \infty$) and zone III ($x > l$, $-\infty < y < \infty$) the basin depth is constant and equals H_1 and H_3, respectively; in between lies the transition zone II ($|x| \leq l$, $-\infty < y < \infty$). The water depth H_2 in this transition zone is assumed to depend only on the coordinate x: $H_2 = H_2(x)$. (The right-hand Cartesian coordinate system lies within the undisturbed free surface with the Ox-axis pointing towards the shore, also shown in Figure 1.3.) Such simple geometry can be a model for oceanic ridges or near coastal zones which include the continental slope and the shelf.

The basin is filled with a continuously and stably stratified fluid. The vertical stratification is supposed to match oceanic conditions, having a seasonal pycnocline, a weakly stratified near bottom layer, and a mixed surface layer. Two profiles of the buoyancy frequency, $N(z)$, are presented schematically in the zones of constant depth, I and III. In general they can be different, and consequently the stratification in zones I and III will then also differ. Such a situation is typical for oceanic near coastal zones; it reflects the various conditions of the formation of the water masses in the shallow and abyssal parts of the ocean. If it is stable, i.e. steady, the existing frontal zone, with horizontal cross-topography density gradients presented in Figure 1.3 by the dashed–dotted line, must be maintained by the geostrophic current along the shelf topography. The influence of such horizontal density gradients on internal wave dynamics will be studied in Chapter 3.

Two different physical processes will be considered:

- the generation of (linear and nonlinear) internal waves over variable bottom topography (in zone II) by the barotropic tidal flux and transformation of these waves in zones I and III during their propagation from the source of generation;
- the scattering of internal waves (linear periodic and nonlinear solitary) by underwater obstacles and frontal zones.

Interaction of the barotropic tides or incident internal waves with the bottom topography generates secondary internal waves that radiate outwards from the source of their generation (see Figure 1.3). The overall objective of the present work consists in defining and describing the generated internal wave fields.

Boundary and initial conditions

There are two basic approaches to the formulation of boundary conditions at the upper liquid boundary. In the more common case for an ideal fluid, we have to satisfy the following dynamic and kinematic boundary conditions at the oceanic surface $z = \xi(x, t)$:

$$P = P_{\mathrm{a}}, \tag{1.31}$$

$$\frac{\partial \xi}{\partial t} + u\frac{\partial \xi}{\partial x} = w. \tag{1.32}$$

As our focus is on internal gravity waves, and as we are only interested in the baroclinic response of the ocean, we will apply the so-called "rigid lid" approximation. The formal justification for its applicability lies in the fact that the amplitudes of the surface elevations due to baroclinic internal waves are very small (a few centimeters), generally a factor 10^2 to 10^3 times smaller than the amplitudes of the largest internal signals. Note that modern satellite-borne altimeters, such as TOPEX/Poseidon, can detect this weak signal from internal waves at the free surface [55], [205]. This has given rise to a series of works on the description of the global field of baroclinic tides [116], [152], [177]. However, the maximum values of baroclinic tidal disturbances are reached at the position of the density jump (layered model) or in the pycnocline. It is known that the "rigid lid" assumption eliminates all surface (barotropic) gravity waves. These are not our concern here; moreover, barotropic signals travel much faster than their baroclinic counterparts, so that wave packets quickly separate. This justifies the elimination of the surface signals that accompany the internal waves.

We will return again to the discussion of the applicability of the "rigid lid" approximation in Section 3.3, where a comparative analysis of solutions will be performed with and without this assumption.

In the framework of the "rigid lid" approximation, the dynamic, (1.31), and kinematic, (1.32), boundary conditions at the undisturbed sea surface $z = 0$ are transformed into

$$w = \tilde{P} = \tilde{\rho} = 0, \quad z = 0. \tag{1.33}$$

In other words, these conditions state that the free surface cannot move vertically (corresponding to the "rigid lid" assumption) and that associated with it there are no pressure and no density perturbations.

Additional ("flux") boundary conditions must be added to (1.33), namely

$$u_z = v_z = 0, \quad z = 0, \tag{1.34}$$

which imply the absence of shear tractions at the sea surface, so circulation processes due to wind forcing are not of concern; furthermore, we also require that

$$\mathsf{T}_z = \mathsf{S}_z = 0, \quad z = 0, \tag{1.35}$$

which expresses the fact that salt and heat fluxes through the surface vanish (no external atmospheric impact on the ocean is considered).

At the bottom, $z = -H(x)$, for an ideal fluid the water velocity must be tangential to the bed (slip condition), implying

$$w = -uH_x, \quad v_n = 0, \quad z = -H(x), \tag{1.36}$$

whilst, for a viscous fluid, the no-slip condition,

$$u = v = w = 0, \quad z = -H(x), \tag{1.37}$$

applies. For temperature, salinity, and density, the natural boundary conditions at the bottom are the postulation of vanishing normal fluxes,

$$\mathsf{T}_n = \mathsf{S}_n = \rho_n = 0 \quad \text{at} \quad z = -H(x), \tag{1.38}$$

where the index n identifies the direction perpendicular to the bottom surface.

At the lateral "liquid" or open boundaries of the computational region, i.e. at $x = \pm L$ (see Figure 1.3), depending on the considered problem, two types of boundary conditions will be imposed (either initial value problems or periodic motions are analyzed).

- Radiation of the internal waves when linear analytic or semianalytic periodic problems are analyzed. The phase speed of the secondary generated (reflected and diffracted) waves in zones I and III must be directed from the source of generation away to infinity (Sommerfeld radiation condition).

- In numerical problems, the "zero" wave and "zero" initial perturbation conditions are applied at $x = \pm L$. These conditions are valid at times prior to the time slice when the wave disturbances, propagating from the transition zone II, reach the outer lateral boundaries of the computational domain.

The final problem regarding the model initialization is to choose the initial fields. In the linear models with the imposed periodicity, initial conditions are not required. In the nonlinear case, "zero" initial fields are used for all unknown functions except for the along-bottom horizontal velocity v. In problems of internal wave generation by the barotropic tidal flux, the initial transverse velocity field v is defined by prescribing the clockwise tidal ellipses aligned in the cross-topography direction with ellipticity factor f/σ, where $\sigma = 2\pi/T$ is the tidal frequency and T is the tidal period. The boundary and initial conditions will be justified in detail in Section 1.6 and in the formulations of individual problems.

1.4 Linear wave equation

A formal solution of the above described problem of internal wave dynamics in regions of bottom nonuniformities can be written in terms of a power series expansion for the dependent variables, where the expansion parameter may, for example, be the dimensionless amplitude. The leading order term of this series describes the linear solution; it usually dominates the higher order contributions when the expansion parameter is small. The governing linearized equations can be obtained by substituting the perturbation series into the original nonlinear system and subsequently omitting all higher order (product) terms.

One can follow also the heuristic approach and assume that for infinitesimal waves the effects related to turbulent mixing are negligibly small and may thus be omitted (the processes can be considered in the adiabatic approximation). Then, instead of the Reynolds equations, we may use the Euler equations, i.e. (1.27) with $\mathbf{F}(\mathbf{v}) = \mathbf{0}$. If the fluid is not subjected to any external forcings, and if the background density ρ_0 depends only on the vertical coordinate z, then for two-dimensional internal waves for which $\partial/\partial y = 0$, and when all mass, momentum, and energy flux terms as well as the radiation are ignored so that the energy reduces to $(d\rho/dt = 0)$, the governing equations read

$$\left.\begin{aligned} u_t + \boxed{uu_x + wu_z} - fv &= -\tilde{P}_x/\rho_0, \\ v_t + \boxed{uv_x + wv_z} + fu &= 0, \\ w_t + \boxed{uw_x + ww_z} &= -\tilde{P}_z/\rho_0 - g\tilde{\rho}/\rho_0, \\ \tilde{\rho}_t + \boxed{u\tilde{\rho}_x + w\tilde{\rho}_z} + w\rho_{0z} &= 0, \\ u_x + w_z &= 0. \end{aligned}\right\} \tag{1.39}$$

The continuity equation can be satisfied identically if the stream function $\psi(x, z, t)$ is introduced such that

$$u = \psi_z, \quad w = -\psi_x. \tag{1.40}$$

Our interest lies in wave motions with small amplitudes; (1.39) can then be simplified by linearizing about a stable ground state. For infinitesimal waves the nonlinear convective acceleration terms, which are marked out by the boxes, become negligible in comparison with the local derivatives. Introducing a time scale $\mathbb{T}$ (the wave period), velocity scale $\mathbb{U}$, and a horizontal scale $\mathbb{L}$ (wavelength λ) of the wave motion, the range of validity of such a linearization can be estimated. Writing the equation in nondimensional form by using these scales, the nonlinear convective terms can be neglected if the following inequality is valid:

$$\frac{1}{\mathbb{T}} \gg \frac{\mathbb{U}}{\mathbb{L}} \quad \text{or} \quad \frac{\mathbb{L}}{\mathbb{T}} \gg \mathbb{U}.$$

Taking into account that $\mathbb{L}/\mathbb{T}$ may be identified with the phase speed of internal waves c_{p}, this last condition means that the linear theory is valid when the orbital velocity is negligible in comparison with the phase speed, i.e. $\mathbb{U} \ll c_{\mathrm{p}}$.

In the linear case (without the terms in boxes) equations (1.39) admit the harmonic periodic solutions

$$\left\{u, v, w, \psi, \tilde{P}, \tilde{\rho}\right\}(x, z, t) = \left\{\overset{*}{u}, \overset{*}{v}, \overset{*}{w}, \overset{*}{\psi}, \overset{*}{\tilde{P}}, \overset{*}{\tilde{\rho}}\right\}(x, z) \exp(\iota \sigma t). \tag{1.41}$$

Inserting this plane wave representation along with the stream function (1.40) and its harmonic representation into equations (1.39) and eliminating by cross-differentiation all variables except $\overset{*}{\psi}$, the following equation for the stream function $\overset{*}{\psi}(x, z)$ emerges:

$$\overset{*}{\psi}_{zz} - \alpha^{-2}(z)\overset{*}{\psi}_{xx} + N^2(z)g^{-1}\overset{*}{\psi}_z = 0, \tag{1.42}$$

in which

$$\alpha^2(z) = (\sigma^2 - f^2)/(N^2(z) - \sigma^2). \tag{1.43}$$

Let us analyze equation (1.42). Provided $N^2(z) > 0$, i.e. for stable fluid stratification and for $\alpha^2(z) > 0$, this equation is a wave equation with variable coefficients depending on z; it describes linear *monochromatic* waves. Via the vertical profile of the buoyancy frequency, the coefficient $\alpha(z)$ depends on the fluid stratification. In general, this dependence is strong. However, $\alpha(z)$ also has a latitudinal dependence expressed by the Coriolis parameter f, and it depends on the wave frequency through σ.

For low-frequency internal waves such as baroclinic tides, the buoyancy frequency is usually larger than the wave frequency through almost the entire depth; the exception is a thin upper mixed surface layer and perhaps a weakly stratified bottom layer.[14] So, in most cases in regions below the critical latitude,[15] where $\sigma > f$, the coefficient $\alpha^2(z)$ in (1.43) is positive, and the differential equation (1.42) is of hyperbolic type. The Riemann method can be used for its solution.

Let us transform this wave equation to canonical form. To this end, let $\alpha^2(z)$ be strictly positive and employ the substitution

$$z_1 = \int_0^z \alpha^{-1}(s)\,ds. \tag{1.44}$$

Then, (1.42) can be transformed to the following form:

$$\overset{*}{\psi}_{xx} - \overset{*}{\psi}_{z_1 z_1} + \overset{*}{\psi}_{z_1}\left(\frac{d\alpha(z)}{dz} - \alpha(z)\frac{N^2(z)}{g}\right) = 0. \tag{1.45}$$

This is no simpler than (1.42), but if we introduce the new unknown function

$$\Psi(x, z_1) = \overset{*}{\psi}(x, z_1)\alpha^{-1/2}[z(z_1)]\exp\left(\int_0^{z(z_1)} \frac{N^2(s)}{2g}\,ds\right), \tag{1.46}$$

the wave equation (1.45) assumes the form of the Klein–Gordon equation with variable coefficient, i.e.

$$\Psi_{xx} - \Psi_{z_1 z_1} + p(z)\Psi = 0, \tag{1.47}$$

where

$$p(z) = \frac{1}{4}\left\{[(\alpha(z)_z - \alpha(z)N(z)^2/g]^2 - 2\alpha(z)[\alpha(z)_z - \alpha(z)N(z)^2/g]_z\right\}. \tag{1.48}$$

Note that this variable transformation can be performed only in the frequency band $f < \sigma < N(z)$, i.e. when $\alpha^2(z) > 0$, for which the inverse transformation $z = z(z_1)$ exists. Indeed, when $\alpha = 0$ the integral in (1.44) is singular, and the integral for z_1 may no longer exist.

The wave equation (1.47) was obtained from the momentum equation (1.27) when the latter was written in the free convection approximation. With the use of the Boussinesq approximation (1.28), in which the density $\rho_0(z)$ is replaced by an

[14] If this occurs, the wave solution is restricted to a wave guide in which $\alpha^2(z) > 0$. Above and below this wave guide the wave is accompanied by "tails" which decay with increasing distance from the wave-guide boundaries.

[15] The value of the critical latitude φ_c is defined by the condition $\sigma = f$, which gives $\varphi_c = \arcsin(\sigma/2\Omega)$. For $\varphi > \varphi_c$, (1.42) changes its type from hyperbolic to elliptic.

average value $\bar{\rho}_0$, equation (1.42) for $\overset{*}{\psi}(x, z)$ has the form

$$\overset{*}{\psi}_{zz} - \alpha^{-2}(z)\overset{*}{\psi}_{xx} = 0, \tag{1.49}$$

and after the substitution of variables (1.44) it takes the form

$$\overset{*}{\psi}_{xx} - \overset{*}{\psi}_{z_1 z_1} + \overset{*}{\psi}_{z_1}\frac{d\alpha}{dz} = 0. \tag{1.50}$$

If we introduce the new function

$$\Psi(x, z_1) = \overset{*}{\psi}(x, z_1)\alpha^{-1/2}[z(z_1)], \tag{1.51}$$

it is simplified to the same form as (1.47), namely

$$\Psi_{xx} - \Psi_{z_1 z_1} + p(z)\Psi = 0, \tag{1.52}$$

but here $p(z)$ has the form

$$p(z) = \tfrac{1}{4}[(\alpha(z)_z)^2 - 2\alpha(z)\alpha(z)_{zz}], \tag{1.53}$$

and in contrast with (1.48) the function $p(z)$ depends only on $\alpha(z)$ and its first and second derivatives.

Note that the term in $N^2(z)/g$ in (1.42), (1.45), and (1.48) appears only in the free convection but not in the Boussinesq approximation (compare with (1.49), (1.50), and (1.52)). Usually, the term involving $N^2(z)/g$ is substantially smaller than the first two terms in (1.42). Nevertheless, in certain situations, this term can be very important, and neglecting it may lead to a loss of solutions [144].

For monotonic stratification with exponential dependence of the density,

$$\rho_0(z) = \rho_0(0)\exp\left(-\frac{N_0^2}{g}z\right), \tag{1.54}$$

where $\rho_0(0)$ is the density at the free surface, the buoyancy frequency is constant, $N(z) = N_0$. Equation (1.53) is simplified in this case to $p(z) \equiv 0$, and the wave equation (1.52) takes the form

$$\Psi_{xx} - \Psi_{z_1 z_1} = 0. \tag{1.55}$$

There are also some other profiles of $N(z)$ for which the function $p(z)$ is simplified. They will be considered below.

Equation (1.55) is the standard wave equation in canonical form that possesses d'Alembert's solutions. Thus, any differentiable function F or G of the arguments $x + z_1$ or $x - z_1$ is a solution of (1.55), so that, owing to linearity,

$$\Psi = F(x + z_1) + G(x - z_1) \tag{1.56}$$

is also a solution. Ψ represents a pair of propagating Riemann waves. The lines $x + z_1$ and $x - z_1$ are the characteristic lines of equation (1.55). For the stratification (1.54) α is constant, and so, according to (1.44), $z = \alpha z_1$, implying that the characteristics take the form

$$\mathscr{X} = x + \alpha^{-1} z \quad \text{and} \quad \mathscr{Z} = x - \alpha^{-1} z.$$

As already mentioned in the Preamble, the behavior of these lines in space and their correlation with the bottom profile are two of the more important factors determining the dynamics of internal gravity waves over variable bottom topography. We will return to the analysis of this result in Chapters 2 and 3.

1.5 Linear boundary value problem and dispersion relation

1.5.1 Formulation of the boundary value problem

The functions $F(x + z_1)$ and $G(x - z_1)$ are two independent solutions of the wave equation (1.55) for an infinite medium, and they represent two obliquely propagating internal waves. However, the ocean is not an infinite system. It is bounded by the free and bottom surfaces, and these two boundaries introduce specific fundamental features in the structure of the internal waves. Because of the reflection of the obliquely propagating waves $F(x + z_1)$ and $G(x - z_1)$ at the upper and lower boundaries, and the successive superposition of the incident and reflected waves, standing vertical internal waves are formed, which, as entities, propagate horizontally. This is described mathematically in the context of a standard boundary value problem formulated for the wave equation.

Boundary conditions for an ideal fluid are presented by relations (1.33) and (1.36). Let us formulate the kinematic boundary conditions in terms of the stream function. At the free surface $z = 0$, according to the "rigid lid" condition (1.33) $w = 0$, and, according to definition (1.40), we have

$$\psi = C_1(t) \quad \text{at} \quad z = 0, \tag{1.57}$$

where, formally, $C_1(t)$ is some function of time.

To derive the second boundary condition, at the bottom, consider the differential

$$d\psi(x, z, t) = \psi_x \, dx + \psi_z \, dz = \left(\psi_x + \psi_z \frac{dz}{dx} \right) dx.$$

At the bottom, $z = -H(x)$, so, according to (1.36) and (1.40), the term in parenthesis vanishes; consequently,

$$\psi = C_2(t) \quad \text{at} \quad z = -H(x), \tag{1.58}$$

where $C_2(t)$ is a further function of time.

The physical meaning of the stream function emerges if the volume flow is evaluated through a vertical section between two horizons. In particular, the total water flux through a vertical cross-section of the ocean from the free surface to the bottom is

$$\int_{-H(x)}^{0} u\, dz = \int_{-H(x)}^{0} \frac{\partial \psi}{\partial z}\, dz = \psi|_{z=0} - \psi|_{z=-H(x)} = C_1(t) - C_2(t).$$

It is clear that a physical meaning can be assigned, not to the algebraic values of the functions $C_1(t)$ and $C_2(t)$, but only to their difference. Moreover, in the absence of sources and sinks of mass this value $C_1(t) - C_2(t)$ (the total horizontal water flux) must be the same for any vertical section. Thus, as the stream function is defined to within an arbitrary additive function of time, we may put $C_1(t) = 0$ at the free surface, and the boundary condition (1.57) may be rewritten as

$$\psi = 0 \quad \text{at} \quad z = 0. \tag{1.59}$$

The second function, $C_2(t)$, must now in general be assumed to be nonzero with a value that ought to be chosen by physical reasoning. As free propagating linear internal waves do not produce any resulting water transport, we may choose for them also the homogeneous boundary condition at the bottom,

$$\psi = 0 \quad \text{at} \quad z = -H(x). \tag{1.60}$$

Together with (1.59), this simply means that the mass transport through any vertical cross-section vanishes. If, however, we consider baroclinic motions together with the barotropic tidal waves (internal wave generation by the tides) we must in this case define $C_2(t)$ as a function with amplitude Ψ_0 and periodic dependence on time, as in (1.41), and we must write

$$\psi = \Psi_0 \exp(\imath\sigma t) \quad \text{at} \quad z = -H(x). \tag{1.61}$$

The simplest solution of equation (1.42) which satisfies the formulated boundary conditions is the vertically uniform barotropic flux. For the stream function $\psi(x, z, t)$, this solution reads

$$\psi(x, z, t) = -z(\Psi_0/H)\exp(\imath\sigma t), \tag{1.62}$$

and the horizontal velocity $u(x, z, t)$ may be readily found. The solution (1.62) is a good approximation for the barotropic tidal flux. It is used in many baroclinic tidal models as an external forcing. In fact, use of the periodic barotropic flux (corresponding to the "nondivergent tidal wave" model) instead of the real long barotropic tidal wave does not introduce any essential errors [9], [11] because the characteristic horizontal scale of the barotropic tidal wave is of the order of a few thousand kilometers. This is much more the characteristic width of oceanic underwater obstacles such as continental slopes or oceanic ridges, which extend to several tens of kilometers.

Under the action of the Coriolis force, the barotropic tidal flux consists of two components which form the tidal ellipse. With the solution (1.62), the component $v(x, z, t)$ along the y-axis is readily found from the linear approximation of (1.39) in the following form:

$$v(x, z, t) = \imath(f/\sigma)(\Psi_0/H)\exp(\imath\sigma t). \tag{1.63}$$

Both functions, (1.62) and (1.63), will be used below for the definition of the boundary and initial conditions.

Thus, summarizing all the above considerations, the dynamics of plane internal waves in a rotating ocean of variable depth in the Boussinesq approximation is described by the boundary value problem (BVP)

$$\left.\begin{aligned} \Psi_{xx} - \Psi_{z_1z_1} + p(z)\Psi = 0, \\ \Psi = 0 \quad \text{at} \quad z_1 = 0, \\ \Psi = \alpha^{-1/2}(-H(x))\Psi_0 \quad \text{at} \quad z_1 = h(x). \end{aligned}\right\} \tag{1.64}$$

Here, the function $p(z)$ is defined in (1.53), and $h(x)$ is the "effective" water depth defined by (1.44): $h(x) = z_1(-H(x))$. The above BVP is written for any stable vertical fluid stratification, and is valid for any bottom profile $z = -H(x)$ that conforms with the linearization assumption.

Note that the differential equation is homogeneous and that the external driving force is introduced into the problem via the boundary conditions. A specific feature is that the bottom boundary condition is defined not at the flat bottom but at the curved surface $z = -H(x)$. Furthermore, the model domain, as mentioned above, consists of three regions. In two of them (regions I and III) the depth is uniform; see Figure 1.3. So, it is reasonable in the first instance to study the properties of the solution outside the bottom obstacle II, in the regions of constant depth, I and III. The depths H_1 and H_3 in these areas are, for simplicity, denoted by $H_i = \text{const.}$, where $i = 1, 3$.

1.5.2 Linear vertical mode analysis

Equations (1.64) with homogeneous boundary conditions ($\Psi_0 = 0$) describe "pure" baroclinic wave motions. For constant basin depth, the BVP reduces to

$$\left.\begin{aligned} \Psi_{xx} - \Psi_{z_1z_1} + p(z)\Psi = 0, \\ \Psi = 0 \quad \text{at} \quad z_1 = 0 \quad \text{and} \quad z_1 = h_i, \end{aligned}\right\} \tag{1.65}$$

where $h_i = z_1(-H_i)$, $i = 1, 3$. One can try to find a periodic solution of (1.65) in the form

$$\Psi(x, z_1) = q(z_1)\exp(\imath kx), \tag{1.66}$$

and then obtain after substitution of (1.66) into (1.65) the eigenvalue problem

$$\left.\begin{aligned} q_{z_1 z_1} + (k^2 - p(z))q = 0, \\ q = 0 \quad \text{at} \quad z_1 = 0 \quad \text{and} \quad z_1 = h_i. \end{aligned}\right\} \tag{1.67}$$

This looks very simple, although in the common case of arbitrary fluid stratification it can only be solved numerically. However, for several specific profiles $N(z)$, there exist also analytical solutions. This is the case for instance when the coefficient $p(z)$ in (1.67) reduces to a constant. Better known and simpler is the case of monotonic stratification with exponential density variation (1.54), $N(z) = N_0 =$ const., $p(z) = 0$, and $z = \alpha z_1$. Then, in physical variables, the eigenvalue problem (1.67) takes the form

$$q_{zz} + k^2 \frac{N_0^2 - \sigma^2}{\sigma^2 - f^2} q = 0,$$
$$q = 0 \quad \text{at} \quad z = 0 \quad \text{and} \quad z = -H_i,$$

and admits the counting number of periodic solution $q_j(z) = A_j \sin(k_j z) + B_j \cos(k_j z)$, $j = 1, 2, 3, \ldots$ Satisfying the boundary conditions leads to $B_j = 0$ and the following discrete set of eigenvalues and eigenfunctions,

$$k_j = \left(\frac{j\pi}{H_i}\right)\left(\frac{\sigma^2 - f^2}{N_0^2 - \sigma^2}\right)^{1/2}, \tag{1.68a}$$

$$q_j(z) = \sin\left(\frac{j\pi z}{H_i}\right), \tag{1.68b}$$

where j is the baroclinic mode number. Equation (1.68a) is the *dispersion relation*, $k_j = k_j(\sigma)$, which can be rewritten in the form $\sigma = \sigma(k_j)$ as follows:

$$\sigma = \left[\frac{(k_j N_0)^2 + (j\pi f/H_i)^2}{k_j^2 + (j\pi/H_i)^2}\right]^{1/2}, \quad i = 1, 3; \quad j = 1, 2, \ldots \tag{1.69}$$

Equation (1.68b) lists the mode functions. They are trigonometric functions and arise in a countably infinite number $j = 1, 2, 3, \ldots$

The vertical structure of the first three baroclinic modes $q_j(z)$ for the monotonic stratification (1.54) are presented in Figure 1.4(a). The dispersion curves (1.69), $\sigma = \sigma(k)$, displayed in Figure 1.4(b), clearly show that the frequencies of the free propagating periodic internal gravity waves are located in the frequency band between the Coriolis parameter f and the maximum of the buoyancy frequency N_0.

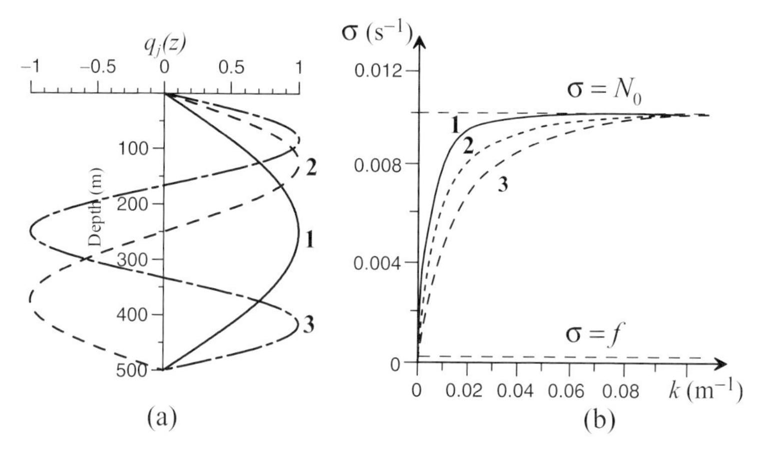

Figure 1.4. (a) First three baroclinic eigenmodes for constant buoyancy frequency N_0 for a 500 m deep ocean. (b) Corresponding dispersion curves plotted against wave number.

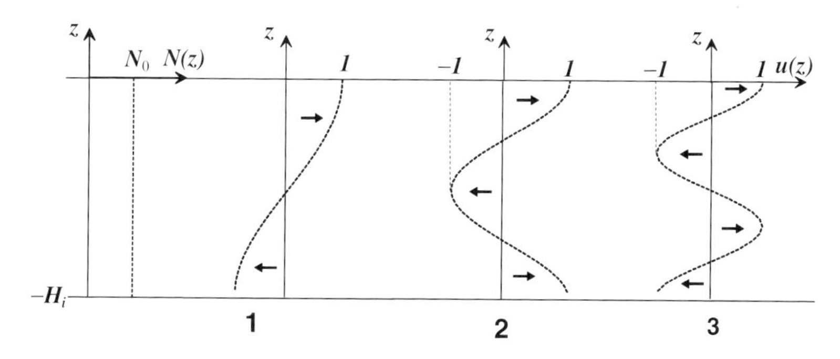

Figure 1.5. Constant buoyancy frequency N_0 and profiles of horizontal velocity distribution for the corresponding first three baroclinic modes. Velocity direction in each mode alternates from one sublayer to the next, and the total mass flux for each mode integrates to zero.

Physically more appealing than the eigenmodes for ψ are those for the horizontal velocity component $u = \psi_z$. For the monotonic stratification with exponential density profile ($p(z) = 0$), this yields $u(z) \propto \cos(j\pi z/H_i)$, where j is again the mode number. Figure 1.5 shows the first three eigenmodes together with the buoyancy frequency distribution for the horizontal velocity component. Each mode evidently represents a virtual layering into two, three, four, ... layers with alternating velocity directions. The number of zeros in the horizontal velocity profiles is, for each mode, equal to the mode number, and the number of layers with alternating velocity directions is equal to $(j + 1)$. These results are obtained for constant N and trigonometric mode functions, but the stated properties can be proven for the linear case for any stable stratification.

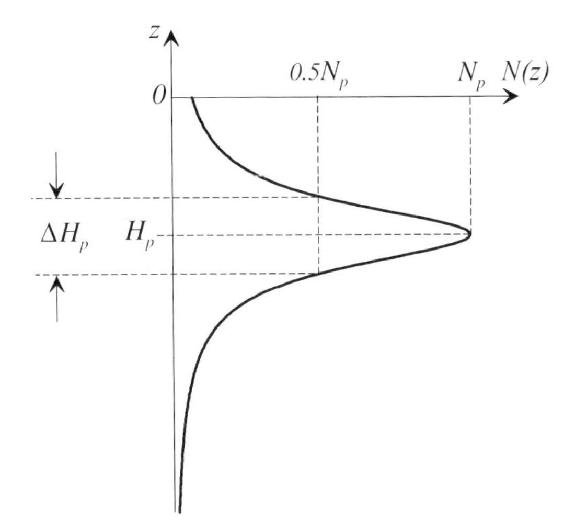

Figure 1.6. Depth distribution of the buoyancy frequency for the profile (1.70). The relations between the parameters c_1, c_2, and c_3, and H_p, ΔH_p, and N_p are given by (1.71).

From an oceanographic point of view, a constant buoyancy frequency is untypical. Normally, stratification in the ocean $\rho_0(z)$ contains a density jump called a pycnocline, located in the upper 50–100 m layer. The following three-parameter family of curves [241] allows a fairly realistic oceanic pycnocline:

$$N^2(z) = \sigma^2 + (\sigma^2 - f^2)[c_1(z + c_2)^2 + c_3]^{-2}, \tag{1.70}$$

in which c_1, c_2, and c_3 are arbitrary constants. Expression (1.70) with $c_1 c_3 > 0$ simulates smooth pycnoclines. The constants c_1, c_2, and c_3 are uniquely expressible by more obvious and more convenient parameters, such as the pycnocline depth, H_p, the maximum value of the buoyancy frequency at that depth, N_p, and the width, ΔH_p, over which the values of the buoyancy frequency are larger than $N_\mathrm{p}/2$; see Figure 1.6. If the values of N_p, H_p, and ΔH_p are known, the dependence $N(z)$ according to (1.70) is determined unambiguously. Using these parameters, the coefficients c_1, c_2, and c_3 are expressible as follows:

$$\left.\begin{aligned} c_1 &= \frac{4(\sigma^2 - f^2)^{1/2}}{\Delta H_\mathrm{p}^2}\left[\frac{1}{(N_\mathrm{p}^2/4 - \sigma^2)^{1/2}} - \frac{1}{(N_\mathrm{p}^2 - \sigma^2)^{1/2}}\right], \\ c_2 &= H_\mathrm{p}, \\ c_3 &= \left(\frac{\sigma^2 - f^2}{N_\mathrm{p}^2 - \sigma^2}\right)^{1/2}. \end{aligned}\right\} \tag{1.71}$$

Several profiles of the buoyancy frequency are shown in Figure 1.7(b). It is readily seen that with the dependence (1.71) it is possible to

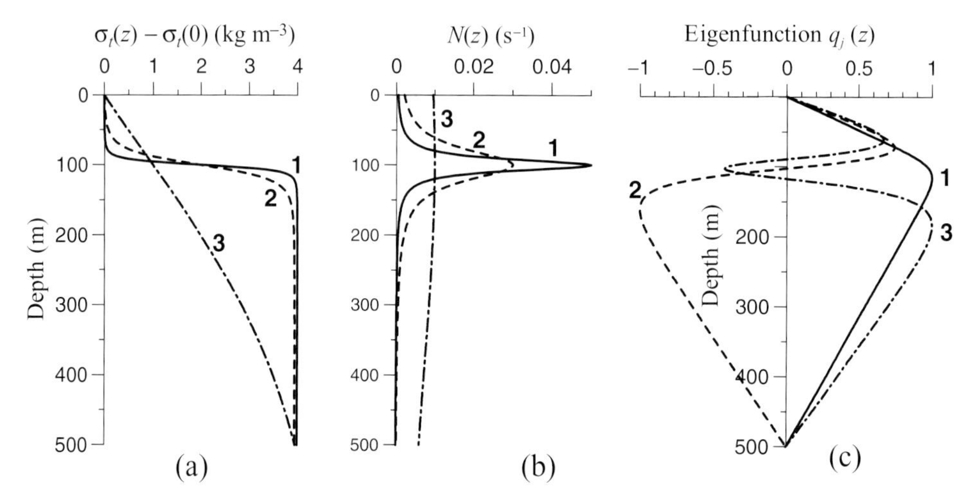

Figure 1.7. (a) Density anomaly and (b) buoyancy frequency profiles for formula (1.70) when $H_\mathrm{p} = 100\,\mathrm{m}$, $N_\mathrm{p} = (1; 3; 5) \times 10^{-2}\,\mathrm{s}^{-1}$, $\Delta H_\mathrm{p} = (20; 55; 10^3)\,\mathrm{m}$, $f = 0$, and $T = 12.4\,\mathrm{h}$. (c) First three eigenfunctions computed for $N_\mathrm{p} = 10^{-2}\,\mathrm{s}^{-1}$ and $\Delta H_\mathrm{p} = 20\,\mathrm{m}$.

model a very wide range of oceanographic density stratifications: from the near two-layer stratification (sharp pycnocline, solid lines 1 in Figures 1.7(a) and (b)) to monotonic stratification (dashed–dotted lines 3 in Figures 1.7(a) and (b)). In passing to the limit $\Delta H_\mathrm{p} \to 0$, one approaches the two-layer density approximation, and as $\Delta H_\mathrm{p} \to \infty$ the monotonic stratification is obtained. With the law of fluid stratification (1.70) the function $p(z)$ in (1.53) for the Boussinesq approximation is

$$p(z) = -c_1 c_3 = \text{const.}, \tag{1.72}$$

and the wave equation (1.52) has the form of the Klein–Gordon equation with constant coefficients, i.e.

$$\Psi_{xx} - \Psi_{z_1 z_1} - c_1 c_3 \Psi = 0. \tag{1.73}$$

With (1.70), the eigenvalue problem (1.67) allows an exact solution, which is straightforward to determine; it reads

$$k_j = \left[\left(\frac{j\pi}{h_i} \right)^2 - c_1 c_3 \right]^{1/2}, \tag{1.74}$$

$$q_j(z_1) = \sin\left(\frac{j\pi z_1}{h_i} \right), \quad i = 1, 3; \quad j = 1, 2, \ldots \tag{1.75}$$

Then, with (1.72) the solution of the BVP (1.65) is expressible as

$$\Psi(x, z_1) = a_j q_j(z_1)\exp(\pm\imath k_j x), \tag{1.76}$$

where j is the mode number and a_j is the free amplitude of the function $\Psi(x, z_1)$. Taking into account that for the law (1.70) $\alpha(z)$ has the form

$$\alpha(z) = c_1(z + c_2)^2 + c_3,$$

we can obtain from (1.44) the following connection between z_1 and z:

$$z_1(z) = \left\{\arctan[(c_1/c_3)^{1/2}(z + c_2)] - \arctan[(c_1/c_3)^{1/2}c_2]\right\}/(c_1c_3)^{1/2}. \tag{1.77}$$

Although for the new variable z_1 the vertical structure of the baroclinic modes is determined by the simple periodic function (1.75), in fact the transformation (1.77) leads to a compression of the eigenfunctions in the z-coordinate in the pycnocline layer (compare Figures 1.4(a) and 1.7(c)). In Figure 1.7(c) the profiles of the first three eigenfunctions $q_j(z)$ $(j = 1, 2, 3)$ are represented. On inspection of the behavior of these curves, we see that the maxima of the functions $q_j(z)$ are not in the pycnocline layer, but are below it. With growing mode number, the positions of these maxima deepen.

Thus, the solution (1.76) describes periodic progressive internal waves for the model stratification (1.70) with baroclinic mode number j that are horizontally propagating in a basin of constant depth. The dispersion relation in this case can formally be written as

$$k_j = \left\{\left(\frac{j\pi}{h_i}\right)^2 - \frac{4(\sigma^2 - f^2)}{\Delta H_{\mathrm{p}}^2(N_{\mathrm{p}}^2 - \sigma^2)^{1/2}} \times \left[\frac{1}{(N_{\mathrm{p}}^2/4 - \sigma^2)^{1/2}} - \frac{1}{(N_{\mathrm{p}}^2 - \sigma^2)^{1/2}}\right]\right\}^{1/2}, \tag{1.78}$$

and follows from (1.71) by substitution for c_1 and c_3 in (1.74).

Realistic buoyancy profiles are determined in the ocean from the temperature and salinity measurements by CTD probes (conductivity, temperature, and depth). With the help of (1.10), these define pointwise buoyancy profiles which can be extended to continuous profiles by spline interpolation. The boundary value problem (1.64) must in this case be solved numerically. The emerging eigenvalue problem is of so-called Sturm–Liouville type, and this property allows one to prove, for arbitrary stable stratification, that there is a countably infinite number of baroclinic modes, which can be ordered according to their phase speeds with mode numbering 1, 2, 3, ... in decreasing magnitude of the phase speed. The eigenfunction for the horizontal velocity belonging to mode number j has $j + 1$ layers with alternating direction of the velocity and j isolated points within the profile at which the velocity changes direction. These are exactly the properties explicitly analyzed for the monotonic stratification in Figure 1.5.

1.6 Nonlinear wave problem

The linear theory of baroclinic wave motions presented in Sections 1.4 and 1.5 is only valid for small-amplitude internal waves. For the derivation of the model equations it was assumed that the external exciting forcing is weak and that the amplitudes of the generated waves are infinitesimal; consequently, it was possible to neglect the nonlinear terms in system (1.39). However, for many regions of the World Ocean that were mentioned in the Preamble as "hot spots" or "storm areas," such conditions are not appropriate. It is necessary to build models in which both realistic input parameters determining the structure of the internal waves (these are, e.g., the external forcing, the law of fluid stratification, and the profile of bottom relief) and the effects related to the nonlinearities of the wave processes are taken into account.

In this section, we derive the basic equations (i.e. the differential equations and the boundary and initial conditions) describing the motion of nonlinear internal waves in a basin of variable depth. These equations will then be used for the development of analytical as well as numerical models.

In nonhydrostatic numerical models, one of the more difficult problems, from a mathematical, as well as a physical, point of view, is the definition of the boundary conditions for the pressure disturbances. To circumvent this difficulty we eliminate the pressure from the governing equations; this is done by taking the curl of the momentum equation and can be achieved by performing the appropriate cross-differentiations with the first three equations in (1.39) if one considers an ideal fluid, or the first three equations in (1.29) if the fluid is viscous. The method is well known, and for a Boussinesq fluid when $\partial/\partial y = 0$, we obtain

$$\left.\begin{aligned} \omega_t + J(\omega, \psi) - f v_z &= g\tilde{\rho}_x/\bar{\rho}_0 + \boxed{A^{\mathrm{H}}\omega_{xx} + (A_z^{\mathrm{V}}\psi_{zz})_z + (A^{\mathrm{V}}\omega_z)_z}, \\ v_t + J(v, \psi) + f\psi_z &= \boxed{A^{\mathrm{H}}v_{xx} + (A^{\mathrm{V}}v_z)_z}, \\ \omega &= \psi_{xx} + \psi_{zz}. \end{aligned}\right\} \tag{1.79}$$

We have used definition (1.40) for the stream function ψ and we also introduced the new unknown function, the vorticity $\omega = u_z - w_x$. The symbol J denotes the Jacobian operator, which is given by $J(a, b) = a_x b_z - a_z b_x$. If an ideal fluid is considered, the terms in the boxes can be ignored. We also implemented representation (1.24) for the density as the sum of a stationary density $\rho_0(z)$ and a perturbation $\tilde{\rho}(x, z, t)$.

The last three equations of system (1.29) (balances of heat, salt, and equation of state) for the viscous fluid in terms of the new variables are transformed to

$$\left.\begin{aligned} \mathsf{S}_t + J(\mathsf{S}, \psi) &= \boxed{K^{\mathrm{H}}\mathsf{S}_{xx} + (K^{\mathrm{V}}\mathsf{S}_z)_z}, \\ \mathsf{T}_t + J(\mathsf{T}, \psi) &= \boxed{K^{\mathrm{H}}\mathsf{T}_{xx} + (K^{\mathrm{V}}\mathsf{T}_z)_z}, \\ \rho &= \rho(\mathsf{S}, \mathsf{T}, P). \end{aligned}\right\} \tag{1.80}$$

The system (1.79), (1.80) is complete and can be used as a basis for investigations of baroclinic wave motions. In many cases, instead of equations (1.80), one can use the single diffusion equation for the density, (1.30), which in terms of the stream function reads as

$$\tilde{\rho}_t + J(\tilde{\rho}, \psi) + \bar{\rho}_0 g^{-1} N^2(z)\psi_x = \boxed{K^{\mathrm{H}}\tilde{\rho}_{xx} + (K^{\mathrm{V}}\tilde{\rho}_z)_z + (K^{\mathrm{V}}\rho_{0z})_z}. \tag{1.81}$$

Such a substitution does not essentially restrict the generality of the approach, and can be applied for the study of many mesoscale wave processes.

Initial and boundary conditions are chosen according to the specific features and demands of the problem under consideration. Here we shall formulate conditions for the generation and evolution of baroclinic tides. We are interested in a baroclinic response of the ocean which is produced by the barotropic tide over the localized bottom topography. It is therefore feasible to prescribe "zero" initial conditions at the initial moment $t = 0$ in the whole area:

$$\left.\begin{aligned} \omega = \psi = \tilde{\rho} = 0, \\ v = -(f/\sigma)\Psi_0/H(x). \end{aligned}\right\} \tag{1.82}$$

The only type of motion which may exist at $t = 0$ is the homogeneous tidal flux along the shelf topography represented by the v-component of the barotropic tide (see (1.63)). Strictly speaking, (1.62) and (1.63) are the exact solutions of the governing system only for a basin of constant depth which is filled with an ideal fluid. However, the expressions for the horizontal velocities in the forms (1.62) and (1.63) are also valid with good accuracy for variable bottom topography and in the viscous case through all the water depth except for a thin bottom boundary layer.

One further aspect of the considered problem should be taken into account for the definition of the boundary conditions. As already mentioned, the source of the internal wave generation is considered in a restricted area (i.e. in the band $-l \leq x \leq l, -\infty < y < \infty$; see Figure 1.3). Outside this strip, i.e. for $x > |l|$, the basin depth is constant, additional sources of wave generation are absent, and the wave field can be represented as a sum of the barotropic tidal flux and the progressive internal waves propagating away from the source of generation.

Taking this into account, the following conditions have to be satisfied at the vertical open boundaries $x = \pm L$ (see Figure 1.3):

$$\left.\begin{aligned} \psi(\pm L, z, t) &= (z/H_i)\Psi_0 \sin(\sigma t), & \tilde{\rho} &= 0, \\ v(\pm L, z, t) &= -(f/(\sigma H_i))\Psi_0 \cos(\sigma t), & \omega &= 0. \end{aligned}\right\} \tag{1.83}$$

Here the "−" ("+") and index $i = 1$ ($i = 3$) correspond to the left (right) boundary. These conditions imply the presence of only barotropic wave motions at the liquid boundaries (horizontal tangents of the streamlines, isopycnals, and vortex lines) with tidal ellipses aligned in the cross-topography direction and ellipticity factor σ/f (ratio of the semiaxes). This assumption can be justified because of the existence of an upper limit for the velocity of the baroclinic disturbances, described by the hyperbolic system (1.79). In the ocean, this limit of the velocity value does not usually exceed 1.5 to 2 m s^{-1}. The trick is to set the model boundaries sufficiently distant from the source of the internal waves ($l \ll L$) that the leading waves generated in region II at $|x| \leq l$ do not reach the boundary before several wave periods have passed. During this time, the fluid motions in the vicinity of the bottom topography are unaffected by the presence of the side boundaries.

At the free surface ($z = 0$), due to the "rigid lid" condition we require

$$\psi = \omega = \tilde{\rho} = v_z = 0 \tag{1.84}$$

for the vorticity ω, the normal derivative of the transverse velocity component v, and the tangential stress τ_{yz} at $z = 0$ (wind forcing is not considered).

A few words should also be said about the conditions for the density. Due to the no-slip condition applied at the rigid boundaries for the wave perturbations of the density, the use of the "zero-field" condition (at the sea surface as well as at the bottom) will be the adequate boundary condition. On the other hand, it is also possible to apply the "zero-flux" condition (zero mass flow through a rigid surface, (1.38)). It was found by numerical runs that the use of these two alternative boundary conditions does not introduce appreciable differences in the characteristics of the wave fields.

Two sets of boundary conditions are used at the bottom, $z = -H(x)$. For a viscous fluid the no-slip conditions are justified:

$$\psi = \Psi_0 \sin(\sigma t), \quad \psi_n = v = 0, \quad \tilde{\rho} \text{ or } \tilde{\rho}_n = 0, \quad \omega = \omega_0. \tag{1.85a}$$

For an ideal fluid we require the slip conditions,

$$\psi = \Psi_0 \sin(\sigma t), \quad \tilde{\rho} \text{ or } \tilde{\rho}_n = 0, \quad \omega = 0. \tag{1.85b}$$

Here, $\partial/\partial n$ denotes the derivative normal to the bottom surface. The value of the vorticity at the bottom, ω_0, is usually calculated from the stream function obtained at the previous time step. The technique of defining ω_0 will be described in Chapter 4. With the above equations for the solution of the nonlinear initial boundary value problem, our introductory analysis on tidally induced internal waves in a stably stratified ocean comes to an end. In the subsequent chapters, both the linear and nonlinear equations will be employed in typical situations when a barotropic external tidal wave encounters a bottom feature of distinct variability.

2

Linear baroclinic tides over variable bottom topography

This chapter is devoted to a study of the dynamics of *infinitesimal waves* in a continuously stratified ocean of variable depth. Small-amplitude baroclinic tides are usually generated when the intensity of the barotropic tidal forcing is small. As mentioned earlier, depending on the wave amplitude, the analysis can be carried out by means of either the linear or nonlinear theory. A quantitative estimation of the efficiency of the tidal generation of internal waves and the discussion of the necessity to use the full nonlinear system of equations to study baroclinic tides is given in refs. [73] and [75]. We return to this latter point in detail in Chapter 4. Here we simply assume that the amplitudes of the considered waves are so small that with sufficient accuracy the advective terms in the governing system (1.39) can be neglected.

Historically, the first linear models of baroclinic tides were developed for a two-layer ocean. In the models in refs. [199] and [265], the internal tide was generated as a result of the interaction of the barotropic tidal wave with a Heaviside-like bottom step, an obstacle, which approximates the transition zone between the deep and shallow parts of the ocean. In these works, reflection of the waves from the coastal line was used as a boundary condition. Because of this, the properties of the solution were basically defined by the resonance of the internal waves on the shelf. Later it became obvious that a more suitable and more realistic condition along a coastal line would require wave absorption [81].

In more realistic models, [197], [198], [200], the layered stratification of the fluid was replaced by a continuous counterpart. However, these models employed a method of solution in which linear combinations of the orthogonal modes on the shelf were pasted together at the shelf boundary with those in the deep water. As a result, an infinite system of algebraic equations for the determination of the mode amplitudes had to be solved. Furthermore, the applicability of this method was limited to piecewise linear profiles of the bottom topography. Nevertheless, in these works it was possible to reproduce the observed occurrence of the wave-beam structure of the baroclinic tides that is usually generated over such supercritical bottom features.

In ref. [198] it was found that in a viscous fluid the decrement of the attenuation of the generated waves is proportional to the square root of the viscosity coefficient. Waves propagating from the shelf break towards the shore generally attenuate before they are reflected from the shore line. This conclusion, together with the results of experimental work [77], allows us to exclude the reflection of a wave from the coast and, in turn, to apply inviscid models paired with Sommerfeld radiation conditions on the liquid boundary of the shallow zone.

The problem of the interaction of the barotropic tidal flux with two-dimensional bottom features in a continuously stratified fluid was solved by Baines [9]–[11], who derived and solved an integral equation with the method of characteristics. This transformation was successful because of the use of a specific law for the vertical fluid stratification. The model allowed Baines to draw the important conclusion that steep, supercritical bottom slopes are the topographic locations of the most effective generation of internal waves. An equivalent approach was developed by Sandström [211], who used the method of characteristics as well as the method of representation of the wave field by a series of orthogonal baroclinic modes. The results obtained by Sandström [211] are in agreement with the conclusions stated by Baines [9], [10].

The more advanced linear slice models of baroclinic tides, [217], [241], allow the incorporation of arbitrary vertical fluid stratification as well as complicated bottom topography. Because of their relative simplicity, slice models are very fruitful for qualitative analyses and interpretation of results obtained from field experiments. Nevertheless, to obtain a more reliable and more realistic prediction of the baroclinic tidal activity at separate positions within the World Ocean, both full nonlinear and three-dimensional effects must be included in the models (we consider these effects in Chapters 4–7).

Linear models share the advantage that the mathematical theory is so well developed that it often allows us to find analytical solutions to given initial and boundary value problems. For instance, in ref. [43] the problem of the generation of baroclinic tides from the prescribed barotropic tidal motion above localized bottom features (e.g. oceanic banks) was treated for an arbitrary profile of fluid stratification. Three-dimensional bottom topography located in the deep part of the ocean was used in this model. The height of the subsurface obstacle was assumed to be small; i.e. the ratio of the height, $H_{\max}$, of the underwater mountain to the total water depth of the ocean, H, was assumed to be a small parameter ($H_{\max}/H \ll 1$). By use of the perturbation method, the energy sink from the surface waves to the internal modes was estimated. This estimation showed that the transfer of tidal energy into internal waves in an open part of the ocean is larger by two orders of magnitude than the energy lost by dissipation. These estimates were corrected and revised in ref. [17].

Reference [17] should be considered in conjunction with the fundamental work of Bell [18], in which the problem of topographic generation of baroclinic tides was solved analytically with the use of Fourier transforms for a vertically unbounded, uniformly stratified fluid ($N(z) = N_0 =$ const.). It was found that internal waves are generated over an obstacle not only at the fundamental tidal frequency, σ, but also at other tidal harmonics. Two important limiting regimes of generation, depending on the magnitudes of the ratios σ/N_0 and $U_0/l\sigma$, were described (l is the horizontal scale of the obstacle and U_0 is the amplitude of tidal velocity). In the quasistatic limit, i.e. when $\sigma/N_0 \to 0$, the time derivatives in the governing equations may be ignored, and the problem is reduced to a slowly varying case of the classical lee wave problem, the parameter U_0/l becomes the characteristic frequency of the waves, and U/lN_0 is the slope of the wave rays. In the other limiting case, when $U_0/l\sigma \to 0$, the problem reduces to the classical problem of a vibrating disturbance in a stratified fluid (the acoustic limit). In this case, the slope of the wave ray is σ/N_0.

This study became a basis for many other analytical investigations of the problem of baroclinic tide generation. In the most recent papers, refs. [14], [118], [220], [225], [226], the theory developed by Bell [17], [18] is modified and used for the estimation of the energy sink from barotropic to baroclinic tides in applications to different shapes of bottom features. For instance, ref. [14] focuses on the investigation of how sensitive Bell's theory is to the slope γ and amplitude H_{max} of the bottom topography. The basic result is that the barotropic–baroclinic energy conversion is proportional to H^2_{max} and is described well by Bell's formula in the subcritical case (for small bottom inclinations when $\gamma \ll \alpha$), but the discrepancy exceeds 50% when the bottom inclination tends to the characteristic lines (in a near-critical case when $\gamma \to \alpha$).

Bell's theory was extended to the case of an ocean of finite depth H with an arbitrary fluid stratification (however, the restriction of weak topography approximation $H_{max}/H \ll 1$ still remains) [220]. The basic finding was that in the presence of an upper reflecting surface the conversion rate from barotropic to baroclinic tides can be substantially smaller than with infinite-ocean predictions; it takes place when the scale of the topography and the wavelength of the generated waves are comparable. A similar result was obtained in ref. [118], where not only analytical but also numerical nonlinear solutions were obtained. In particular, it was shown that the linear theory underestimates the conversion rate near steep bottom features, yet, even at the critical slope, the difference does not exceed 20%.

For supercritical topographies (knife-edge ridge, top-hat ridge, topographic step), the analytical solution of the baroclinic tidal problem was found by St. Laurent *et al.* [226]. It can be seen in this work that the energy flux from abrupt topographies can significantly exceed that from gentle ones.

Obviously, the solution found in refs. [14], [43], [118], and [220] can be valid only for "small obstacles," i.e. when $H_{\max}/H \ll 1$. In Section 2.1 it will be shown that a more successful approach is to use as a small parameter (instead of $H_{\max}/H$) the "effective" height of the bottom roughness, which accounts not only for the obstacle height but also for the pertinent vertical fluid stratification. Such a reformulation makes it possible to consider small bottom roughnesses as well as large obstacles for which the height is comparable to the total oceanic depth. The use of the characteristic variables (1.44) instead of the original Cartesian space and time variables considerably expands the range of applicability of the analytical methods in problems of the generation of internal waves. This inference will be drawn in Section 2.3 on the basis of a comparative analysis of results obtained analytically with the perturbation theory on the basis of the semianalytical model presented in Section 2.2 and developed for arbitrary bottom features and arbitrary fluid stratification.

In Section 2.5 some aspects of the generation of baroclinic tides in the vicinity of "steep" underwater obstacles will be considered when the steepness of the bottom topography is comparable to the inclinations of the characteristic lines of the hyperbolic wave equation (1.5) ($\alpha \approx \gamma$). A theory of this kind was constructed for the first time in ref. [10] with the intention of studying the generation process of waves in a fluid with specific laws of stratification. Here, we derive relations which are valid in more general situations. In particular, it is shown that an increase of the bottom steepness results in an accompanying increase of the effective generation of high baroclinic modes, the superposition of which leads to the formation of a "wave beam."

2.1 Analytical solution for "small" bottom features

In this section we consider the problem of the generation and scattering of internal waves by isolated bottom obstacles. The geometric configuration here is similar to that presented in Figure 1.3: from the far end of zone I, perpendicular to the elongated bottom feature in zone II, two-dimensional harmonic, barotropic, or baroclinic waves propagate into zone II. Interaction of these waves with the obstacle leads to the generation in zones I and III of reflected and transmitted waves; we are interested in the construction of the internal wave fields generated above the underwater sill and beyond it in terms of the form of the bottom relief, the law of stratification, and the parameters of the incoming waves. Mathematically, the problem is formulated in the characteristic variables (x, z_1) leading to the BVP (1.64) for which the function $p(z)$ is defined in (1.53).

Before solving the problem, two preliminary remarks should be made. First, as was mentioned in Section 1.5, the stream function is defined up to an additive constant. For this reason, instead of applying the zero boundary condition (1.59) at the free surface and the nonzero condition (1.61) for the barotropic tidal wave at the bottom, we can apply the zero boundary condition (1.60) at the bottom and the nonzero condition at the free surface. In any case, the free constant should be selected so as to simplify the mathematical formulas that generate the solution.

So, with the new boundary conditions instead of (1.64), the equations

$$\left.\begin{aligned} \Psi_{xx} - \Psi_{z_1 z_1} + p(z)\Psi = 0, \\ \Psi = \alpha^{-1/2}(0)\Psi_0 \quad \text{at} \quad z_1 = 0, \\ \Psi = 0 \quad \text{at} \quad z_1 = h(x) \end{aligned}\right\} \tag{2.1}$$

must be solved. Ψ_0 is the amplitude of the vertically integrated water flux that is related to the incoming wave. When the internal wave is generated by a barotropic tide, this value is uniquely defined by the amplitude of the incident wave. If the incident wave of the wave scattering problem is baroclinic, we can put $\Psi_0 = 0$.

The second remark concerns the geometrical configuration of the considered area. We restrict our attention to wave fields generated over oceanic ridges. So, we take $H_1 = H_3 = H_0$, and $H_2(x) = H_0 - H_{\max} r(x)$, where $H_{\max}$ is the ridge height, the function $r(x)$ defines its shape ($|r(x)| \leq 1$ at $|x| < l$, and $r(x) \equiv 0$ at $|x| > l$). Such a restriction is dictated by the necessity to apply Fourier transforms[1] and the perturbation method, which will be used below. Of course, to some extent, this prevents the application of results of the linear model to a wide variety of oceanic conditions, but, on the other hand, such an approach allows us to investigate the generation process with the help of analytic techniques, which is more important.

In this section we will obtain the solution of problem (2.1) for a symmetric oceanic ridge (Figure 2.1) of "small" height with the law of stratification (1.70) according to ref. [248], and then, in the subsequent sections, we will generalize these results to obstacles of any form and arbitrary height.

To describe the wave field generated by the barotropic tidal flux or the incident internal wave over the oceanic ridge, we follow the procedure developed by Cox and Sandström [43]. The main difference is that we solve equations (2.1) instead of (1.42).

[1] For Fourier transforms to exist, $r(x)$ does not need to have compact support; exponential decay as $|x| \to \infty$ is sufficient.

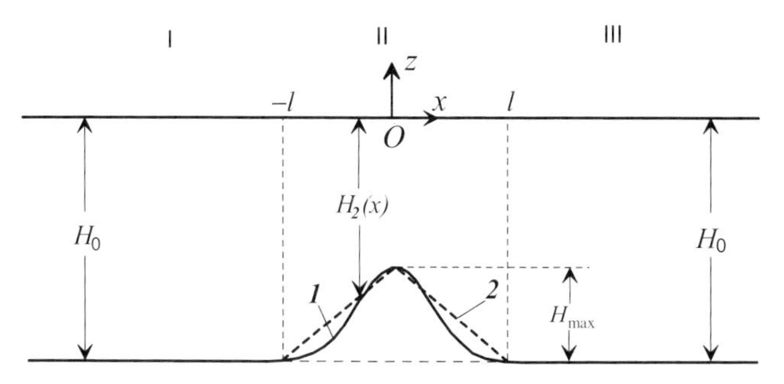

Figure 2.1. Sketch of the geometry with two profiles of an oceanic ridge. The shapes of the underwater obstacles 1 and 2 are given by (2.18) and (2.32), respectively.

In terms of the new variables (x, z_1), the "depth" $h(x)$ of the basin can be written as

$$h(x) = \int_0^{-H(x)} \frac{ds}{\alpha(s)} = \int_0^{-H_0} \frac{ds}{\alpha(s)} - \int_{H_{max}r(x)-H_0}^{-H_0} \frac{ds}{\alpha(s)},$$

where $\alpha(s)$ is defined in (1.43). The last integral can be represented as follows:

$$\int_{H_{max}r(x)-H_0}^{-H_0} \frac{ds}{\alpha(s)} = \int_{H_{max}-H_0}^{-H_0} \frac{ds}{\alpha(s)} \left(\int_{H_{max}r(x)-H_0}^{-H_0} \frac{ds}{\alpha(s)} \Big/ \int_{H_{max}-H_0}^{-H_0} \frac{ds}{\alpha(s)} \right),$$

where the first factor is the "height" of the sill, h_{max}, in canonical variables,

$$h_{max} = \int_{H_{max}-H_0}^{-H_0} \frac{ds}{\alpha(s)},$$

and the factor in brackets is its shape $r_1(x)$ in the new variables:

$$r_1(x) = \left(\int_{H_{max}r(x)-H_0}^{-H_0} \frac{ds}{\alpha(s)} \Big/ \int_{H_{max}-H_0}^{-H_0} \frac{ds}{\alpha(s)} \right).$$

So, in the space of the canonical variables the basin depth is given by

$$h(x) = h_0 - h_{max} r_1(x) = h_0 \left(1 - r_1(x) \frac{h_{max}}{h_0}\right), \tag{2.2}$$

where

$$h_0 = \int_0^{-H_0} \frac{ds}{\alpha(s)}, \quad h_{max} r_1(x) = \int_{H_{max} r(x) - H_0}^{-H_0} \frac{ds}{\alpha(s)},$$

and where

$$r_1(x) = 0 \quad \text{if} \quad |x| > l.$$

Here, the maximum height of the ridge, H_{max}, is transformed into h_{max}. So, as suggested by (2.2), a small parameter can be introduced if it is further assumed that the height of the ridge in the new variables is small, so that

$$\varepsilon_0 = h_{max}/h_0 \ll 1. \tag{2.3}$$

Note that condition (2.3) does not mean that, in physical coordinates, the bottom obstacle necessarily has to be small. Actually, h_{max} is the integral of the buoyancy frequency profile $N(z)$ between the bottom and the depth of the mountain peak; h_0 has the same meaning for the total oceanic depth. Thus, the value of $\varepsilon_0 = h_{max}/h_0$ actually not only reflects the fact how high the sill may be in comparison with the total depth, but is also equal to a measure of stratification of the deep layers. Usually, oceanic waters are primarily stratified near the free surface, whilst the abyssal layers are very weakly stratified. Thus, the condition $\varepsilon_0 \ll 1$ is probably valid for the majority of ocean ridges and banks, the tops of which are located below the seasonal pycnocline (namely, several hundred meters below the free surface). For instance, for the model pycnocline (1.70), (1.71) with the parameters $N_p = 5 \times 10^{-3}\,\text{s}^{-1}$, $H_p = 100\,\text{m}$, $\Delta H_p = 60\,\text{m}$, an oceanic ridge at middle latitudes with height $H_{max} = 3.7$ km in a basin with total depth $H_0 = 4$ km yields the value $\varepsilon_0 \approx 0.09$, whereas $H_{max}/H_0 = 0.925$, which is not small.

This example shows that the analytical solution obtained by means of the perturbation method with the small parameter ε_0 is formally valid not only for "small" ridges (when $H_{max} \ll H_0$, as was assumed in ref. [43]) but also for "large" ridges with realistic oceanic heights such that $H_{max} = O(H_0)$. The only condition that must be satisfied is that the top of the ridge must lie below the pycnocline. This is a fortunate consequence of the transformation (1.44). The range of applicability of the presented method will be estimated in Section 2.3.

The boundary condition in (2.1) at the bottom reads

$$\Psi = 0 \quad \text{at} \quad z_1 = h_0(1 - \varepsilon_0 r_1(x)),$$

where $r_1(x)$ is the shape of the bottom topography in the new variables (x, z_1). Let us use the special stratification given by (1.70)–(1.72). Then, problem (2.1) can be rewritten as

$$\left.\begin{aligned} \Psi_{xx} - \Psi_{z_1 z_1} - c_1 c_3 \Psi = 0, \\ \Psi(x, z_1) = \alpha^{-1/2}(0)\Psi_0 \quad \text{at} \quad z_1 = 0, \\ \Psi(x, z_1) = 0 \quad \text{at} \quad z_1 = h_0[1 - \varepsilon_0 r_1(x)], \end{aligned}\right\} \tag{2.4}$$

where c_1 and c_3 are defined in (1.71). This problem describes the generation of internal waves by a barotropic tidal flux oscillating over an underwater ridge, as well as the scattering of harmonic internal waves incident on the underwater obstacle. Note that the small parameter ε_0 appears in the problem not in the differential equation but in the bottom boundary condition.

Suppose (see ref. [43]) that $\Psi(x, z_1)$ has the asymptotic expansion[2]

$$\Psi(x, z_1) = \sum_{i=0}^{\infty} \varepsilon_0^i \Psi^{(i)}(x, z_1) = \Psi^{(0)}(x, z_1) + \varepsilon_0 \Psi^{(1)}(x, z_1) + O(\varepsilon_0^2). \tag{2.5}$$

Then, the boundary condition $\Psi(x, z_1) = 0$ at the bottom $z_1 = h_0[1 - \varepsilon_0 r_1(x)]$ can be written as

$$\begin{aligned} \Psi(x, h(x)) = \Psi^{(0)}\,(x, h_0[1 - \varepsilon_0 r_1(x)]) + \varepsilon_0 \Psi^{(1)}\,(x, h_0[1 - \varepsilon_0 r_1(x)]) \\ + \varepsilon_0^2 \Psi^{(2)}\,(x, h_0[1 - \varepsilon_0 r_1(x)]) + o(\varepsilon_0^2) = 0. \end{aligned}$$

Taking into account the fact that the parameter ε_0 is small, the functions $\Psi^{(i)}(x, h_0[1 - \varepsilon_0 r_1(x)])$ in the preceding equation may be expanded in Taylor series near $z_1 = h_0$. This leads to the following boundary condition at the bottom:

$$\Psi(x, h(x)) = \Psi^{(0)}(x, h_0) + \varepsilon_0 \left(\Psi^{(1)}(x, h_0) - h_0 r_1(x) \frac{\partial \Psi^{(0)}}{\partial z_1}\bigg|_{z_1 = h_0} \right) + O(\varepsilon_0^2) = 0. \tag{2.6}$$

Substituting (2.5) into (2.4) and equating terms of the same order of ε_0 (also taking into account equation (2.6)), the following sequence of boundary value problems is obtained.

To *zeroth order*,

$$\left.\begin{aligned} \Psi^{(0)}_{xx} - \Psi^{(0)}_{z_1 z_1} - c_1 c_3 \Psi^{(0)} = 0, \\ \Psi^{(0)}(x, 0) = \alpha^{-1/2}(0)\Psi_0, \\ \Psi^{(0)}(x, h_0) = 0. \end{aligned}\right\} \tag{2.7}$$

[2] In fact, in ref. [43] the expansion was performed in terms of the nondimensional height of the bottom roughness, H_{max}/H_0, whereas here we use the small parameter h_{max}/h_0.

To *first order*,

$$\left.\begin{aligned} \Psi^{(1)}_{xx} - \Psi^{(1)}_{z_1 z_1} - c_1 c_3 \Psi^{(1)} &= 0, \\ \Psi^{(1)}(x, 0) &= 0, \\ \Psi^{(1)}(x, h_0) &= h_0 r_1(x) \left.\frac{\partial \Psi^{(0)}}{\partial z_1}\right|_{z_1 = h_0}. \end{aligned}\right\} \tag{2.8}$$

Notice that the lowest-order problem involves only $\Psi^{(0)}$ as an unknown, whilst the first-order problem is driven by the solution of the lowest-order problem through the boundary term at $z_1 = h_0$. Furthermore, both problems are formulated in such a way that the boundary conditions are applied at $z_1 = 0$ and $z_1 = h_0$. So, the perturbation method has shifted the bottom boundary to a straight line. This shift makes an analytic solution possible. Both problems are used below for the analysis of the generation of internal waves, as well as for the scattering of internal waves approaching from infinity.

2.1.1 Generation of internal waves by an oscillating tidal flux

Let us consider the generation of internal waves by an oscillating barotropic tidal flux.

Zeroth-order solution

First we must find a particular solution of problem (2.7), which, according to the basic aim of the present study, represents the barotropic tidal wave as a forcing factor (scattering of incident waves will be considered in Section 2.1.2).

For the model pycnocline (1.70), the BVP (2.7) has the following analytic solution:

$$\Psi^{(0)}(x, z_1) = \alpha^{-1/2}(0)\Psi_0 \frac{\sin[(c_1 c_3)^{1/2}(h_0 - z_1)]}{\sin[(c_1 c_3)^{1/2} h_0]} := \Phi(z_1), \tag{2.9}$$

which, after the replacement of z_1 by z with the use of (1.77), takes the very simple form

$$\Psi^{(0)}(x, z) = \alpha^{-1/2}(0)\Psi_0(1 + z/H_0). \tag{2.10}$$

Thus, the solution of the zeroth-order problem is an undisturbed barotropic flux with constant velocity $U^{(0)} = \Psi_0/H_0$ from the top to the bottom that is modeling the long barotropic tidal wave. This should not be surprising because the BVP (2.7) is the flow in a constant depth layer that obviously cannot generate any disturbance.

First-order solution

Now let us find the solution of BVP (2.8). The homogeneous equation, the first equation in (2.8), with inhomogeneous boundary conditions, the second and third equations in (2.8), is solved by throwing this inhomogeneity from the boundary condition into the differential equation. So, when (2.9) is substituted into the third equation in (2.8), it is seen that a new stream function $\tilde{\psi}(x, z_1)$ can be defined

$$\left.\begin{aligned} \tilde{\psi}(x, z_1) &= \Psi^{(1)}(x, z_1) - Bf_1(z_1)r_1(x), \\ B &= \frac{h_0\Psi_0 \cdot (c_1c_3)^{1/2}}{\alpha^{1/2}(0)\sin[(c_1c_3)^{1/2}h_0]}, \\ f_1(z_1) &= \frac{\sin[(c_1c_3)^{1/2}z_1]}{\sin[(c_1c_3)^{1/2}h_0]}, \end{aligned}\right\} \tag{2.11}$$

which makes the boundary condition $(2.8)_3$ homogeneous. It reduces (2.8) to the following BVP involving the function $\tilde{\psi}(x, z_1)$:

$$\left.\begin{aligned} \tilde{\psi}_{xx} - \tilde{\psi}_{z_1z_1} - c_1c_3\tilde{\psi} &= -Bf_1(z_1)r_{1xx}(x), \\ \tilde{\psi}(x, 0) = \tilde{\psi}(x, h_0) &= 0, \end{aligned}\right\} \tag{2.12}$$

in which the inhomogeneity is now removed from the boundary conditions and transferred to the differential equation.

Problem (2.12) is solved with the help of a Fourier transform, which we use with the definition

$$\hat{\psi}(k, z_1) = \int_{-\infty}^{\infty} \tilde{\psi}(x, z_1)\exp(\imath kx)\,dx.$$

With this, (2.12) can be shown to lead to the Fourier transformed BVP

$$\left.\begin{aligned} \hat{\psi}_{z_1z_1} + (k^2 + c_1c_3)\hat{\psi} &= \sin[(c_1c_3)^{1/2}z_1]F(k), \\ \hat{\psi}(k, 0) = \hat{\psi}(k, h_0) &= 0, \end{aligned}\right\} \tag{2.13}$$

where

$$F(k) = \frac{Bk^2}{2\pi\sin[(c_1c_3)^{1/2}h_0]}\int_{-\infty}^{\infty} r_1(x)\cdot\exp(\imath kx)\,dx.$$

The general solution of (2.13) consists of (i) its particular solution plus (ii) the solution of the homogeneous problem [189]

$$\hat{\psi}(k, z_1) = \underbrace{\text{particular solution}}_{\text{(i)}} + \underbrace{\sum_{j=1}^{\infty} C_j \hat{\psi}_j(k, z_1)}_{\text{(ii)}}.$$

Here $\hat{\psi}_j(k, z_1)$ are the eigenfunctions of the homogeneous BVP (2.13) and C_j are arbitrary constant values. In fact, $\hat{\psi}_j(k, z_1)$ are monochromatic internal waves like those presented by formula (1.76), which propagate horizontally from $+\infty$ to $-\infty$ if $k > 0$ and in the opposite direction if $k < 0$. Taking into account the Sommerfeld radiation condition – an absence of waves approaching from infinity – we can set $C_j = 0$. This means that the solution of (2.13) consists only of the particular solution of the inhomogeneous equation (2.13), which has the form

$$\hat{\psi}(k, \zeta_1) = F(k) \int_0^{h_0} \sin[(c_1 c_3)^{1/2} z_1] \, \mathcal{G}(z_1, \zeta_1) \, dz_1.$$

$\mathcal{G}(z_1, \zeta_1)$ is the Green function[3] for the BVP (2.13),

$$\mathcal{G}(z_1, \zeta_1) = \frac{1}{\delta \sin(\delta h_0)} \begin{cases} \sin[\delta(\zeta_1 - h_0)] \sin(\delta z_1), & \zeta \le z_1 \le 0, \\ \sin[\delta(z_1 - h_0)] \sin(\delta \zeta_1), & h_0 \le z_1 \le \zeta_1, \end{cases} \tag{2.14}$$

where $\delta = (k^2 + c_1 c_3)^{1/2}$. We find, after performing the integration and substitution of z_1 instead of ζ_1,

$$\hat{\psi}(k, z_1) = \frac{F(k)}{k^2} \sin[(c_1 c_3)^{1/2} h_0][f_1(z_1) - f_2(k, z_1)], \tag{2.15}$$

where

$$f_2(k, z_1) = \frac{\sin(\delta z_1)}{\sin(\delta h_0)} = \frac{\sin[(c_1 c_3 + k^2)^{1/2} z_1]}{\sin[(c_1 c_3 + k^2)^{1/2} h_0]}. \tag{2.16}$$

Let us investigate the function $f_2(k, z_1)$:

(1) When $k = 0$,

$$f_2(0, z_1) = \frac{\sin[(c_1 c_3)^{1/2} z_1]}{\sin[(c_1 c_3)^{1/2} h_0]}.$$

(2) When $k = \pm k_j = \pm[(j\pi/h_0)^2 - c_1 c_3]^{1/2}$ $(j = 1, 2, \ldots)$, $f_2(k, z_1)$ has a vanishing denominator or poles of the first order.

[3] For details of the construction of the solution of (2.13) by the Green function, see ref. [80], p. 38.

This means that $f_2(k, z_1)$ is a meromorphic function.[4] According to the Cauchy theorem of residues, such a function can be represented as a series of rational functions [41],

$$f_2(k, z_1) = f_2(0, z_1) + \sum_{j=1}^{\infty} A_j \left(\frac{1}{k - k_j} + \frac{1}{k_j} \right),$$

where $A_j = \operatorname*{Res}_{k=k_j} f_2(k, z_1)$ are the residues of function $f_2(k, z_1)$ at the points $k = k_j$, $j = 1, 2, \ldots$ Thus, $f_2(k, z_1)$ can be expanded into elementary fractions as follows:

$$f_2(k, z_1) = f_1(z_1) + \sum_{j=1}^{\infty} (-1)^j \cdot \frac{j\pi}{h_0^2 k_j^2} q_j(z_1) \frac{2k^2}{k^2 - k_j^2}, \tag{2.17}$$

where $f_1(z_1)$ is given in (2.11), k_j is the horizontal wavenumber of the jth mode, and $q_j(z_1)$ is the vertical structure function for the stratification (1.70) as defined in (1.74) and (1.75).

If we now introduce in the Fourier integral representation of $F(k)$ the specific profile (the solid line in Figure 2.1),

$$r_1(x) = \begin{cases} [1 + \cos(\pi x / l)]/2, & |x| \le l, \\ 0, & |x| > l, \end{cases} \tag{2.18}$$

then, after substitution of the representations for f_1 and f_2 into (2.15) and subsequently performing the inverse Fourier transform

$$\tilde{\psi}(x, z_1) = \frac{1}{2\pi} \int_{-\infty}^{\infty} \hat{\psi}(k, z_1) \exp(-\imath kx)\, dk,$$

we find the solution of problem (2.12) in the form

$$\tilde{\psi}(x, z_1) = \frac{\Psi_0 \pi^2 (c_1 c_3)^{1/2} h_0}{\alpha_0^{1/2} \sin[(c_1 c_3)^{1/2} h_0]} \sum_{j=1}^{\infty} \frac{j(-1)^{j+1}}{k_j^2} q_j(z_1) I_j, \tag{2.19}$$

where

$$I_j = \int_{-\infty}^{\infty} \frac{k \sin(kl)}{(k^2 - k_j^2)[(kl)^2 - \pi^2]} \cdot \exp(-\imath kx)\, dk,$$

[4] A function that is analytic, except for a countably infinite number of poles.

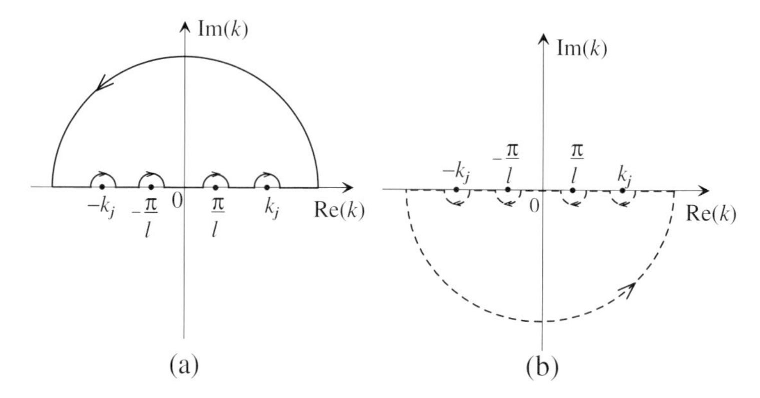

Figure 2.2. Integration contours in the complex plane $k = \mathrm{Re}(k) + \imath \mathrm{Im}(k)$, chosen in conformity with the radiation conditions.

or, with $\sin(kl) = \left[\exp(\imath kl) - \exp(-\imath kl)\right]/2\imath$,

$$I_j = \underbrace{\frac{1}{2\imath}\int_{-\infty}^{\infty} \frac{k\exp[-\imath k(x-l)]}{(k^2-k_j^2)[(kl)^2-\pi^2]}\,dk}_{\mathscr{I}_1} - \underbrace{\frac{1}{2\imath}\int_{-\infty}^{\infty} \frac{k\exp[-\imath k(x+l)]}{(k^2-k_j^2)[(kl)^2-\pi^2]}\,dk}_{\mathscr{I}_2}.$$

The last two integrals are calculated with the use of the residue theorem (upon imposing the Jordan lemma [41]). In so doing, the Sommerfeld radiation condition should be satisfied; in other words, the generated waves must propagate away from the bottom obstacle and radiate outward. According to this principle, four cases must be considered: two for the first integral $\mathscr{I}_1$ and two for the second integral $\mathscr{I}_2$. For the integral $\mathscr{I}_1$ to satisfy the radiation condition, the integrating contour shown in Figure 2.2(a) (waves propagate to the left) is used when $x < l$, and that in Figure 2.2(b) (waves propagate to the right) is considered when $x > l$. Similar reasoning is valid for the second integral $\mathscr{I}_2$ when the regions $x < -l$ and $x > -l$ are considered.

The final equation can be written in the form

$$I_j = \begin{cases} \imath D_j \sin(k_j l)\exp(\imath k_j x), & x < -l, \\ -D_j[\cos(\pi x/l) + \cos(k_j x)\exp(-\imath k_j l)], & |x| \le l, \\ \imath D_j \sin(k_j l)\exp(-\imath k_j x), & x > l, \end{cases} \tag{2.20}$$

where

$$D_j = \pi/[\pi^2 - (k_j l)^2].$$

The function $\tilde{\psi}(x, z_1)$ is found when (2.20) is inserted into (2.19). Taking into account (2.11), this gives us the first-order solution $\Psi^{(1)}$ of the BVP (2.8).

Having determined the first two terms, $\Psi^{(0)}$ and $\Psi^{(1)}$, of the expansion (2.5) of the BVP (2.4), the stream function $\Psi(x, z_1)$ can be written with accuracy $O(\varepsilon_0^2)$. Indeed, with (2.9), (2.11), (2.19), and (2.20), we have

$$\Psi(x, z) = \Phi(z_1) + \varepsilon_0 \sum_{j=1}^{\infty} q_j(z_1) \times \begin{cases} b_j \exp\left\{\imath[k_j x + \pi(j + 0.5)]\right\}, & x < -l, \\ (b_j + a_j)\cos(k_j x)\exp[\imath\pi(j + 0.5)], & |x| \le l, \\ a_j \exp\left\{\imath[-k_j x + \pi(j + 0.5)]\right\}, & x > l, \end{cases} \tag{2.21}$$

where $\Phi(z_1)$ is defined by (2.9), $q_j(z_1)$ are eigenfunctions given by (1.75) (when $h_i = h_0$), and the amplitudes a_j and b_j are given by

$$a_j = -b_j = -B\pi^3 k_j^{-2} \sin(k_j l)/[(k_j l)^2 - \pi^2]. \tag{2.22}$$

Equation (2.21) describes the generation of internal waves by a barotropic tidal flux over the ridge (2.18). Obviously, the first term in (2.21) is the undisturbed barotropic flux, and only the sum that is premultiplied by ε_0 represents the excited baroclinic response.

Let us discuss the solution (2.21) when $|x| > l$. It is clear that the generated wave field on this side of the underwater ridge consists of a sum of radiated baroclinic modes which propagate left and right from the source of generation; they move on the background of the barotropic tidal flux given by (2.9).

The amplitudes (2.22) of the generated modes depend on the intensity of the barotropic tidal flux, height, and width of the underwater obstacle and on the parameters of the stratification; see the definition of B in (2.11). The dependence of $a_j(l)$ on the width l of the obstacle reveals a resonating character of the generating mechanism. It is clear that $a_j(l) = 0$ when $l = n\pi k_j^{-1}$, $n = 2, 3, 4, \ldots$, but at $n = 1$ the function $a_j(l)$ assumes a maximum, as can easily be corroborated by employing the rule of Bernoulli and Hôpital when evaluating the singular expression $0/0$. This means that the jth baroclinic mode is not generated whenever the width of the ridge is equal to an integer number of the wavelength, λ_j, of this mode (i.e. when $2l = n\lambda_j$, $n = 2, 3, 4, \ldots$).

The exact positions of the maxima of the function $a_j(l)$ can be determined by solving the transcendental equation $da_j(l)/dl = 0$. However, for rough estimations of these positions, one may simply use the expression $l = (n + 1/2)\pi/k_j$, $n = 1, 2, 3, \ldots$ This assumes that the lobes have half-sinusoidal form. Moreover, with

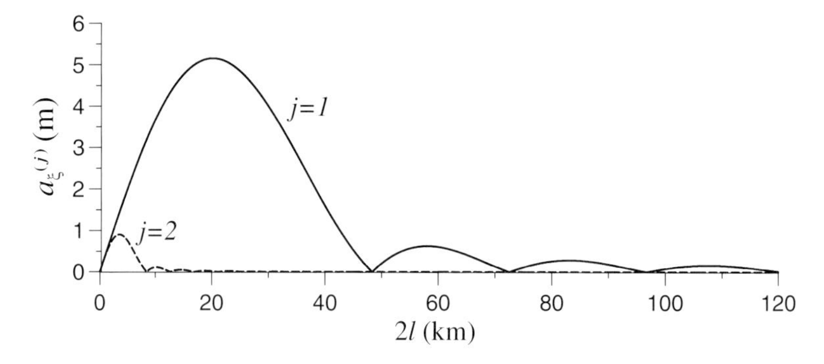

Figure 2.3. Amplitudes of the vertical displacement ($a_\xi^{(j)} = a_j(l)k_j/\sigma$) of the first two baroclinic modes generated by the barotropic tidal flux at $\Psi_0 = 20$ m^2 s^{-1} over the oceanic ridge (2.18) with height $H_{\max} = 2$ km in an ocean with total depth $H_0 = 4$ km versus the width $2l$ of the ridge. Parameters of the pycnocline (1.70) were $N_p = 0.02$ s^{-1}, $H_p = 100$ m, $\Delta H_p = 50$ m.

increasing l, the maxima of $a_j(l)$ decrease asymptotically for large l proportional to l^{-2}.

Note also that, for fixed values of l, the amplitudes a_j for large j are proportional to j^{-4}. This follows from (2.22) if l is held fixed and finite while k_j is becoming large, implying $a_j(l) \sim k_j^{-4}$, and so equally also $a_j(l) \sim j^{-4}$. Consequently, the amplitudes of high baroclinic modes are usually negligible in comparison with the first baroclinic mode (see Figure 2.3), a fact that is in accordance with the data of many field observations: *in the ocean the first baroclinic mode usually predominates*.

The exception to this rule is the baroclinic response for wave interaction with very steep bottom obstacles (at $l < 5$ km in our case) when amplitudes of all generated baroclinic modes are comparable. This fact reflects the peculiarity of the generation of internal waves near steep bottom topographies where the superposition of a large number of baroclinic modes leads to the formation of a wave beam near-critical and supercritical slopes. This mechanism will be discussed further in Section 2.5.

In the region $|x| \leq l$, i.e. above the bottom topographic feature, the solution (2.21) can be characterized as a sum of standing internal waves which are generated by the superposition of two systems of waves propagating in opposite directions. These waves are exposed to a multiple reflection from the topographic boundaries, which creates a system of standing waves. The dependence of the amplitudes of these waves on the width of the obstacle, the number of modes, and the parameters of the stratification is similar to that for the waves propagating beyond the ridge.

The above solution has been constructed analytically, and this construction was possible because the bottom obstruction had the special mathematical form (2.18). This shape is not the only possible form allowing construction of a simple analytical solution. In principle, such a closed form solution can be found for any bottom profile for which the integral $F(k)$ can be calculated analytically.

2.1.2 Scattering of internal waves by a bottom obstacle

Wave scattering at an infinitesimal topography has been studied by the perturbation method in refs. [171] and [203]. In this section we investigate the problem of wave scattering following ref. [248].

We consider a monochromatic internal wave propagating in a basin of constant depth which encounters a localized bottom feature (a ridge or canyon); a certain portion of the wave energy is reflected back from the obstacle, the other penetrates through the region, interacts with the obstacle, and propagates beyond it; the question is, how much of the total energy is reflected and transmitted, respectively? Furthermore, we ask what form the wave solution in front of and behind the bottom feature will assume and how the wave field will depend on the basic controlling parameters such as the amplitude and modal structure of the incoming wave, the shape, height, and width of the bottom topographic features, and the parameters of stratification.

Mathematically, this problem is very similar to that considered in Section 2.1.1. The only difference is that, to lowest order, (2.7) should be solved subject to homogeneous conditions with $\Psi_0 = 0$, instead of the inhomogeneous boundary conditions at the free surface, because the incident linear internal wave produces no residual flux. Thus, the first two lowest-order BVPs are as follows.

To *zeroth order*,

$$\left.\begin{aligned} \Psi^{(0)}_{xx} - \Psi^{(0)}_{z_1 z_1} - c_1 c_3 \Psi^{(0)} &= 0, \\ \Psi^{(0)}(x, 0) &= 0, \\ \Psi^{(0)}(x, h_0) &= 0. \end{aligned}\right\} \tag{2.23}$$

To *first order*,

$$\left.\begin{aligned} \Psi^{(1)}_{xx} - \Psi^{(1)}_{z_1 z_1} - c_1 c_3 \Psi^{(1)} &= 0, \\ \Psi^{(1)}(x, 0) &= 0, \\ \Psi^{(1)}(x, h_0) &= h_0 r_1(x) \left.\frac{\partial \Psi^{(0)}}{\partial z_1}\right|_{z_1 = h_0}. \end{aligned}\right\} \tag{2.24}$$

First, we should find the solution of the zeroth-order problem (2.23), which, in fact, triggers the first-order BVP via the boundary condition at the bottom. With zero boundary conditions, the nontrivial solutions of problem (2.23) can only be propagating internal waves. Such a solution of problem (2.23) was constructed in detail in Section 1.5. It is given in (1.76), and its vertical structure and wavenumbers are defined by (1.74) and (1.75), respectively.

Thus, we may consider the progressing internal wave with mode number m and amplitude a_m^{inc},

$$\Psi^{(0)}(x, z_1) = a_m^{\text{inc}} q_m(z_1) \exp(-\imath k_m x), \tag{2.25}$$

propagating in the positive x-direction and interacting with the bottom topography located in the interval $-l < x < l$. Considering (2.25) as the basic zeroth-order solution, let us determine the small perturbations brought into the system by the presence of a small bottom roughness ("small" in the sense of (2.3)). These perturbations are found from the solution of the BVP (2.24).

As in Section 2.1.1, we wish to move the inhomogeneity from the boundary condition into the differential equation. To this end, we introduce a new function $\tilde{\psi}(x, z_1)$ according to

$$\tilde{\psi}(x, z_1) = \Psi^{(1)}(x, z_1) - z_1 r_1(x) h_0 \left(\frac{\partial \Psi^{(0)}}{\partial z_1} \right)_{z_1 = h_0}. \tag{2.26}$$

Then the BVP (2.24) can be rewritten as

$$\left. \begin{aligned} &\tilde{\psi}_{xx} - \tilde{\psi}_{z_1 z_1} - c_1 c_3 \tilde{\psi} \\ &\quad = z_1 \Gamma \{ [r_1(x) \exp(-\imath k_m x)]_{xx} - c_1 c_3 r_1(x) \exp(-\imath k_m x) \}, \\ &\qquad \tilde{\psi}(x, 0) = 0, \quad \tilde{\psi}(x, h_0) = 0, \end{aligned} \right\} \tag{2.27}$$

where

$$\Gamma = (-1)^m a_m^{\text{inc}} \pi m / h_0.$$

Applying Fourier transforms to (2.27), we deduce the two-point BVP,

$$\left. \begin{aligned} \hat{\psi}_{z_1 z_1} + (k^2 + c_1 c_3) \hat{\psi} &= z_1 F_I(k), \\ \hat{\psi}(k, 0) = \hat{\psi}(k, h_0) &= 0, \end{aligned} \right\} \tag{2.28}$$

in which

$$F_I(k) = \frac{\Gamma}{2\pi} \left(k^2 + c_1 c_3 \right) \int_{-\infty}^{\infty} r_1(x) \exp[\imath (k - k_m) x] \, dx.$$

The Green function for (2.28) is given by (2.14). Proceeding then as in Section 2.1.1, we can write down the solution of (2.28) through use of the Green function as

$$\hat{\psi}(x, z_1) = \frac{F_I(k)}{k^2 + c_1 c_3} \left[\frac{z_1}{h_0} - f_2(z_1, k) \right], \tag{2.29}$$

where $f_2(z_1, k)$ is defined by (2.16). The inverse Fourier transform of (2.29) provides the integral representation of the solution $\tilde{\psi}(x, z_1)$, namely

$$\tilde{\psi}(k, z_1) = \frac{\Gamma}{2\pi} \int_{-\infty}^{\infty} \frac{F_I(k)}{k^2 + c_1 c_3} \left[\frac{z_1}{h_0} - f_2(z_1, k) \right] \exp(-\imath kx)\, dk. \tag{2.30}$$

Replacing the function $f_2(z_1, k)$ by its decomposition into elementary fractions, (2.17), the first-order solution to problem (2.24) reads as

$$\left.\begin{aligned} \Psi^{(1)}(x, z_1) &= -\Gamma \Bigg\{ f_1(z_1) r_1(x) \exp(-\imath k_m x) \\ &+ \sum_{j=1}^{\infty} \frac{j(-1)^j}{[(j\pi)^2 - c_1 c_3 h_0^2]} q_j(z_1) I_j \Bigg\}, \\ I_j = \int_{-\infty}^{\infty} \frac{k^2}{k^2 - k_j^2} & \left\{ \int_{-\infty}^{\infty} r_1(x) \exp[\imath(k - k_m)x]\, dx \right\} \exp(-\imath kx)\, dk. \end{aligned}\right\} \tag{2.31}$$

The calculation of the integrals I_j for particular shapes of the bottom topography $r_1(x)$ is performed by contour integration and use of the residue theorem and Jordan's lemma. The integration contours are chosen so as to satisfy the radiation conditions; i.e. beyond the bottom obstruction all generated modes (except the incoming wave) must propagate away from the source of generation.

We deduce below the final form of the solution away from the ridge, when its form is given by expression (2.18) and, alternatively, by

$$r_1(x) = (1 - |x|/l) \quad \text{at} \quad |x| < l \tag{2.32}$$

(lines 1 and 2 in Figure 2.1). In both cases, the solution $\Psi^{(1)}$ for $|x| > l$, in areas I and III, can be represented as a sum of radiated modes propagating away from the source of generation; it is given by

$$\Psi^{(1)}(x, z_1) = \sum_{j=1}^{\infty} q_j(z_1) \times \begin{cases} b_j \exp\left\{\imath \left[k_j x + \pi \left(m + j + \frac{1}{2}\right)\right]\right\}, & x < -l, \\ a_j \exp\left\{\imath \left[-k_j x + \pi \left(m + j + \frac{1}{2}\right)\right]\right\}, & x > l. \end{cases} \tag{2.33}$$

Thus, wave perturbations $\Psi^{(1)}(x, z_1)$ produced by the interaction of an incident internal wave $\Psi^{(0)}(x, z_1)$ with a small bottom roughness consist outside its boundaries (at $|x| > l$) of a sum of radiated baroclinic modes. The vertical structure $q_j(z_1)$ of the generated modes (2.33) was already analyzed in Section 1.5 (see also Figure 1.7(c)).

Taking into account the asymptotic expansion (2.5) of the stream function, and the zeroth- and first-order solutions (2.25) and (2.33), the stream function, accurate to $O(\varepsilon_0^2)$, may be written as

$$\Psi(x, z_1) = a_m^{\text{inc}} q_m(z_1) \exp(-\imath k_m x) + \varepsilon_0 \sum_{j=1}^{\infty} q_j(z_1)$$
$$\times \begin{cases} b_j \exp\left\{\imath\left[k_j x + \pi\left(m + j + \frac{1}{2}\right)\right]\right\}, & x < -l, \\ a_j \exp\left\{\imath\left[-k_j x + \pi\left(m + j + \frac{1}{2}\right)\right]\right\}, & x > l. \end{cases} \tag{2.34}$$

The amplitudes of the generated waves depend on many model parameters and are given by the function $a_j(l)$. For profile (2.18) they are

$$\left.\begin{aligned} b_j(l) &= \frac{\pi^4 a_m^{\text{inc}} m j}{h_0^2} \frac{\sin[(k_m + k_j)l]}{k_j(k_m + k_j)[(k_m + k_j)^2 l^2 - \pi^2]}, \\ a_j(l) &= -\frac{\pi^4 a_m^{\text{inc}} m j}{h_0^2} \frac{\sin[(k_m - k_j)l]}{k_j(k_m - k_j)[(k_m - k_j)^2 l^2 - \pi^2]}. \end{aligned}\right\} \tag{2.35}$$

The positive and negative signs of the wavenumber k_j of the jth generated mode correspond to the waves generated at $x < -l$ and $x > l$, respectively. Although the dependence of $a_j(l)$ looks symmetrical with respect to the wavenumber of the waves generated to the left and right of the ridge, it is not; in fact, the reflected waves (for $x < -l$) are usually much weaker than the waves transmitted through and behind the obstacle (at $x > l$). This is a consequence of the initial anisotropy of the system when the energy flux of the incoming wave initially has a prescribed direction (from left to right, in our case). Thus, the internal modes, excited at $x < -l$, propagate upstream, whilst waves excited at $x > l$ propagate downstream (relative to the incoming wave). Note that, unlike the considered case, amplitudes of internal waves generated by a barotropic tidal flux (2.22) have equal amplitudes on both sides of the obstacle.

The quasiperiodic dependence of the amplitudes a_j and b_j on the width of the ridge is a consequence of the resonating character of the internal wave interaction with the bottom obstacle (the singularities at $j = m$ are removable, when the identifier of the generated baroclinic mode, j, coincides with that of the incoming mode, m, and when $(k_m \pm k_j)l = \pm\pi$). The amplitude of the jth ($j \neq m$) mode is equal to zero when the half-width of the ridge equals $l = n\pi(k_m \pm k_j)^{-1}$, $n = \pm 2, \pm 3, \ldots$ (n is the number of the extremum of the curve $a_j(l)$).

Between two zeros of the function $a_j(l)$ there is a local maximum. Its position on the x-axis can be found by solving the equation $da_j(l)/dl = 0$. With an increase of l, the values of the local maxima decrease, and at large l we have the asymptotic behavior $a_j(l) \sim l^{-2}$.

The above solution was obtained for the special bottom profile (2.18), which permits analytical calculation of the integrals I_j. To understand how general the derived inferences concerning the resonance character of internal wave dynamics over bottom features are, profiles of several different ridges were studied. For instance, repeating the above procedure for ridge (2.32), i.e. line 2 in Figure 2.1, we find the following dependence of the amplitudes of the generated waves on the width of the ridge:

$$\left.\begin{aligned} b_j &= jm\pi^2 \frac{\sin^2[(k_m + k_j)l/2]}{h_0^2[(k_m + k_j)l/2]^2} a_m^{\text{inc}}, \\ a_j &= jm\pi^2 \frac{\sin^2[(k_m - k_j)l/2]}{h_0^2[(k_m - k_j)l/2]^2} a_m^{\text{inc}}. \end{aligned}\right\} \tag{2.36}$$

It is seen that, similarly to (2.34), the dependence of (2.36) on l is also quasiperiodic. However, contrary to (2.34), where the periodicity $a_j = a_j(l)$ was defined by the function $\sin[(k_m - k_j)l]$, here the quasiperiodic character of the solution is set by the expression $\sin^2[(k_m - k_j)l/2]$. The asymptotic decrease of the local maxima of the function $a_j(l)$ with increasing width of the ridge in (2.36) is also different when compared with the similar law in (2.35). Moreover, at large values of l, the maxima $a_j(l)$ are proportional to l^{-1} instead of l^{-2} as for (2.35). However, these distinctions do not contradict the basic conclusion on the resonating behavior of the internal wave over the bottom topography.

2.2 Numerical model for large bottom obstacles

The linear theory described in Section 2.1 was based on the application of the perturbation method. It is valid only for underwater obstacles that are "small" (in the sense of (2.3)). The assumption $\varepsilon_0 \ll 1$ provided us with the opportunity to find analytical solutions of the problems of generation and scattering of internal waves. Unfortunately, assumption (2.3) is not valid for all underwater oceanic ridges and banks. A further restriction of the developed theory is that it cannot be applied directly to slope-shelf areas, in which the basic part of the tidal energy is transformed from the barotropic to the baroclinic modes. This reasoning is motivation for us to develop an algorithm which is valid without any restriction on the height of the bottom protuberance. Such a model will be constructed on the basis of refs. [241] and [249].

The formulation of the problem is the same as in Section 2.1.

A barotropic or baroclinic wave, propagating in a basin of constant depth (in region I, Figure 1.3) interacts with the bottom topography (area II), localized at $|x| \leq l$; the task is to find the characteristics of the resulting internal wave field, generated over the bottom obstacle and beyond it.

As was shown in Section 1.4, the governing system describing the dynamics of infinitesimal internal waves in an ocean of variable depth is reduced to equation (1.52) for the function $\Psi(x, z_1)$. So, the BVP describing the dynamics of internal waves in an ocean of variable depth reads as (see (1.64))

$$\left.\begin{aligned} \Psi_{xx} - \Psi_{z_1 z_1} + p(z)\Psi &= 0, \\ \Psi = 0 \quad \text{at} \quad z_1 &= 0, \\ \Psi = \alpha^{-1/2}(-H(x))\Psi_0 \quad \text{at} \quad z_1 &= h(x). \end{aligned}\right\} \tag{2.37}$$

Before we start to construct the numerical procedure for an ocean of variable depth, let us discuss some peculiarities of the solution for basins of constant depth. Infinitesimal waves in a horizontally homogeneous medium of constant depth do not interact with each other; i.e. in the linear approach, baroclinic modes propagate independently. The solution of the BVP (2.37) in the areas of constant depth I and III can be presented as a superposition of external forcing (barotropic flow or incident baroclinic wave) and radiated internal waves, generated by the forcing and propagating from the source of generation (i.e. from region II) as follows:

$$\Psi(x, z_1) = \Phi_1(x, z_1) + \sum_{j=1}^{\infty} b_j q_{1j}(z_1) \exp[\imath(k_{1j}x + \chi_j)], \quad x < -l, \tag{2.38}$$

$$\Psi(x, z_1) = \Phi_3(x, z_1) + \sum_{j=1}^{\infty} a_j q_{3j}(z_1) \exp[\imath(-k_{3j}x + \phi_j)], \quad x > l. \tag{2.39}$$

Here, a_j, ϕ_j $(j = 1, 2, 3, \ldots)$ are the amplitudes and phases of the generated modes with number j, propagating in the region $x > l$ to the right with horizontal wavenumber k_{3j}. Similarly, b_j, χ_j, k_{1j} $(j = 1, 2, 3, \ldots)$ are the amplitude, phase, and wavenumber, respectively, of the jth mode in the area $x < -l$, propagating to the left. The functions $q_{ij}(z)$ $(i = 1, 3;\ j = 1, 2, 3, \ldots)$ are found from the BVP (1.67). Representations (2.38) and (2.39) are written in such a way as to satisfy the Sommerfeld radiation condition, which states that the generated wave must only propagate away from the source of generation, not towards it.

When internal waves are generated by the barotropic tidal flow, the functions $\Phi_i(x, z_1)$ $(i = 1; 3)$ for the external forcing Ψ_0 were found in (2.9) and (2.10) as follows:

$$\left.\begin{aligned} \Phi_1(x, z_1) &= -\Psi_0 \frac{z(z_1)}{\alpha^{1/2}[z(z_1)]H_1}, \\ \Phi_3(x, z_1) &= -\Psi_0 \frac{z(z_1)}{\alpha^{1/2}[z(z_1)]H_3}. \end{aligned}\right\} \tag{2.40}$$

Alternatively, for the scattering of a baroclinic wave (a single baroclinic mode with number m) the external forcing is defined only in zone I (the incoming wave) so that

$$\left.\begin{aligned}\Phi_1(x, z_1) &= a_m^{\text{inc}} q_{1m}(z_1) \exp[\imath(k_{1m}x)],\\ \Phi_3(x, z_1) &= 0,\end{aligned}\right\} \tag{2.41}$$

in which a_m^{inc} is the amplitude of the mth mode.

For the definition of the unknown amplitudes a_j, b_j and phases ϕ_j, χ_j, it is necessary to find a relation between the two expansions (2.38) and (2.39). This can be done with the aid of the solution of the BVP (2.37) in area II. The differential equation in this BVP is of hyperbolic type, and the two families of characteristic lines

$$x + z_1 = \mathscr{X}, \quad x - z_1 = \mathscr{Z} \tag{2.42}$$

transform it to the equivalent canonical form $\Psi_{\mathscr{X}\mathscr{Z}} + \frac{1}{4}p(\mathscr{X}, \mathscr{Z})\Psi = 0$. In the last case its characteristic variables can be simply found from (2.42):

$$\zeta = \tfrac{1}{2}(\mathscr{X} + \mathscr{Z}), \quad \eta = \tfrac{1}{2}(\mathscr{X} - \mathscr{Z}).$$

To construct the solution of the BVP (2.37) in area II, we can use the Riemann method [42]. This method uses the fact that the solution of a hyperbolic equation can be found in a region $\mathscr{G}$ bounded by two characteristic lines (2.42) and any plane noncharacteristic curve $\bar{\mathscr{G}}$, provided the function and also its derivatives are defined at this boundary $\bar{\mathscr{G}}$. In our case, it is reasonable to choose the free surface as the external boundary $\bar{\mathscr{G}}$ of the physical region to be analyzed. Consider any point $A(x, z_1)$ in the region occupied by water, see Figure 2.4. The lines AB and AC are the characteristic lines (2.42) through A which intersect the free surface at the points $B(x - z_1, 0)$ and $C(x + z_1, 0)$.

The philosophy of the Riemann approach is now apparent. Central to the method is the construction of the linear operator

$$\mathscr{M}[\Psi] = \Psi_{xx} - \Psi_{z_1 z_1} + p(z)\Psi$$

and its adjoint[5]

$$\mathscr{M}^*[\mathcal{R}] = \mathcal{R}_{xx} - \mathcal{R}_{z_1 z_1} + p(z)\mathcal{R},$$

[5] In the present case, the operator $\mathscr{M}^*$ coincides with $\mathscr{M}$ because the initial differential equation (2.37) is without first derivatives.

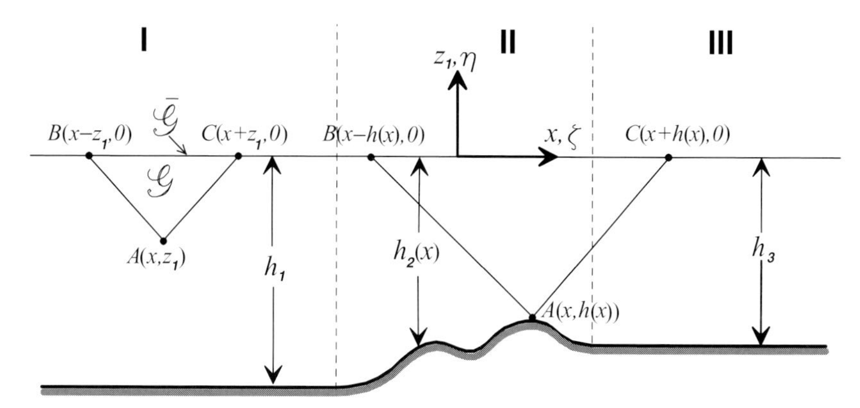

Figure 2.4. Schematic representation of the computational domain $\mathscr{G}$ (left triangle ABC) for the definition of the stream function $\Psi(x, z_1)$ at any point $A(x, z_1)$; the right triangle is the same but for the functional integral equation (2.49) when point $A(x, h(x))$ is located at the bottom.

which is determined by the compatibility condition

$$\mathcal{R}\mathscr{M}[\Psi] - \Psi\mathscr{M}^*[\mathcal{R}] = \frac{\partial}{\partial x}\left(\mathcal{R}\frac{\partial \Psi}{\partial x} - \Psi\frac{\partial \mathcal{R}}{\partial x}\right) + \frac{\partial}{\partial z_1}\left(\Psi\frac{\partial \mathcal{R}}{\partial z_1} - \mathcal{R}\frac{\partial \Psi}{\partial z_1}\right). \tag{2.43}$$

By integrating equation (2.43) over the region $\mathscr{G}$ (triangle ABC in Figure 2.4) and applying Stokes' theorem to its right side, we have

$$\iint_{BAC} (\mathcal{R}\mathscr{M}[\Psi] - \Psi\mathscr{M}^*[\mathcal{R}])\, d\xi\, d\eta$$
$$= \frac{1}{2}\int_{B\to A\to C\to B}\left[\mathcal{R}\left(\frac{\partial \Psi}{\partial \zeta}d\eta + \frac{\partial \Psi}{\partial \eta}d\zeta\right) - \Psi\left(\frac{\partial \mathcal{R}}{\partial \zeta}d\eta + \frac{\partial \mathcal{R}}{\partial \eta}d\zeta\right)\right]. \tag{2.44}$$

In this equation, ζ, η are the variables (coordinates) in the domain $\mathscr{G}$. Then we can find the solution Ψ of the BVP (2.37) by expressing it in terms of its Cauchy data and in terms of a simple solution to the adjoint differential equation, namely the Gursa problem

$$\left.\begin{aligned} \mathscr{M}^*[\mathcal{R}] &= 0, \\ \mathcal{R} = 1 &\quad \text{along } AC, \\ \mathcal{R} = 1 &\quad \text{along } BA. \end{aligned}\right\} \tag{2.45}$$

Riemann noticed that it is often easier to solve the adjoint boundary value problem (2.45) for the function $\mathcal{R}(x, z_1; \zeta, \eta)$, called the Riemann function, and, once one has solved it, the solution of the BVP (2.37) is immediate. Taking into account

(2.45), (2.44) is simplified to

$$0 = \underbrace{\int_{BA} \left[\mathcal{R} \left(\frac{\partial \Psi}{\partial \zeta} d\eta + \frac{\partial \Psi}{\partial \eta} d\zeta \right) - \Psi \left(\frac{\partial \mathcal{R}}{\partial \zeta} d\eta + \frac{\partial \mathcal{R}}{\partial \eta} d\zeta \right) \right]}_{I_1}$$

$$+ \underbrace{\int_{AC} \left[\mathcal{R} \left(\frac{\partial \Psi}{\partial \zeta} d\eta + \frac{\partial \Psi}{\partial \eta} d\zeta \right) - \Psi \left(\frac{\partial \mathcal{R}}{\partial \zeta} d\eta + \frac{\partial \mathcal{R}}{\partial \eta} d\zeta \right) \right]}_{I_2}$$

$$+ \underbrace{\int_{CB} \left[\mathcal{R} \left(\frac{\partial \Psi}{\partial \zeta} d\eta + \frac{\partial \Psi}{\partial \eta} d\zeta \right) - \Psi \left(\frac{\partial \mathcal{R}}{\partial \zeta} d\eta + \frac{\partial \mathcal{R}}{\partial \eta} d\zeta \right) \right]}_{I_3}. \tag{2.46}$$

With a little scrutiny of the integrals I_1 and I_2, and with the help of the boundary conditions for the Riemann function (2.45), we find that

$$I_1 \equiv -\Psi_A + \Psi_B,$$
$$I_2 \equiv -\Psi_A + \Psi_C.$$

Then, the solution of the BVP (2.37) at the point $A(x, z_1)$ of the two-dimensional region $\mathcal{G}$, restricted by the contour BC and the two characteristics AB and AC, can be written in integral form as

$$\Psi(x, z_1) = \frac{\Psi_B + \Psi_C}{2}$$
$$+ \frac{1}{2} \int_{CB} \left[\mathcal{R} \left(\frac{\partial \Psi}{\partial \zeta} d\eta + \frac{\partial \Psi}{\partial \eta} d\zeta \right) - \Psi \left(\frac{\partial \mathcal{R}}{\partial \zeta} d\eta + \frac{\partial \mathcal{R}}{\partial \eta} d\zeta \right) \right]. \tag{2.47}$$

In this equation, x and z_1 are the parameters (coordinates) identifying the point $A(x, z_1)$. For the moment, we will assume that $\mathcal{R}(x, z_1; \zeta, \eta)$ is a known function; more details in finding this solution will be discussed at the end of this section. However, (2.47) is valid provided that no one characteristic line crosses the contour of integration $\bar{\mathcal{G}}$ more than once.

Inspection of (2.47) shows that the solution at point $A(x, z_1)$ can be found if the function Ψ and also its derivatives Ψ_x and Ψ_{z_1} are known at the surface segment BC.

Let us simplify formula (2.47). The contour $\bar{\mathcal{G}}$ (see Figure 2.4) in our case is the free surface $z_1 = 0$, where $\Psi = 0$ (because of the boundary condition). Moreover, this means that $\Psi_x = 0$, and hence $\Psi_\zeta = 0$. As a consequence, formula (2.47) takes now the simpler form

$$\Psi(x, z_1) = \frac{1}{2} \int\limits_{x-z_1}^{x+z_1} \mathcal{R}(x, z_1; \zeta, 0) \left[\frac{\partial \Psi(\zeta, \eta)}{\partial \eta} \right]_{\eta=0} d\zeta. \tag{2.48}$$

This formula says that the value of the stream function at any point (x, z_1) of the oceanic region can be expressed in terms of the horizontal velocity $[\partial \Psi(\zeta, \eta)/\partial \eta]_{\eta=0}$ at the free surface alone (the only unknown function).

The next important step is to satisfy the boundary condition at the bottom. As (2.48) is also valid for the bottom, its application to any point $A(x, h(x))$ at the bottom surface $z_1 = h(x)$ together with the boundary condition leads to the following integral functional equation:

$$\int\limits_{x-h(x)}^{x+h(x)} \mathcal{R}(x, h(x); \zeta, 0) \left[\frac{\partial \Psi(\zeta, \eta)}{\partial \eta} \right]_{\eta=0} d\zeta = Q(x), \tag{2.49}$$

where the function $Q(x)$ depends on the considered phenomenon and is known: (i) it is equal to zero for problem (1.65) i.e. scattering of an incident internal wave on the bottom topography; and (ii) it differs from zero, e.g. if internal waves are generated by a barotropic flux over a topographic obstruction, i.e. BVP (1.64). So, in accordance with this, the function $Q(x)$ takes the form

$$Q(x) = \begin{cases} \dfrac{2\Psi_0}{\alpha^{1/2}(-H(x))}, & \text{wave generation;} \\ 0, & \text{wave scattering.} \end{cases} \tag{2.50}$$

In summary, we have now obtained the two series (2.38) and (2.39), which represent the solution of the problem (2.37) in zones I and III (with unknown wave amplitudes a_j and b_j), and we have the functional equation (2.49) at our disposal which must be used to find the necessary relations between the amplitudes of the waves radiated from the source of generation. In the following, we describe (step by step) the procedure for obtaining all the necessary relations. However, before doing this we introduce a new function

$$S_1(\zeta) := \left[\frac{\partial \Psi(\zeta, \eta)}{\partial \eta} \right]_{\eta=0}, \tag{2.51}$$

which will make our statements more concise.

The following description in Steps 1 to 6 is somewhat technical and may be omitted at first.

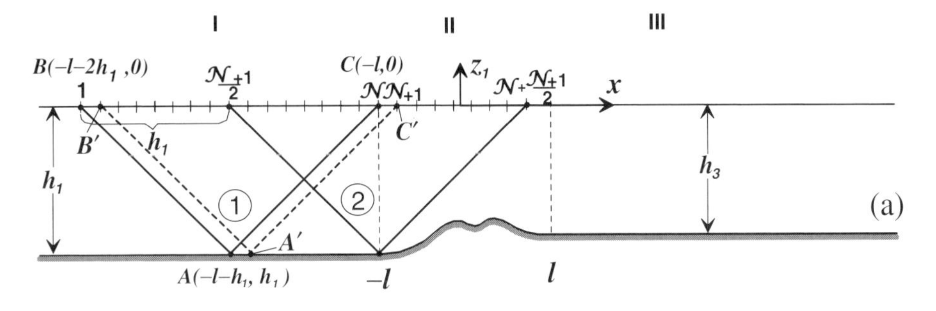

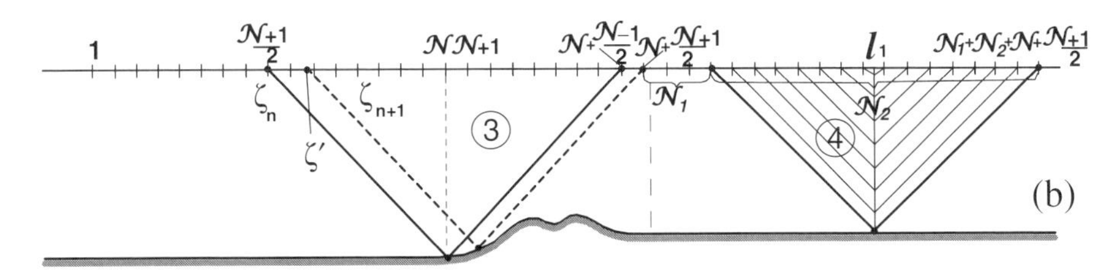

Figure 2.5. Schematic representation of the numerical procedure for the coupling of the wave solutions in zones I and III via the functional equation (2.49). Characteristics in (a) the constant depth region and (b) the variable depth zone.

Step 1: Introduction of the grid

First, we draw the characteristic line $x + z_1 = -l$ emerging from point $C(-l, 0)$, located at the free surface (at the position of the left boundary point of the bottom topography) and pointing to the left into region I, intersecting with the bottom at point $A(-l - h_1, h_1)$; then we draw the second characteristic line $x - z_1 = -l - 2h_1$ from point $A(-l - h_1, h_1)$ to its intersection with the free surface at point $B(-l - 2h_1, 0)$ (see Figure 2.5(a)). On the horizontal axis, we introduce a grid with spatial step $\Delta x = 2h_1/(\mathcal{N} - 1)$, where $\mathcal{N}$ is the number of nodes on the segment BC, so that the points $\zeta_1 = -l - 2h_1$ and $\zeta_{\mathcal{N}} = -l$ are located at the edges of the segment BC (Figure 2.5(a)). According to (2.38), the horizontal velocity $S_1(\zeta)$ defined by (2.51) at the free surface at any node n of the grid is defined by

$$S_1(\zeta_n) = D_1(\zeta_n) + \sum_{j=1}^{\infty} B_j C_{1,j}(\zeta_n), \tag{2.52}$$

where

$$D_1(\zeta_n) = \left. \frac{\partial \Phi_1(\zeta, \eta)}{\partial \eta} \right|_{\substack{\zeta=\zeta_n \\ \eta=0}},$$

$$C_{1,j}(\zeta_n) = \exp(\imath k_{1j} \zeta_n),$$

$$B_j = b_j \exp(\imath \chi_j) \left(\frac{dq_{1j}}{d\eta} \right)_{\eta=0}.$$

Index 1 corresponds to region I, n denotes the number of units of the grid, and j is the number of the baroclinic mode.

Step 2: Finding the recurrence relation

Let us shift the characteristic triangle ABC one spatial step to the right (see the dashed lines in Figure 2.5(a)) and let us try to express how the function $S_1(\zeta)$ in node $n = \mathcal{N} + 1$ can be expressed via functions $S_1(\zeta_n)$ in nodes $n = 1, 2, 3, \ldots, \mathcal{N}$ (located in the area where the bottom is flat, $x < -l$). This can be achieved using the functional equation (2.49): we simply replace the integral in (2.49) by any approximate quadrature formula and apply it to the segment $B'C'$ (the segment $[\zeta_2, \zeta_{\mathcal{N}+1}]$). Mathematically, the integral equation (2.49) can, in this case, be approximated by

$$\begin{aligned}\Delta x[\overbrace{\mathcal{R}(x, h_1; \zeta_2, 0)}^{=1} S_1(\zeta_2) + 2\mathcal{R}(x, h_1; \zeta_3, 0)S_1(\zeta_3) + \cdots \\ + 2\mathcal{R}(x, h_1; \zeta_{\mathcal{N}}, 0)S_1(\zeta_{\mathcal{N}}) \\ + \underbrace{\mathcal{R}(x, h_1; \zeta_{\mathcal{N}+1}, 0)}_{=1} S_1(\zeta_{\mathcal{N}+1})]/2 = Q(x),\end{aligned} \tag{2.53}$$

where $x = -l - h_1 + \Delta x$ at the bottom (point A′ in Figure 2.5(a)) corresponds to the value $\zeta_{(\mathcal{N}+1)/2+1}$ at the free surface. The Newton–Cotes formula of the first order was used here to replace integral (2.49) by its approximate representation.

In the first and last terms of equation (2.53), the Riemann function, according to (2.45), is equal to unity. Taking this into account, the function $S_1(\zeta)$ at point $\mathcal{N} + 1$, $S_1(\zeta_{\mathcal{N}+1})$, is expressible via $S_1(\zeta_2)$, $S_1(\zeta_3), \ldots, S_1(\zeta_{\mathcal{N}})$ as follows:

$$\begin{aligned}S_1(\zeta_{\mathcal{N}+1}) = \frac{2Q(x)}{\Delta x} - [S_1(\zeta_2) + 2\mathcal{R}(x, h_1; \zeta_3, 0)S_1(\zeta_3) + \cdots \\ + 2\mathcal{R}(x, h_1; \zeta_{\mathcal{N}}, 0)S_1(\zeta_{\mathcal{N}})].\end{aligned} \tag{2.54}$$

This formula provides a relation between the function S_1 evaluated at $\zeta_{\mathcal{N}+1}$ and the horizontal velocities S_1 in the grid points $2, 3, 4, \ldots, \mathcal{N}$. Note that $S_1(\zeta_2), S_1(\zeta_3), \ldots, S_1(\zeta_{\mathcal{N}})$ in (2.54) could be replaced by (2.52) in order to show explicitly the unknown amplitudes B_j of the internal waves in zone III, which must be found. If this is done, (2.54) takes the form

$$S_1(\zeta_{\mathcal{N}+1}) = \mathscr{D}_1(\zeta_2, \zeta_3, \ldots, \zeta_{\mathcal{N}}) + \sum_{j=1}^{\infty} B_j \mathscr{C}_{1,j}(\zeta_2, \zeta_3, \ldots, \zeta_{\mathcal{N}}), \tag{2.55}$$

where $\mathscr{D}_1(\zeta)$ and $\mathscr{C}_{1,j}(\zeta)$ are piecewise linear functions defined at grid point $\mathcal{N}$.

Step 3: Upstream procedure

So, an expression for S_1 at point $\mathcal{N}+1$ has been found via information at the previous $\mathcal{N}$ nodes. If we now repeat the above procedure for the characteristic triangle shifted by two steps ($2\Delta x$) to the right and write formulas similar to (2.53) and (2.54) but for $S_1(\zeta_{\mathcal{N}+2})$ in those formulas, then, instead of an explicit expression for $S_1(\zeta_{\mathcal{N}+1})$, a representation for $S_1(\zeta_{\mathcal{N}+2})$ is obtained. In particular, because of (2.55), an expression $S_1(\zeta_{\mathcal{N}+2})$ via the first $\mathcal{N}$ nodes is obtained. Following the procedure (2.53)–(2.55) and shifting the characteristic triangle step by step from the left to the right, we can express the horizontal velocity $S_1(\zeta_n)$ at any point ζ_n (over the bottom topography and behind it in zone III for $x > l$) as a function of the first $\mathcal{N}$ nodes,

$$S_1(\zeta_n) = \mathscr{D}_1(\zeta_n) + \sum_{j=1}^{\infty} B_j \mathscr{C}_{1,j}(\zeta_n), \tag{2.56}$$

where $\mathscr{D}_1(\zeta)$ and $\mathscr{C}_{1,j}(\zeta)$ are known piecewise functions defined by application of the approximate analog (2.53) of the integral equation (2.49), and B_j are unknown amplitudes.

It should be noted that nontrivial difficulties with the application of the outlined procedure arise when the characteristic lines cross the bottom in region II where the basin depth is not constant. In this case, a characteristic line at a nodal point of the free surface may originate from a surface point that may lie between two nodal points. Such a situation is presented in Figure 2.5(b) by the dashed line. The intersection point ζ' lies between nodes ζ_n and ζ_{n+1}. In this case, interpolation at point ζ' via the neighboring points must be employed. Yet another problem, namely multiple intersections of a characteristic line with the bottom surface, is possible; this will be discussed in Section 2.5. This defines the so-called "supercritical case."

Step 4: Orthogonalization

Having established formula (2.56) for the horizontal velocity $S_1(\zeta_n)$ in zone III expressed via the amplitude B_j, we can construct the vertical profile of the stream function at any vertical cross-section $x = l_1$ by use of formula (2.48):

$$\Psi(l_1, z_1) = \frac{1}{2} \int_{l_1 - z_1}^{l_1 + z_1} \mathcal{R}(l_1, z_1; \zeta, 0) \left[\frac{\partial \Psi(\zeta, \eta)}{\partial \eta} \right]_{\eta=0} d\zeta$$

$$= \mathcal{D}_1(l_1) + \sum_{j=1}^{\infty} B_j \mathcal{C}_{1,j}(l_1, z_1). \tag{2.57}$$

Here, the functions $\mathcal{C}_{1,j}$ and $\mathcal{D}_1$ are obtained from the functions $\mathscr{C}_{1,j}$ and $\mathscr{D}_1$ after discretizing the integral (2.57) and accordingly substituting (2.56). Alternatively, the stream function $\Psi(l_1, z_1)$ is defined by (2.39) by

$$\Psi(l_1, z_1) = \Phi_3(l_1, z_1) + \sum_{j=1}^{\infty} A_j q_{3j}(z_1) \exp(-\imath k_{3j} l_1), \tag{2.58}$$

where $A_j = a_j \exp(\imath\phi_j)$. The functions $q_{3j}(z_1)$ are the eigenfunctions of the BVP (1.67), and hence they are orthogonal to each other on the vertical segment $[0, h_3]$. Equating the right-hand sides of (2.57) and (2.58) and performing the procedure of orthogonalization, i.e. multiplying the emerging equation successively by $q_{3n}(z_1)$ with $n = 1, 2, 3, \ldots$ and integrating over z_1 from 0 to h_3, the following system of linear algebraic equations for the amplitudes A_j and B_j is found:

$$A_n \exp(-\imath k_{3,n} l_1) = \sum_{j=1}^{\infty} B_j \hat{C}_{1j,n}(l_1) + Q_{1n}(l_1), \tag{2.59}$$

where $n = 1, 2, 3, \ldots$, and

$$\hat{C}_{1j,n}(l_1) = \int_0^{h_3} \mathcal{C}_{1,j}(l_1, z_1) q_{3n}(z_1)\, dz_1,$$

$$Q_{1n}(l_1) = \int_0^{h_3} [\mathcal{D}_1(l_1) - \Phi_3(l_1, z_1)]\, q_{3n}(z_1)\, dz_1.$$

Step 5: Downstream procedure and orthogonalization

The procedure outlined in Steps 1 to 4 to find the solution is repeated by shifting the characteristic triangle from zone III to zone I. Then the stream function at the vertical $x = -l_1$ (to the left from the bottom topography) has the form

$$\Psi(-l_1, z_1) = \int_{-l_1 - z_1}^{-l_1 + z_1} \mathcal{R}(-l_1, h_1; \zeta, 0) \left[\frac{\partial \Psi(\zeta, \eta)}{\partial \eta}\right]_{\eta=0} d\zeta$$

$$= \mathcal{D}_3(-l_1) + \sum_{j=1}^{\infty} A_j \mathcal{C}_{3,j}(-l_1, z_1). \tag{2.60a}$$

According to (2.38), $\Psi(-l_1, z_1)$ can also be expressed as a sum of modes:

$$\Psi(-l_1, z_1) = \Phi_1(-l_1, z_1) + \sum_{j=1}^{\infty} B_j q_{1j}(z_1) \exp(-\imath k_{1j} l_1). \tag{2.60b}$$

Equating (2.60a) and (2.60b) and performing an orthogonalization equivalent to that performed above yields the system of equations

$$B_n \exp(-\imath k_{1,n} l_1) = \sum_{j=1}^{\infty} A_j \hat{C}_{3j,n}(-l_1) + Q_{3n}(-l_1), \tag{2.61}$$

where $n = 1, 2, 3, \ldots$, and

$$\hat{C}_{3j,n}(-l_1) = \int_0^{h_1} C_{3,j}(-l_1, z_1) q_{1n}(z_1)\, dz_1,$$

$$Q_3(-l_1) = \int_0^{h_1} [\mathcal{D}_3(-l_1) - \Phi_1(-l_1, z_1)]\, q_{1n}(z_1)\, dz_1.$$

Step 6: Truncation

Problem (2.37) has thus been reduced to the simultaneous solution of two systems of linear algebraic equations (2.59) and (2.61) for the unknown amplitudes of the generated internal waves A_j and B_j. The specific feature of the two systems is that they consist of an infinite number of equations with an infinite number of unknown amplitudes. This is a consequence of the representation of the solution in zones I and III as infinite series, (2.38) and (2.39). To find the solution of such an infinite system in a general case is impossible; therefore we use a procedure of truncation, a method which is commonly used in such cases; see, e.g., ref. [211].

Instead of the infinite system let us consider only the first $\mathcal{J}_0$ equations of (2.59) and (2.61) with $2\mathcal{J}_0$ unknowns. Thus, we truncate the summation over j at $j = \mathcal{J}_0$ and set $n = 1, 2, \ldots, \mathcal{J}_0$. Such a procedure of truncation introduces into the system a certain error in the values of the unknown amplitudes. The idea of restricting consideration to only the first $\mathcal{J}_0$ modes is based on the assumption that the major part of the energy of the barotropic flux or the incident baroclinic wave is transferred to the first $\mathcal{J}_0$ baroclinic modes. Of course, this is not valid in all circumstances. If the inclination of the bottom topography is close to critical (or even supercritical), this truncation procedure is no longer valid because it loses the physical basis: not only the lowest but also higher modes are effectively generated in such cases. This problem of wave dynamics near steep bottom topographies is discussed in Section 2.5. It is, however, hoped that most continental slopes and shelves of the World Ocean are subcritical, and thus the truncation procedure will be meaningful.

Thus, the amplitudes of the internal waves with numbers larger than $\mathcal{J}_0$ are assumed to be negligible. Of course, in each particular case the chosen number of equations depends on the steepness of the bottom topography. Naturally, with the growth of bottom steepness, the required number of equations increases.

The description given in Steps 1 to 6 of the semianalytical procedure of finding the solution to problem (2.37) requires that the Riemann function $\mathcal{R}(x, z_1; \zeta, \eta)$ is known. In the usual case, for an arbitrary distribution of the buoyancy frequency $N(z)$, the solution to the Gursa problem (2.45) can be found only numerically. However, there are several special cases of fluid stratification when (2.45) has an analytical solution. These are connected with the buoyancy frequency law given by formula (1.70), for which $p(z) = c_1 c_3$, and the solution to the adjoint Gursa problem (2.45) is simply achieved by setting a new variable $\mathscr{Y} = [(x - \zeta)^2 - (z_1 - \eta)^2]^{1/2}$ in the Riemann function $\mathcal{R}$. Such a manipulation leads to the equation

$$\mathcal{R}_{\mathscr{Y}\mathscr{Y}} + \frac{1}{\mathscr{Y}}\mathcal{R}_{\mathscr{Y}} - c_1 c_3 \mathcal{R} = 0,$$

and the Riemann function is equal to the Bessel function of the zeroth order, J_0,

$$\mathcal{R}(x, z_1; \zeta, \eta) = J_0\{c_1 c_3[(x - \zeta)^2 - (z_1 - \eta)^2]\},$$

which will be used below in some cases.

2.3 Wave dynamics over oceanic ridges: applicability of the perturbation method

At the beginning of this chapter we discussed the peculiarities of the generation of internal waves and wave scattering over localized bottom topographic features of small height. "Smallness" was introduced with the help of the parameter ε_0 in (2.3). However, the condition $\varepsilon_0 \ll 1$ does not mean that the topographic height $H_{\max}$ must really be much less than the total basin depth H_0. This is a consequence of the structure of the eigenfunctions q_j of the standard boundary value problem (1.67), which encounter large vertical changes in a layer of relatively large density changes; in the ocean abyss, where the vertical stratification of the fluid is relatively weak, such changes are less pronounced (see Figure 1.7). As a consequence, the height $h_{\max}$ of an underwater obstacle in the variables (x, z_1) is usually much smaller than the total depth h_0; this occurs when the mountain top lies below the pycnocline.

Some interesting results were obtained in Section 2.1 using the perturbation method. One of the basic conclusions was that the perturbation theory, which is based on the assumption $\varepsilon_0 \ll 1$, can be applied for most oceanic ridges of realistic height. To ascertain that this is true and, furthermore, to find the range of applicability of the underlying model, we constructed, in Section 2.2, the semianalytical method, which is valid without the mentioned restriction of small ε_0. This computational procedure will be applied in the following for ridges of arbitrary height, and the obtained results will be compared with analytical solutions.

2.3.1 Generation of internal waves

First, we apply the developed semianalytical procedure to the study of the generation of internal waves by a barotropic tidal flux over an underwater obstacle as in ref. [241]. We consider the ridge (2.18) with the intention of comparing the numerical results with the analytical solution (2.21).

As was already shown in Section 2.1, the generation mechanism has resonance character (see, for instance, formula (2.22) and Figure 2.3). This means that the amplitude of any jth mode is equal to zero, whenever the width of the ridge, $2l$, equals an integer multiple of the wavelength of this mode. In view of the fact that this result has been obtained for $\varepsilon_0 \ll 1$, one can assume that this conclusion, with some corrections, must also be valid for large obstacles. The search for the dependence of the amplitudes of the generated internal waves on the horizontal and vertical scales of the underwater ridge – width $2l$ and height $H_{\max}$ – was carried out for the following parameter values:

$$\left.\begin{array}{c} H_1 = H_3 = 4000\,\mathrm{m}, \quad N_\mathrm{p} = 5 \times 10^{-3}\,\mathrm{s}^{-1}, \quad H_\mathrm{p} = 100\,\mathrm{m}, \\ \Delta H_\mathrm{p} = 50\,\mathrm{m}, \quad T = 12.4\,\mathrm{h}, \quad \varphi = 30^\circ. \end{array}\right\} \tag{2.62}$$

The value of the amplitude of the vertically integrated water flux, $\Psi_0 = 200\,\mathrm{m}^2\,\mathrm{s}^{-1}$, corresponds to a barotropic speed in areas I and III equal to $5\,\mathrm{cm}\,\mathrm{s}^{-1}$.

The dependences of $a_\xi^{(1)}(l)$, obtained numerically for different heights of the ridge, are shown in Figure 2.6. The lines 1, 2, and 3 are constructed for $\varepsilon_0 = (42.1; 8.4; 1.6) \times 10^{-3}$ corresponding to $H_{\max} = (3.5; 2.5; 1)$ km. It is seen that the behavior of the functions $a_\xi^{(1)}(l)$, found numerically using the computational procedure described in Section 2.2, is qualitatively similar to the behavior of the dependences (2.22) obtained analytically and displayed in Figure 2.3: the amplitude $a_\xi^{(1)}(l)$ of the generated wave is a quasiperiodic function of the width of the ridge $2l$; moreover, the absolute values of the local maxima decrease asymptotically as $\sim l^{-2}$ when the width of the ridge increases (as was the case for the perturbation method in Section 2.1.1).

One should indicate also the evident discrepancies between the two solutions. If one compares curves (1)–(3) in Figure 2.6 with those in Figure 2.3, it is evident that the positions of the maxima and zeros of the functions $a_\xi^{(1)}(l)$ change with a change of the obstacle height $H_{\max}$, contrary to (2.22). Furthermore, whilst the positions of the zeros of the functions $a_\xi^{(j)}(l)$ were defined by the condition $\sin(k_j l) = 0$, independently of $h_{\max}$, here a distance Δl between the neighboring two zeros exists and decreases with increasing $H_{\max}$.

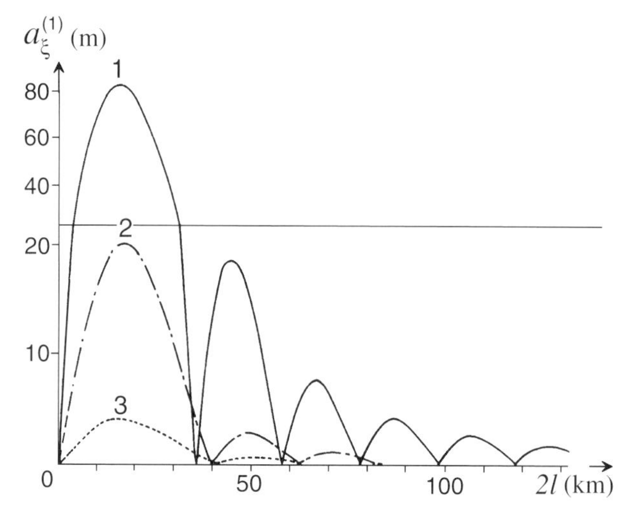

Figure 2.6. Amplitude of the first baroclinic mode generated by a barotropic tidal flux over the ridge (2.18) for different values of the height H_{max} versus the width of the ridge $2l$. The solid line (1) corresponds to $H_{max} = 3.5$ km ($\varepsilon_0 = 0.0421$), the dashed–dotted line (2) corresponds to $H_{max} = 2.5$ km ($\varepsilon_0 = 0.0084$), and the dotted line (3) corresponds to $H_{max} = 1$ km ($\varepsilon_0 = 0.0016$). All other parameters are given by (2.62).

This behavior becomes clear if one tries to understand the physical meaning of these maxima and zeros. To this end, let us calculate the time which an internal wave takes to travel through the bottom roughness from $x = -l$ to $x = l$. It can be estimated by the integral

$$T_1(2l) = \int_{-l}^{l} \frac{dx}{\tilde{c}_p(x)}, \tag{2.63}$$

where $\tilde{c}_p(x) = \sigma/\tilde{k}_1(x)$ and $\tilde{k}_1(x) = [(\pi/h(x))^2 - c_1c_3)]^{1/2}$ are the local phase speed and local wavenumber of the first baroclinic mode, respectively. If the depth of the basin is constant, $\tilde{k}_1(x) = k_1$, and this wavenumber coincides with the horizontal wavenumber of the first baroclinic mode of the BVP (1.67), then the integral (2.63) describes the time which the wave takes to pass the distance $2l$.

In the general case, when the basin depth is variable, it is not correct to consider the value $\tilde{\lambda}_1(x) = 2\pi/\tilde{k}_1(x)$ as the wavelength. However, if the steepness of an obstacle is small (as compared with the steepness of the characteristic lines), it is possible to consider the values $\tilde{\lambda}_1(x)$ and $\tilde{k}_1(x)$ as the local wavelength and local wavenumber on the interval $x \pm \Delta x$. If so, $\tilde{c}_p(x)$ has the meaning of a local phase speed, and the integral (2.63) calculates the time of the wave passage from one boundary of the underwater ridge to the other.

Table 2.1. *Errors with which the perturbation method defines the maxima and positions of zeros of the function $a_\xi^{(1)}(l)$.*

Note that n is the index number of the extremum.

	$\delta_1(\%)$			$\delta_2(\%)$		
	H_{max} (km)			H_{max} (km)		
n	1.0	2.5	3.5	1.0	2.5	3.5
1	1.2	4.7	13.7	1.7	8.7	31.1
2	3.6	14.2	34.5	0.8	4.4	25.5
3	8.9	18.3	25.5	1.8	7.9	28.7
4	5.7	16.2	22.0	2.5	10.2	35.5
5	6.2	12.4	19.4	3.1	15.4	42.6

Scrutiny of the integral (2.63), calculated for the extrema of the curves of Figure 2.6, shows that the values of $T_1(2l)$ for all three curves are with good accuracy (i.e. an error less than 1%) divisible by half the wave period, $T/2$. Thus, one can conclude that the two solutions (analytical and numerical) qualitatively enjoy similar behavior, namely they reflect a resonance mechanism of the generation of internal waves.

The deduced disagreement in shifting the extrema of the functions $a_\xi^{(j)}(l)$ when H_{max} increases is explained by the fact that internal waves become shorter over the bottom topography when the height of the bottom feature increases. So it takes a slightly longer time to propagate from $x = -l$ to $x = l$ (see (2.63)).

It is obvious using chracteristics and the Riemann function that the numerical solution is valid for any height of the bottom features, whereas the solution found by means of the perturbation method can formally be applied only when the condition $\varepsilon_0 \ll 1$ is satisfied. Let us compare the two solutions in order to define the range of applicability of the perturbation method in problems of the generation of internal waves in areas of small bottom roughnesses. To this end, we compare the dependences $a_\xi^{(1)}(l)$ obtained by the two different methods for different heights, H_{max}.

As discussed above, the behavior of the amplitude curves found analytically and numerically is very similar; compare Figures 2.3 and 2.6. So, for a representative analysis it makes sense to compare only the values of the local maxima and the positions of the zeros of the function $a_\xi^{(1)}(l)$ obtained by the two methods. The results of this comparison for the first baroclinic mode are given in Table 2.1. Calculations were performed for typical oceanic parameters, (2.62). Table 2.1 gives the values of the error $\delta_1 = [(a_\xi^{(1)\mathrm{nu}} - a_\xi^{(1)\mathrm{an}})/a_\xi^{(1)\mathrm{an}}] \times 100\%$ for three heights of the ridge $H_{max} = (1; 2.5; 3.5)$ km. The entries show the error with which the perturbation method allows calculation of the magnitude of the nth maximum of the function $a_\xi^{(1)}(l)$. Here, $a_\xi^{(1)\mathrm{nu}}$ is obtained numerically (Figure 2.6), whilst $a_\xi^{(1)\mathrm{an}}$ is obtained analytically (Figure 2.3). Table 2.1 also gives, for the same ridge heights, the values of the error $\delta_2 = \Delta l_n \lambda_1^{-1} \cdot 100\%$, where Δl_n is the distance between the positions of the nth zeros of the functions $a_\xi^{(1)\mathrm{nu}}(l)$ and $a_\xi^{(1)\mathrm{an}}(l)$, $\lambda_1 = 2\pi/k_1$. This value allows an estimate of the error in the definition of the local minimum of the curve $a_\xi^{(1)}(l)$.

From Table 2.1, it follows that for the considered parameters (2.62) and for heights of the ridges $H_{\max} \leq 1$ km the error introduced into the model by the application of the perturbation method in the definition of the maxima of $a_\xi^{(1)}(l)$ does not exceed 10%. For the first (absolute) maximum, this error equals 1.2%. The positions of the zeros of the function $a_\xi^{(1)}(l)$ are also relatively well described. For example, the position of the first minimum is determined with an error of 1.7%.

With the growth of the height of the ridge the difference in values of the amplitudes $a_\xi^{(1)\mathrm{nu}}$ and $a_\xi^{(1)\mathrm{an}}$, obtained by the two methods, increases. For instance, the maximum $a_\xi^{(1)\mathrm{an}}$ exceeds the corresponding value of the function $a_\xi^{(1)\mathrm{nu}}(l)$ by 14–35% when $H_{\max} = 3.5$ km, and the error in the definition of the positions of the minima becomes as large as 43%.

Summarizing the above results, one can conclude that the perturbation theory gives reasonably good results both qualitatively and quantitatively for the first five modes in the range of ridge heights $0 < H_{\max} \leq 3.5$ km at a total oceanic depth of $H_0 = 4$ km. This is a reasonably good result for the perturbation theory. Although the smallness parameter ε_0 for such obstacles was less than 0.05, the real nondimensional height $H_{\max}/H_0$ reached 0.875. Thus, the conclusion from this comparison is that the perturbation theory can be successfully applied for most oceanic ridges and banks.

The question "What is the dependence of the wave amplitudes upon the ridge height $H_{\max}$?" if we increase it (from 3.5 km to 4.0 km in our case) remains unanswered. The perturbation theory predicts a linear growth of the wave amplitude with an increase of $h_{\max}$, but the exact dependence is shown in Figure 2.7. It has been numerically constructed for the underwater ridge (2.18) with width $2l = 67$ km in such a way that the horizontal scale is linear for $h_{\max}/h_0$ and nonlinear for $H_{\max}$. As may be seen from Figure 2.7, the dependence $a_\xi^{(1)}(H_{\max})$ is nonmonotonic. It is almost linear at the beginning (at $H_{\max} < 3.5$ km or $\varepsilon_0 < 0.05$), as predicted by the perturbation theory, but at large values of the ridge heights, $H_{\max} > 3.5$ km, it exhibits a quasiperiodic character: the local maxima of $a_\xi^{(1)}(H_{\max})$ at $H_{\max} = (3750; 3825; 3865)$ m, etc., are alternating with points of local minima where the amplitude is equal to zero, at $H_{\max} = (3765; 3847; 3875)$ m, etc. With an increase of $H_{\max}$, the values of the local maxima grow, and the distance between the two following zeros decreases.

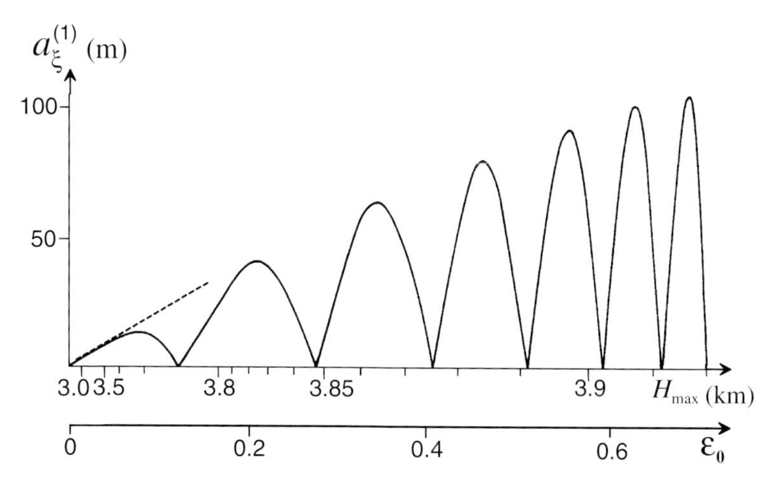

Figure 2.7. Amplitude of the first baroclinic mode generated by a barotropic tidal flux over the ridge (2.18) with width $2l = 67$ km versus H_{max}. All other parameters are as given in (2.62). Along with the height of the ridge, the axis $(0, \varepsilon_0)$ is also shown. The analytical solution obtained with the perturbation method is shown by the dashed line.

The analysis of the values of the integral (2.63), calculated for the points in Figure 2.7 where the function $a_\xi^{(1)}(H_{max})$ has its maxima, shows that, with good accuracy (discrepancy does not exceed 1%), $T_1(2l)$ is equal to half the tidal period of the barotropic flow. Thus, the resonant mechanism of the generation of internal waves found by the perturbation method for obstacles of small height is valid as it is for large obstacles.

Thus, we may state as a brief intermediate conclusion the following.

The perturbation method, developed originally in ref. [43] for "small" bottom roughnesses, for which $H_{max}/H_0 \ll 1$, can actually be used in a much wider range, if in the expansion, instead of the small parameter H_{max}/H_0, the relative height of an obstacle, $\varepsilon_0 = h_{max}/h_0$, and the variables (x, z_1) are used. For the majority of oceanic ridges and banks, for which the top is below the seasonal pycnocline, the condition $\varepsilon_0 \ll 1$ is valid.

2.3.2 Internal wave scattering

We now assume that, instead of the barotropic tide, an internal wave approaches an underwater ridge. The analytical solution, obtained in Section 2.1 for the problem of internal wave scattering, reveals a strong resonant dependence of amplitudes of reflected and transmitted waves from model parameters (see (2.35) and (2.36)). Formally, this solution is valid only for "small" bottom features.

For large bottom topographies the scattering in two dimensions was first studied by Baines [7], [8], who reduced the problem to solving a pair of Fredholm integral equations. In his formulation, the ocean has no upper surface, and an internal wave propagates from above onto an ocean bottom. Baines showed that the correct implementation of radiation conditions leads to the splitting of the incident wave into two reflected waves propagating along the characteristic lines of the wave equation. This process is not described by earlier theories; see, e.g., ref. [145]. In more recent studies by Müller and Liu [169], [170], a finite-depth ocean is considered, and the wave propagates in from the side. As distinct from the analytical solution of Larsen [131], Müller and Liu applied a numerical procedure to calculate reflected and transmitted waves for a large variety of bottom shapes (subcritical, supercritical). They described the redistribution of the incident wave energy among several scattered baroclinic modes and found that supercritical topographies are much more effective for such a redistribution.

Here we study the scattering of internal waves by an underwater ridge on the basis of the results of ref. [247]. The problem is considered for the first five incident baroclinic modes and for the following values of the controlling parameters:

$$\left.\begin{array}{c} H_1 = H_3 = 1000\,\mathrm{m}, \quad N_\mathrm{p} = 5\times 10^{-3}\,\mathrm{s}^{-1}, \quad H_\mathrm{p} = 100\,\mathrm{m}, \\ \Delta H_\mathrm{p} = 50\,\mathrm{m}, \quad T = 12.4\,\mathrm{h}, \quad \varphi = 30^\circ, \\ 0 \le H_{\max} \le 950\,\mathrm{m}, \quad 0 \le 2l \le 100\,\mathrm{km}. \end{array}\right\} \tag{2.64}$$

The shape of the bottom topography was defined by formula (2.18) for which, provided ε_0 is small, the analytical solution is given by (2.34) and (2.35). Thus, the range of applicability of the perturbation method can be estimated by comparing the two different solutions.

The influence of the height $H_{\max}$ of the underwater ridge on the structure of the scattered internal waves can be estimated qualitatively by inspecting Figure 2.8. It is clearly seen that at $H_{\max} = 850$ m (Figure 2.8(a)) the internal wave penetrates behind the obstacle without any essential distortion of its structure, but at $H_{\max} =$ 925 m these distortions are substantial (Figures 2.8(c) and (d)). They are the result of the energy transfer from the incident internal wave to the other higher baroclinic modes.

Quantitatively, the dependences of the amplitudes of the internal waves generated behind the ridge ($2l = 60$ km) in zone III (normalized by the amplitude of the incident wave) upon the height of the ridge are shown in Figure 2.9 for the first three odd-numbered incident baroclinic modes. The first horizontal axis is linear in the parameter $\varepsilon_0 = h_{\max}/h_0$; the second axis is written for the physical variable $H_{\max}$.

Scrutiny of the dependences $a_\xi^{(j)}(H_{\max})$ allows several inferences to be drawn. First, at small and intermediate heights $H_{\max}$, the modes that are effectively excited by the baroclinic mode m behind the obstacle are the neighboring baroclinic modes with mode numbers $m+1$ and $m-1$ closest to the number m of the incoming wave.

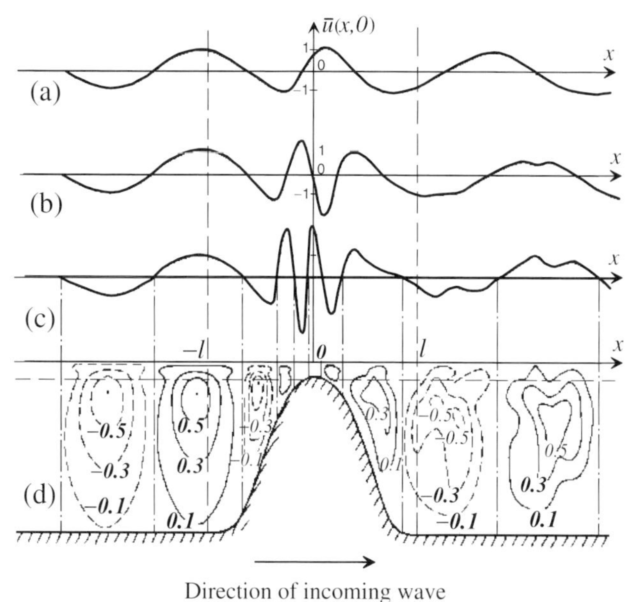

Figure 2.8. Profiles of normalized horizontal velocity $\bar{u}(x, 0)$ of the first baroclinic mode interacting with the underwater ridge (2.18) of heights (a) 850 m, (b) 900 m, and (c) 925 m. All other input parameters are as stated in (2.64). (d) Field of stream function $\psi(x, z)$ calculated for $H_{\max} = 925$ m.

For instance, mode $m = 3$ at $\varepsilon_0 < 0.1$ primarily generates second- and fourth-order modes. The amplitudes of all other modes are negligible.

An increase in $H_{\max}$ (or, alternatively, of ε_0), however, leads to the excitement by large obstacles not only of the modes $m + 1$ and $m - 1$, but also of the next nearest neighbors, $m + 2$ and $m - 2$, and perhaps even those further away. This principle, considered here for the third incident mode, is also valid for the other incident waves that we analyzed. Moreover, dependences $\bar{a}_j(H_{\max})$ for all generated waves are not monotonic, but have local maxima and minima. Thus, next-to-next nearest neighbors, and even modes further away from the incoming mode, may be more excited than the nearest modes. This is because of the above-mentioned resonance mechanism of internal wave scattering. We will discuss this point in more detail below. The next important inference may be drawn by comparing the numerical and analytical solutions. It is clearly seen that the dashed and solid lines in Figure 2.9 at small values of ε_0 ($\varepsilon_0 < 0.2$ for $m = 1$, $\varepsilon_0 < 0.1$ for $m = 3$ and $\varepsilon_0 < 0.05$ for $m = 5$) coincide quite well. This means that the perturbation method can be applied in this range for a valid analysis without reservation. Moreover, by comparing the two bottom axes, we find also that the range of applicability of such an approach is much wider than one would naively expect: so, scattering of internal waves by some real bottom topographies can indeed be considered analytically. Finally, the

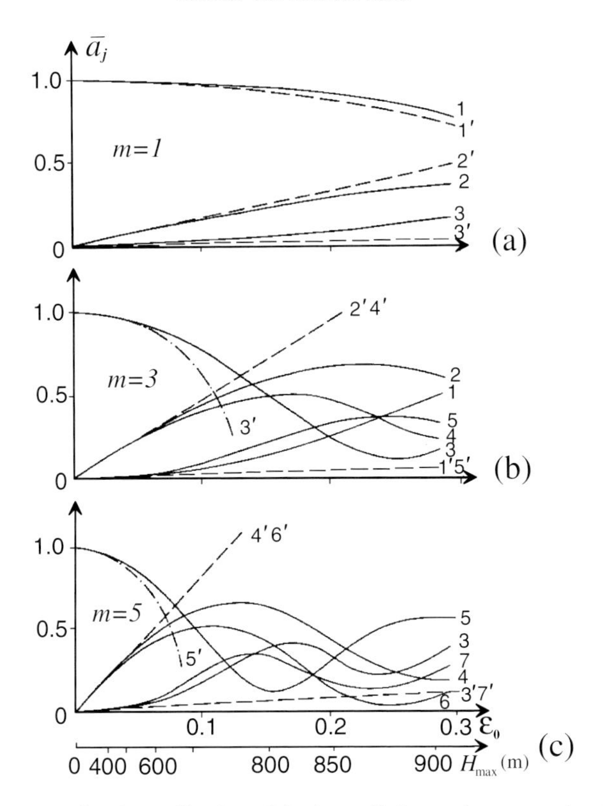

Figure 2.9. Normalized amplitudes of the baroclinic modes transmitted from zone I to zone III during the interaction of the (a) first, (b) third, and (c) fifth mode with the underwater ridge (2.18) versus its height H_{max} (or parameter ε_0). Solid lines correspond to the numerical solution; dashed lines show the analytical solution (2.34) and (2.35). All input parameters are given by (2.64). The labels 1, 1′, ... identify the mode number with results from numerical (1, ...) and analytical (1′, ...) solutions.

dependence of the scattering mechanism upon the number of the baroclinic mode m of the incident wave shows that with growing m the range of the applicability of the perturbation method decreases (compare Figures 2.9(a), (b), and (c)).

Figure 2.10 presents analogous dependences, but for the reflected waves. In the range $\varepsilon_0 < 0.1$, the two (numerical and analytical) models yield results with a high degree of coincidence. However, one should also note the evident distinction between the qualitative structure of the wave fields in zones I and III. This difference concerns the modal structure of the reflected and transmitted waves. Regardless of which incoming baroclinic mode is scattered, the maximum amplitude of the reflected wave is that of the first baroclinic mode, followed by the second, third, and so on. This is different for the transmitted wave, and gives rise to different structures of the reflected and transmitted waves.

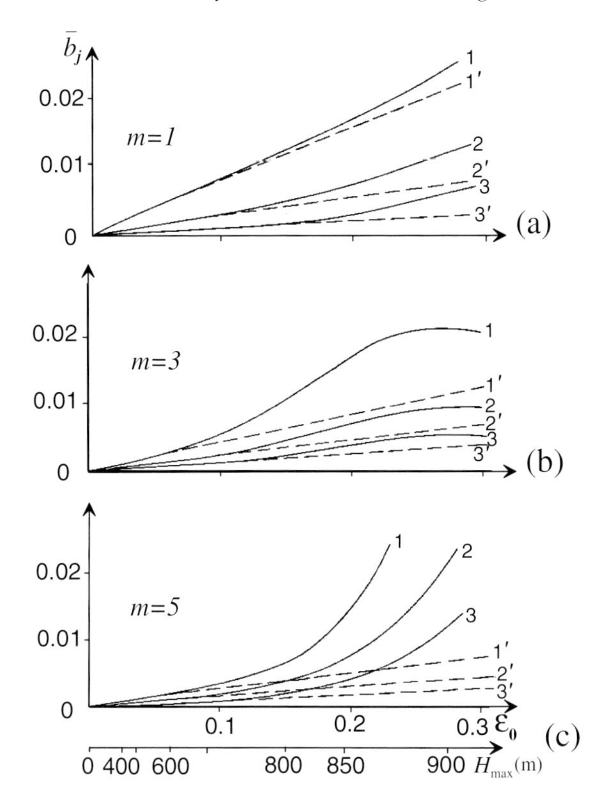

Figure 2.10. Amplitudes of the baroclinic modes reflected in zone I during the interaction of (a) a first, (b) a third, and (c) a fifth mode with the underwater ridge (2.18) versus its height, $H_{\max}$ (or parameter ε_0). Solid lines correspond to numerical solutions; dashed lines show the analytical solutions (2.34) and (2.35). All input parameters are given by (2.64). The labels 1, 1′, ... identify the mode number with results from numerical (1, ...) and analytical (1′, ...) solutions.

The dependences $\bar{a}_j(l)$ for the second, third, and fourth baroclinic modes, generated behind the ridge (2.18) during the scattering of the first baroclinic mode (see Figure 2.11), also reveal a satisfactory coincidence between analytical and numerical solutions in a wide range of obstacle width. If $\varepsilon_0 < 0.1$ ($H_{\max} < 700$ m), a remarkable discrepancy occurs only in close proximity to the origin of the coordinates when $2l < 3$ km (see inset in Figure 2.11). High baroclinic modes are effectively generated by narrow obstacles when the inclination of the bottom topography $\gamma = dH/dx$ is comparable to the inclination of the characteristic lines, $\alpha(z) = [(\sigma^2 - f^2)/(N^2(z) - \sigma^2)]^{1/2}$. This circumstance creates conditions that tend to form a wave beam, a phenomenon that will be discussed in greater detail in Section 2.5.

As already mentioned in Section 2.3.1, the mechanism of the generation of internal waves over bottom topography has a resonance character. This property is

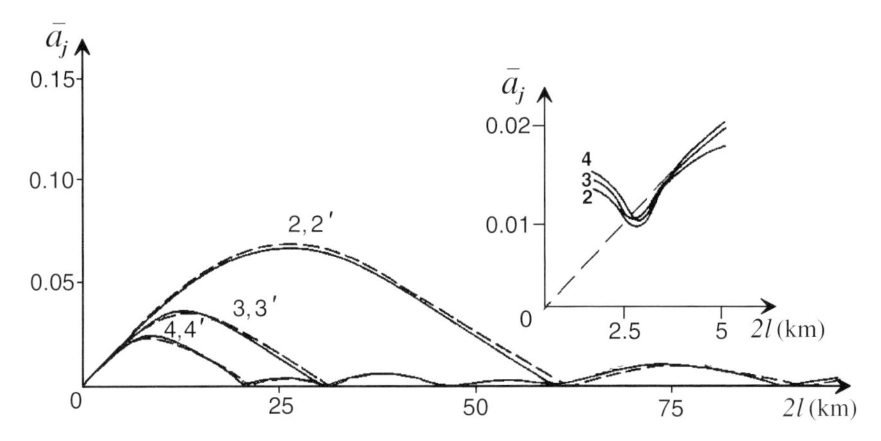

Figure 2.11. Normalized amplitudes of the second, third, and fourth baroclinic modes transmitted in zone III during the interaction of the first baroclinic mode with the ridge (2.18) ($H_{\max} = 500$ m) versus its width $2l$. All the other parameters are as prescribed in (2.64). Solid lines correspond to the numerical solutions; dashed lines show the analytical solutions (2.34) and (2.35). The inset shows the left lower corner of the main figure in greater detail.

also inherent in the process of internal wave scattering, as is obvious from inspection of Figure 2.12, in which the dependences of the normalized amplitudes of the transmitted waves in zone III are plotted against the height $H_{\max}$ of the obstacles for the triangular ridge (2.32) and a wide range of nondimensional obstacle heights $\varepsilon_0 = h_{\max}/h_0$ $(0 < \varepsilon_0 < 0.9)$. The analytical solutions for the amplitudes of the scattered waves for this bottom profile are given by formula (2.36). This solution is linear in ε_0, and it is valid only for small values of ε_0. Regardless of this fact, the numerical solution reveals a conspicuous quasiperiodic dependence $\bar{a}_j(\varepsilon_0)$ also at large values of ε_0. A strong transfer of energy of the incident mth baroclinic mode to the neighboring $m-1, m+1, m-2, m+2, \ldots$ modes near the local maxima (the spikes in Figure 2.12) is evident. At the same time, at points of local minima the energy transfer is minimal: the amplitude of the transmitted internal wave behind the obstacle, for which the number j coincides with the number m of the incoming wave, is equal to the amplitude of the incident wave; simultaneously, the amplitudes of all other waves are equal to zero. Thus, one may conclude that there are several sets of controlling parameters when an incident wave passes an obstacle without any transfer of energy to other baroclinic modes ("zero scattering"). The physical explanation of this phenomenon can be found in the theory described in Section 2.3.1. Estimation of the traveling time obtained with the use of formula (2.63) shows that local extrema correspond to those numbers which are divisible by the wave period.

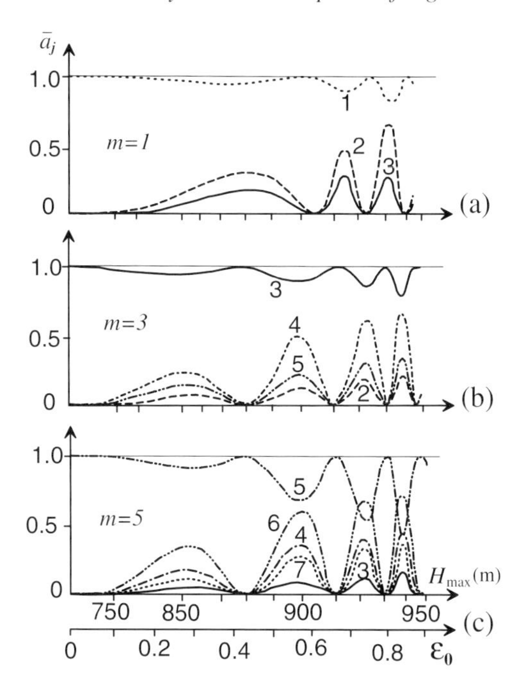

Figure 2.12. Normalized amplitudes of the baroclinic modes transmitted to zone III during the interaction of the (a) first, (b) third, and (c) fifth incoming baroclinic modes with the triangular ridge (2.18) of width $2l = 55$ km versus its height H_{max}. All other parameters are as prescribed in (2.64). Numbers of lines correspond to the number of radiated baroclinic modes.

2.4 Wave dynamics in slope-shelf regions

2.4.1 Generation of baroclinic tides

The mechanism of the generation of internal waves by an oscillating barotropic tidal flux over the continental slope possesses essentially the same characteristic features as that considered above by underwater ridges: the wave excitement tends to increase with the growth of the height and steepness of the underwater obstacle [249]. However, this intensification is not monotonic. It is dictated by the presence of the extrema (minima, maxima) inherent in the manifestation of local resonances (see Figures 2.6 and 2.7).

The above-mentioned mechanism of generation of internal waves in a slope-shelf area is dictated by the specific geometry of the bottom topography defining the transition zone from the deep water to the shallow water. We choose the continental slope profile

$$h_2(x) = h_1 - \frac{(h_1 - h_3)}{2}\left(1 + \sin\frac{\pi x}{2l}\right), \quad -l \leq x \leq l, \tag{2.65}$$

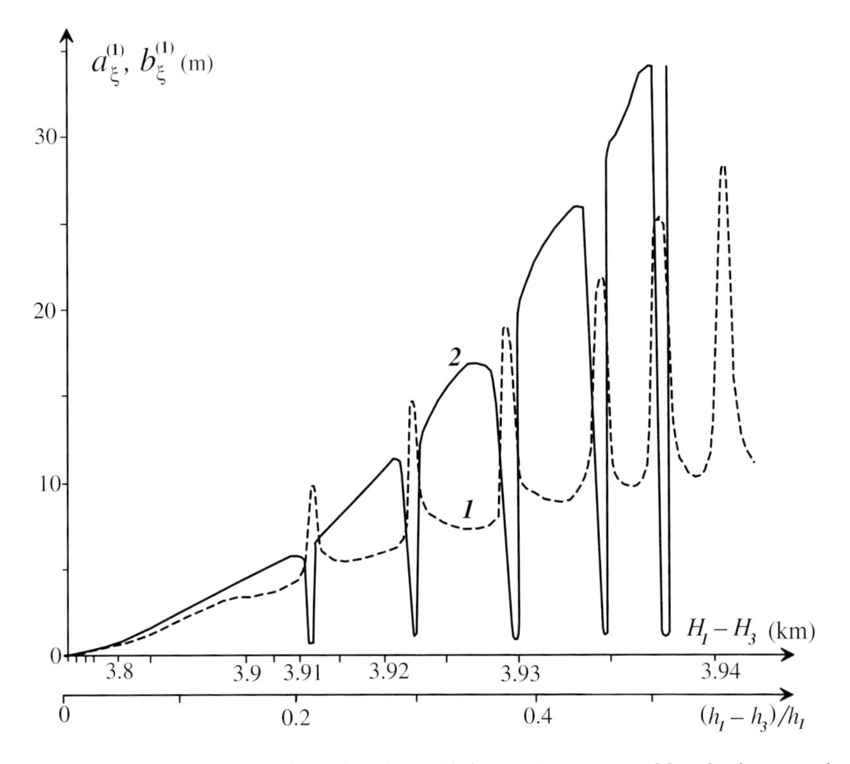

Figure 2.13. Amplitudes of the first baroclinic mode generated by the barotropic tidal flux during the interaction with the continental slope (2.65) in the deep-water zone I (line 1) and in the shelf zone III (line 2) plotted against the depth change, $H_1 - H_3$. The width of the transition zone was 50 km, and all other input parameters are as prescribed in (2.66).

and use the numerical procedure described in Section 2.2. Here $h_1 = z_1(-H_1)$ and $h_3 = z_1(-H_3)$. Calculations were performed for the following values of the input parameters:

$$\left.\begin{array}{c} H_1 = 4\,\text{km}, \quad 60\,\text{m} \le H_3 \le 4\,\text{km}, \quad N_\text{p} = 10^{-2}\,\text{s}^{-1}, \\ H_\text{p} = 60\,\text{m}, \quad \Delta H_\text{p} = 50\,\text{m}, \quad T = 12.4\,\text{h}, \quad \varphi = 30^\circ, \\ 10\,\text{km} \le 2l \le 60\,\text{km}, \quad \Psi_0 = 40\,\text{m}^2\,\text{s}^{-1}. \end{array}\right\} \tag{2.66}$$

Note that the magnitude $\Psi_0 = 40\,\text{m}^2\,\text{s}^{-1}$ corresponds to a horizontal velocity equal to 1 cm s^{-1} in the deep part of the basin.

Scrutiny of Figures 2.13 and 2.14 shows that the dependences $a_\xi^{(1)}(l)$, $b_\xi^{(1)}(l)$, $a_1(H_1 - H_3)$, and $b_\xi^{(1)}(H_1 - H_3)$ are conspicuously nonmonotonous in character exhibiting a sequence of intermediate extrema. These are clearly seen in all curves in the form of narrow and sharp spikes or depressions on the background of an average monotonic increase and decrease, respectively, of the amplitudes with increasing

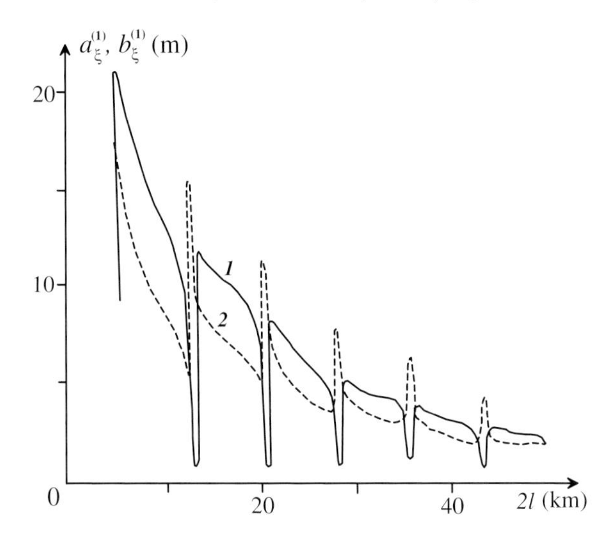

Figure 2.14. Amplitude of the first baroclinic mode generated in the deep part of the ocean (dashed lines) and on the shelf (solid line) plotted against the width of the continental slope $2l$. The depth H_3 in the shallow-water zone III was 100 m, and all other input parameters are from (2.66).

width $2l$ (which is typical for all curves in Figures 2.13 and 2.14). The nature of these extrema was discussed in Section 2.3.1. Analysis of the integral (2.63), calculated for $2l$ corresponding to the mentioned extrema of the functions $a_\xi^{(1)}(l)$, $b_\xi^{(1)}(l)$, $a_\xi^{(1)}(H_1 - H_3)$, and $b_\xi^{(1)}(H_1 - H_3)$, showed that the corresponding values of $T_1(2l)$ are, with good accuracy (i.e. an error of less than 1%), equal to an integer multiple of half the wave period. Thus, the conducted analysis establishes the existence of resonance conditions for the generation of internal waves over the slope-shelf topography. The extrema of the amplitudes correspond to those widths of the transition zone for which the time of the wave propagation from edge to edge of the obstacle is equal to an integer multiple of half the wave period.

The next interesting finding is that the wave amplitudes in the shelf zone are usually larger than those in the deep part of the basin (excluding the area in close proximity to resonance). This is a consequence of the fact that a wave guide on the shelf is narrower than in the deep part of the basin. The energy density in the shallow water must indeed be larger if one assumes equal values of the horizontal energy flux radiated from the obstacle by waves in opposite directions. We shall discuss this point in more detail below.

It also ought to be noted that dependences similar to those considered above for the first baroclinic mode were also found for higher order modes, with the difference, however, that the amplitudes of the higher modes are substantially (i.e. more than ten times) smaller than for the first baroclinic mode.

2.4.2 Transformation of baroclinic tides

The interaction of internal waves with a slope-shelf topography has many features in common with the mechanism of internal wave scattering by oceanic ridges considered above. It concerns the splitting of the incident wave into a group of neighboring baroclinic modes, as well as the manifestation of resonance peculiarities in wave diffraction. However, there are also some obvious differences, which we will discuss here in greater detail.

First we note that the kinematic characteristics of internal waves, i.e. wavelength and vertical structure, strongly depend on the total basin depth. Moreover, the amplitudes of internal waves propagating in a basin of variable depth must change as the wave proceeds. These changes obey the law of energy conservation if the backscattered energy flux is negligible. Such conditions only prevail if the ocean depth varies slowly with position. This fact is illustrated in Figure 2.15, where profiles of the horizontal velocity of the first baroclinic mode propagating from the deep part of the basin onto the shelf are presented for three different positions of the pycnocline (deep, intermediate, and shallow).

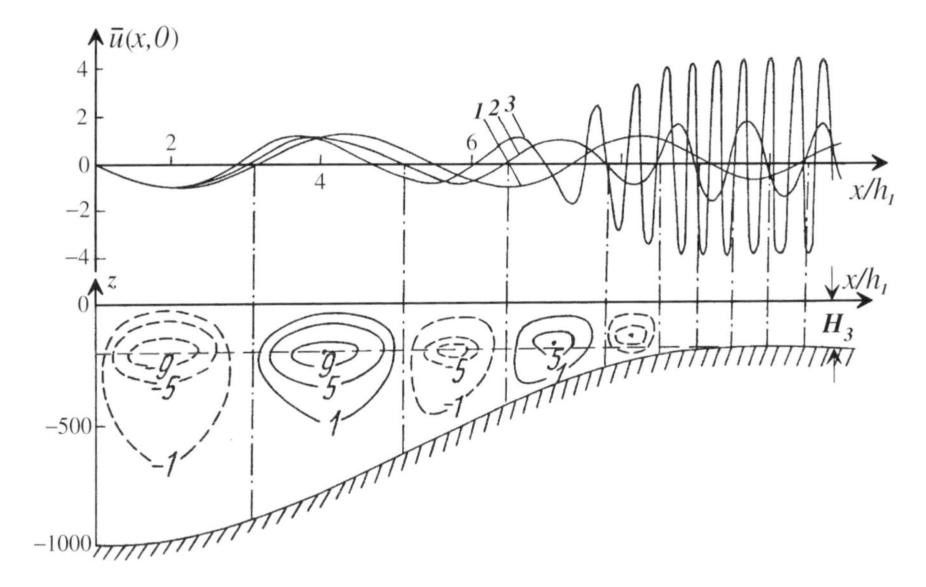

Figure 2.15. Normalized profiles of the horizontal velocity $\bar{u}(x, 0)$ of the first baroclinic mode propagating from the deep part of the ocean onto the shelf for three magnitudes of the pycnocline depth: $H_p = 100$ m (line 1), 200 m (line 2), and 300 m (line 3). The bottom profile is defined by (2.65), the water depth, H_3, on the shelf was equal to 200 m, $2l = 30$ km, and all other input parameters are as prescribed in (2.64). The closed contour lines are the streamlines for the stream function $\psi(x, z)$ ($\text{m}^2\,\text{s}^{-1}$) for line 2.

It may be seen that the wavelength becomes smaller in the shallow water but that the amplitude increases. For example, at $H_p = 100$ m (line 1), the wavelength in zone III is 1.46 times smaller than in the deep part, and the amplitude of $\bar{u}(x, 0)$ increases by 6%; for $H_p = 300$ m (line 3), the amplitude increase equals 320%, and the wavelength decreases by a factor of 10.9.

Figure 2.15 shows that the density of the wave energy increases from the deep part of the ocean towards the shore. This occurs because the inclination of the bottom topography is small, the reflected waves are negligible, and so practically all of the wave energy penetrates onto the shelf. Under these conditions of a narrowing wave guide, the local density of the wave energy must correspondingly increase. The evident growth of the amplitude of the shoaling wave as seen in Figure 2.15 is the obvious consequence of this process.

Note also the evident difference between the three profiles in Figure 2.15. They were all obtained for similar conditions, except the depth of the pycnocline, and this sole difference led to three obviously different waves penetrating into the shelf, waves that may be distinguished by their wavelengths and amplitudes. To make this point clearer, let us consider the conservation law of energy for internal waves.

In the linear case, the vertical velocity w and wave displacement ξ are connected by the relation $w = \xi_t$. Substituting this relation into the equation for the density, (1.39), and integrating the resulting equation with respect to time, yields

$$\tilde{\rho} = \xi \bar{\rho}_0 N^2(z)/g.$$

Then, by scalar multiplication of the equation of momentum balance (1.39) (which can be rewritten in vector form as $\mathbf{v}_t + f \cdot (\mathbf{k} \times \mathbf{v}) = -\nabla \tilde{P}/\bar{\rho}_0 + g\mathbf{k}\tilde{\rho}/\bar{\rho}_0$) with the vector $\mathbf{v}$ and using the evident relation $(\mathbf{k} \times \mathbf{v}) \cdot \mathbf{v} = 0$, we find

$$\bar{\rho}_0 \mathbf{v}_t \cdot \mathbf{v} = -\nabla \tilde{P} \cdot \mathbf{v} + \tilde{\rho} g \mathbf{k} \cdot \mathbf{v}.$$

Note that $\mathbf{v}_t \cdot \mathbf{v} = \frac{1}{2}(\mathbf{v} \cdot \mathbf{v})_t$, $g\mathbf{k} \cdot \mathbf{v} = -g\xi_t$, and hence $\tilde{\rho} g \mathbf{k} \cdot \mathbf{v} = -\xi \bar{\rho}_0 N^2 \xi_t = -\frac{1}{2}\bar{\rho}_0(\xi^2 N^2)_t$. Moreover, in view of the continuity equation, it is readily seen that $\nabla \tilde{P} \cdot \mathbf{v} = \nabla \cdot (\tilde{P}\mathbf{v})$. Finally, after some transformations we find the law of energy conservation for infinitesimal internal waves in the form

$$\frac{\partial}{\partial t}\left[\frac{1}{2}(\mathbf{v} \cdot \mathbf{v} + N^2 \xi^2)\right] + \nabla \cdot (\tilde{P}\mathbf{v}) = 0, \tag{2.67}$$

in which $\frac{1}{2}(\mathbf{v} \cdot \mathbf{v})$ and $\frac{1}{2}(N^2 \xi^2)$ are the densities of the kinetic and potential energy, respectively. From (2.67) it follows that the change of the total energy within a fluid volume is equal to the energy flux $-\tilde{P}\mathbf{v} \cdot \mathbf{n}$ through the liquid boundary of this volume (with exterior unit normal $\mathbf{n}$).

Let us find the energy flux transported by the elementary baroclinic mode (1.66), propagating horizontally into a basin of constant depth. From the linear equations

(1.39) it follows that for two-dimensional waves $\tilde{P}_x = -\bar{\rho}_0 u[\imath\sigma + f^2/(\imath\sigma)]$ and $\tilde{P}_z = \imath w \bar{\rho}_0[N^2 - \sigma^2]/\sigma$. Using the definition of the stream function and the obvious boundary condition $\tilde{P}_{z=0} = 0$, it readily follows that

$$\tilde{P}(x, z) = \frac{\imath\bar{\rho}_0}{\sigma} \int_z^0 [N^2(z) - \sigma^2] \frac{\partial \psi}{\partial x} dz. \tag{2.68}$$

Substituting expressions (1.76) and (1.77) (which were found for the stratification (1.70)) into (2.68), the pressure can be explicitly found to be given by

$$\tilde{P}(x, z) = -\frac{\rho_0(\sigma^2 - f^2)}{2\sigma k_j^2 c_3^{1/2}} \exp(\imath k_j x)$$
$$\times \left[\frac{j\pi}{h(\delta_0^2 + 1)^{1/2}} - \delta_1 \cos(\delta_2 z_1 + \arctan \delta_0) + \delta_2 \cos(\delta_1 z_1 + \arctan \delta_0) \right], \tag{2.69}$$

where h is the depth of the basin given in the variables (x, z_1) and defined by (1.77), $\delta_0^2 = c_1 c_2^2/c_3, \delta_1 = (c_1 c_3)^{1/2} + j\pi/h, \delta_2 = (c_1 c_3)^{1/2} - j\pi/h$, and the constants c_1, c_2, c_3, and the wavenumbers k_j of the jth mode are defined by (1.71) and (1.74), respectively.

It is evident that the flux of the energy of the jth mode expressed by the third term in (2.67) is given by

$$\mathbf{J} = \left[\mathbf{i}(\tilde{P}u^* + \tilde{P}^* u) + \mathbf{k}(\tilde{P}w^* + \tilde{P}^* w)\right] = (\mathbf{i}J_x + \mathbf{k}J_z), \tag{2.70}$$

where the asterisk denotes complex conjugation.

On substituting (2.69) into (2.70) (velocities u, w are defined in (1.40) and the definition of the stream function is employed) and integrating the resulting expression from $z = 0$ to $z = -H$, we may show that the total energy flux transported horizontally by the jth baroclinic mode in a basin of constant depth through the vertical cross-section is given by

$$J_x = \bar{\rho}_0(\sigma^2 - f^2) h a_j^2 k_j / 4\sigma. \tag{2.71}$$

It is evident that, if there is no reflection from the bottom, the value J_x must be constant in every vertical cross-section of the propagating internal waves. This idea can be used for the calculation of the amplitude of the internal wave propagating in a horizontally nonuniform basin provided the water depth changes slowly. The energy balance approach will be used later in Section 4.2 to control the accuracy of the numerical solution procedure to be developed: the total incoming flux of the energy approaching the obstacle must be exactly equal to the total flux of the energy transported from the source of generation by the radiated waves.

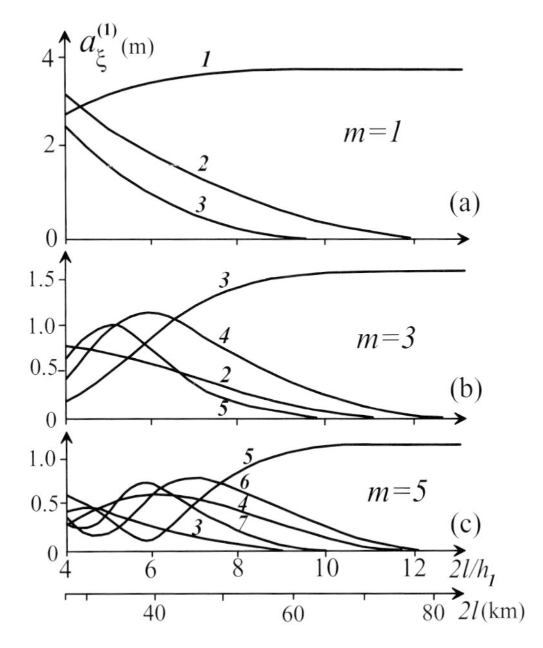

Figure 2.16. Amplitudes of the baroclinic modes transmitted to zone III during the scattering of a (a) first, (b) third, and (c) fifth mode wave over the slope-shelf bottom topography (2.65) versus the width of the transition zone, $2l$. The water depth, H_3, on the shelf is equal to 100 m, and all other input parameters are as prescribed in (2.64). Numbers on the curves correspond to the numbers of the generated mode.

With these preliminary remarks, let us carry out the analysis for the diffraction of internal waves in a slope-shelf region. Figure 2.16 shows the dependence of the amplitudes $a_\xi^{(1)}$ of the baroclinic modes generated within the shallow-water zone III on the width $2l$ for the interaction of a first, third, or fifth mode with the continental slope (2.65). It is seen that, when the continental slope is very wide (in other words, when it is very gently sloping), the incident wave (i.e. any baroclinic mode) penetrates into the shallow water without any essential transfer of energy to some other baroclinic mode. On the shelf it maintains the characteristics inherent in the same baroclinic mode as the incident wave. The single change which can arise is an increase of the wave amplitude and an accompanying decrease of the wavelength.

A typical example of wave propagation over a gently sloping topography was presented in Figure 2.15. The characteristics of the wave in zone III (wavelength and vertical structure) in such a case can be found from the standard BVP (1.67). The wave amplitude is calculated with the aid of the law of energy conservation. According to this law the horizontal energy flux, transferable by the internal wave, is constant if the bottom reflection is negligible, and the amplitude of the internal wave during its shoaling must increase. For instance, in the considered case the

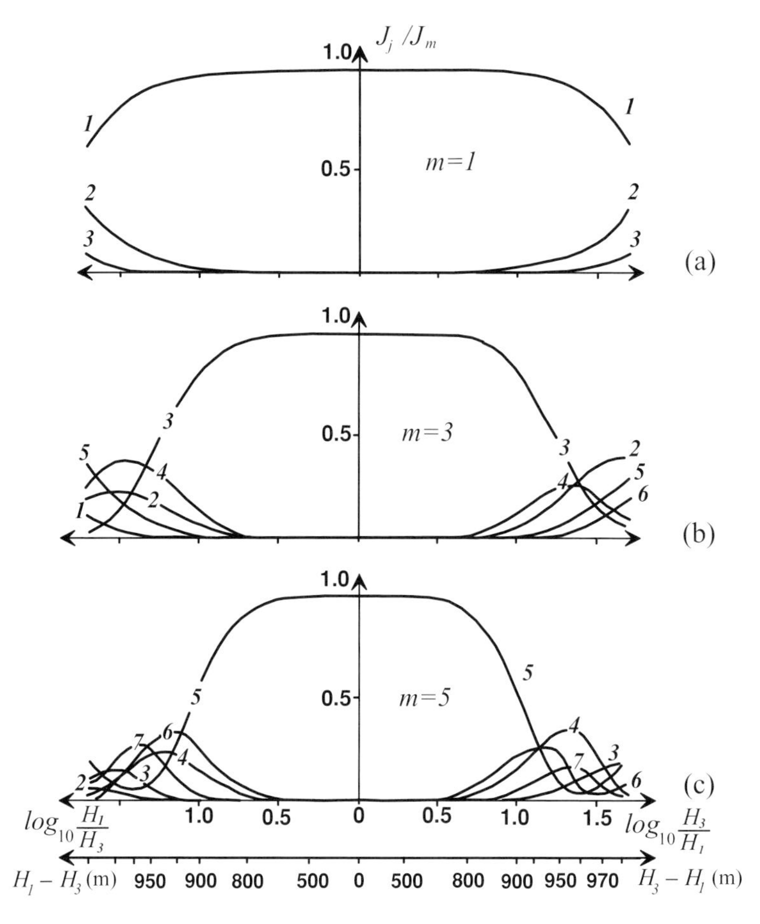

Figure 2.17. Normalized energy flux of the jth baroclinic mode transmitted to zone III as a result of scattering of the (a) first, (b) third, and (c) fifth mode over the slope-shelf bottom topography (2.65) versus the depth difference $H_1 - H_3$ when the wave propagates onto the shelf (right side of the figure) or versus $H_3 - H_1$ when the wave propagates into the other direction. The width $2l$ of the transition zone II is equal to 10 km, and all other input parameters are as given in (2.64).

first, third, and fifth incoming modes with initial amplitude equal to 1 m in the deep part of the ocean have, in the shallow water, amplitudes of 3.9 m, 1.6 m, and 1.2 m, respectively.

A decrease of the slope width, corresponding to an increase of the bottom steepness, leads to a more efficient transfer of energy to the neighboring modes: with a growth of the bottom steepness the modal composition of the wave field is enriched; the number of modes and the amplitudes of the generated waves increase.

The mechanism of splitting a baroclinic mode over the bottom topography into a group of neighboring modes can be clearly seen in Figure 2.17. Here, the energy flux

J_j of each specific jth mode in zone III, normalized by the flux of the incident wave J_m (m is the wavenumber), is given as a function of the depth difference $H_1 - H_3$ (or as a function of $\log_{10}(H_1/H_3)$) when the wave propagates from the deep part of the basin towards the shore, or as a function of $H_3 - H_1$ ($\log_{10}(H_3/H_1)$) when the wave propagates in the opposite direction. Such graphs illustrate how much of the "pure" energy is transferred from the incident wave to the other modes.

It is seen from these graphs that the effective transfer of energy is observed only when the depth difference between the two regions is relatively large ($\log_{10}(H_1/H_3) > 0.5$). It occurs because the basic changes of the eigenfunctions of the standard BVP under conditions of a strongly stratified density jump are located near the free surface (see Figure 1.7). The bottom variations, which are primarily located below the pycnocline, do not introduce any essential changes in the vertical structure of the propagated waves. This is why over gently sloping and relatively shallow obstacles (with their tops located below the pycnocline) an internal wave propagates essentially as a single mode.

The situation is quite different when the bottom topography is steep and high. Under such conditions the wave is efficiently scattered and a transfer to other modes occurs. The amplitudes of the reflected waves also increase. This tendency can be qualitatively estimated with the aid of Figure 2.18. Here, the total flux

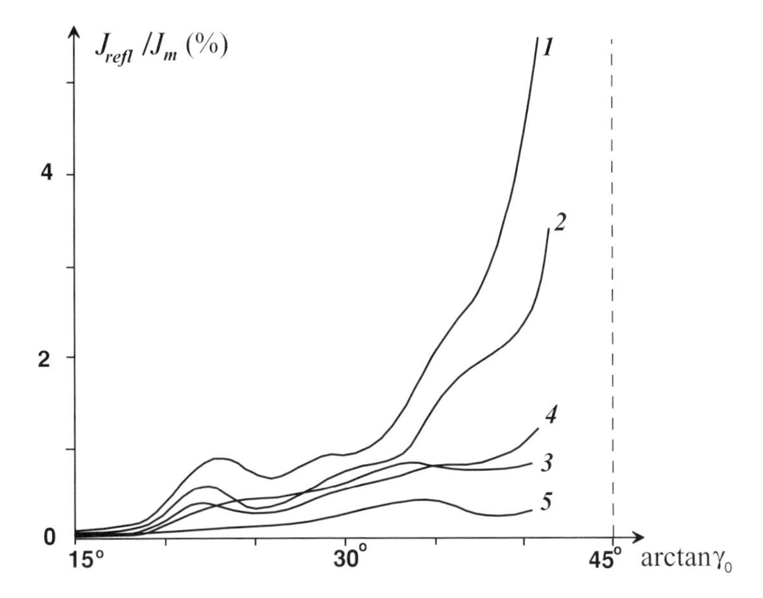

Figure 2.18. Normalized total energy flux of all baroclinic modes reflected from the linear continental slope versus the bottom inclination. The numbers identifying the curves correspond to the mode number of the incoming wave. The water depth H_3 in the shallow water equals 100 m, and all other input parameters are as prescribed in (2.64).

of the energy of the reflected waves (all baroclinic modes propagating in zone I from the transition area) is plotted for the continental slope, linear in the variables (x, z_1), versus arctan γ_0 ($\gamma_0 = dh_2/dx$), the angle of its inclination. The basic conclusion to be drawn from the presented curves is that the reflected waves are relatively weak when arctan $\gamma_0 < 20^\circ$. In such a case, the reflected waves can be excluded from consideration with adequate accuracy. Alternatively, if the bottom inclination is close to the critical value, the reflected waves are relatively strong. In this case of near "critical" obstacles, it is very important to account for the high baroclinic modes, in the problem of wave generation by the barotropic flux as well as for wave scattering. This latter problem is considered below in more detail.

2.5 Internal waves near steep bottom topography

The preceding results on internal wave dynamics were obtained for relatively gently sloping bottom topographies when the inclination of the characteristic lines, $dz/dx = \pm\alpha$ (see (1.43)), well exceeds the bottom inclination, $\gamma = dH/dx$. Under such conditions, only the lowest baroclinic modes are generated by the barotropic tidal flux. For the problem of wave scattering, the incident internal wave usually penetrates behind a gently sloping obstacle without any essential deformation of its structure. Only a small part of the energy of the incident wave is transferred to the neighboring modes. The reflected waves are usually negligible.

Such a situation is typical for many sites of the World Ocean in which internal waves of tidal periods are considered. However, in many shelf zones this rule is violated, for instance near the continental slope of Portugal in the Bay of Biscay. In some areas of the continental slope the condition $\alpha > \gamma$ for semidiurnal baroclinic tides is not valid.

In the supercritical case, when $\alpha < \gamma$, the characteristic lines cross the bottom surface more than once. Mathematically, this leads to more complicated calculations of internal wave fields because of the presence of "shadow zones" at the lateral slopes of steep obstacles. However, in some cases the problem of wave generation (or wave scattering) at abrupt bottom topography can be solved analytically. For instance, an analytical solution, which was found in Section 2.1 for "small" bottom features in the framework of a weak topography approximation (when the bottom boundary condition is applied at a flat surface instead of the actual bumpy bottom), formally has no restriction on the bottom steepness. The next possibility is to consider the specific shape of the bottom obstacle such as a knife-edge barrier, a Gaussian ridge, an abrupt step, or a top-hat ridge. Since the earliest papers of Larsen [131] and Rattray [199], several investigations have been carried out for steep topographies (see e.g. refs. [14], [225], and [226]).

Important results concerning the transfer of energy from the barotropic tide into internal gravity waves at supercritical topographies were obtained in these references. It was found that the steep topographies produce much more baroclinic tidal energy in comparison with weak subcritical obstacles. For instance, the knife-edge barrier as an internal wave generator is twice as effective as a ridge as Witch Agnesi ridge ($H(x) = H_0 - H_{max}/[1 + (x/l)^2]$), and the principal part of the energy (50%) is contained in the first baroclinic mode [226]. Energy flux estimates performed by Gustafsson [86] and Sjöberg and Stigebrandt [219] for a step topography were further, more accurately, analyzed by St. Laurent *et al.* [226]. In particular, it was shown that the representation of the real bottom topography as a number of independent steps is erroneous and presents remarkable errors in the estimation of the wave amplitudes.

Overcoming this situation for the general case of arbitrary bottom topography (for large obstacle heights as well as for arbitrary topography shapes) is possible in the framework of the method presented in Section 2.2. However, additional conditions and relations must be found which account for wave movements in the "shadow zones." The mathematical construction of the wave solution for "steep" bottom features was successful, for a special law of fluid stratification that is monotonically decreasing with depth (a profile of the buoyancy frequency (1.70) with $c_3 = 0$), thanks to the simple form assumed by the wave equation (1.50) in this case [10], [211].

For a realistic ocean stratification, the technique developed in refs. [10] and [211] is not appropriate. In a very common case, i.e. for an arbitrary $N(z)$ profile, apart from the functional equation (2.49), additional relations should be found which account for the multiple intersections of the characteristic lines with the bottom surface. We attempt to find such relations later in this section (we follow the approach presented in ref. [250]). First, let us recognize what happens with the wave fields when the steepness of the bottom topography is still subcritical but tends to a critical value (to 45° in the characteristic variables). To this end we consider a continental slope, linear in the variables (x, z_1),

$$h_2(x) = 0.5\,[h_1 + h_3 - (h_1 - h_3)x/l]\,. \qquad (2.72)$$

Here $h_1 = z_1(-H_1)$ and $h_3 = z_1(-H_3)$. In physical variables (x, z), the slope has the form

$$H_2(x) = -c_2 - (c_3/c_1)^{1/2}\tan[(\delta_1 x/l + \delta_2)/2],$$

where δ_1 and δ_2 are constants.

The numerical procedure described in Section 2.2 is still valid here because the considered bottom inclinations are subcritical. So we apply it for the problem of wave generation and apply BVP (2.1) with the following values of the input

Table 2.2. *Amplitudes* $a_\xi^{(j)}$ *(in meters) of the first ten internal modes in zone III with* $H_1 = 100$ m *and various values of angle* $\arctan \gamma_0$.

	Number of baroclinic mode, j									
$\arctan \gamma_0$	1	2	3	4	5	6	7	8	9	10
10°	5.8	0.8	0.3	0.1	1.2	0.5	0.7	0.4	0.3	0.3
25°	6.4	1.2	2.2	6.0	3.8	6.9	4.8	4.0	2.1	0.5
45°	7.5	7.9	8.6	21.7	13.2	12.0	6.4	25.4	15.7	6.2

parameters:

$$\left.\begin{array}{c} 75\,\text{m} \le H_3 \le 300\,\text{m}, \quad H_1 = 4\,\text{km}, \quad 8\,\text{km} \le 2l \le 60\,\text{km}, \\ \varphi = 30^\circ, \quad N_\text{p} = 10^{-2}\text{s}^{-1}, \\ H_\text{p} = 30\,\text{m}, \quad \Delta H_\text{p} = 50\,\text{m}, \quad T = 12.4\,\text{h}. \end{array}\right\} \qquad (2.73)$$

The magnitude of the stream function is chosen to be $\Psi_0 = 40\,\text{m}^2\,\text{s}^{-1}$, which corresponds to a horizontal velocity equal to 1 cm s^{-1} in the deep part of the basin.

The form of the obstacle (2.72) was chosen to be linear in the variables (x, z_1), the purpose being to investigate features of a wave field in a situation when the bottom steepness tends to the critical value ($\arctan \gamma_0 \to 45^\circ$, the angle of the characteristic lines in the variables (x, z_1)) in the whole water column above the slope when the width $2l$ of the transition zone II decreases.

The tendency of modification of the wave fields with growing bottom inclination ($\arctan \gamma_0 \to 45^\circ$) can be estimated from Table 2.2. It is seen that, at small values of γ_0 ($\arctan \gamma_0 < 10^\circ$), the spatial structure of the baroclinic field is determined by the first baroclinic mode. The contribution of higher modes is not so pronounced, and with reasonable accuracy they can be neglected.

With growing bottom steepness ($\arctan \gamma_0 \to 45^\circ$) the amplitudes of the higher modes increase and can even exceed the amplitudes of the lowest modes. In such a situation it is important for a correct description to account for the high baroclinic modes of the wave field: superposition of a large number of high modes with comparable amplitudes results essentially in the heterogeneity of the wave field. Figure 2.19 makes this statement clear. Here the field of the density of the wave energy $E(x, z) = \frac{1}{4}\bar{\rho}_0(\mathbf{v} \cdot \mathbf{v} + N^2\xi^2)$ averaged over one tidal period is presented at $H_1 = 100\,\text{m}$, $2l = 8\,\text{km}$ for three values of the bottom inclinations of the slope (2.72).

It is seen that the presence of narrow elongated zones with a high level of the energy density is inherent in all patterns. Comparison of Figures 2.19(a), (b), and (c) shows that the density of the wave energy E in these zones increases with the growth of the bottom steepness. The single variable parameter which is responsible

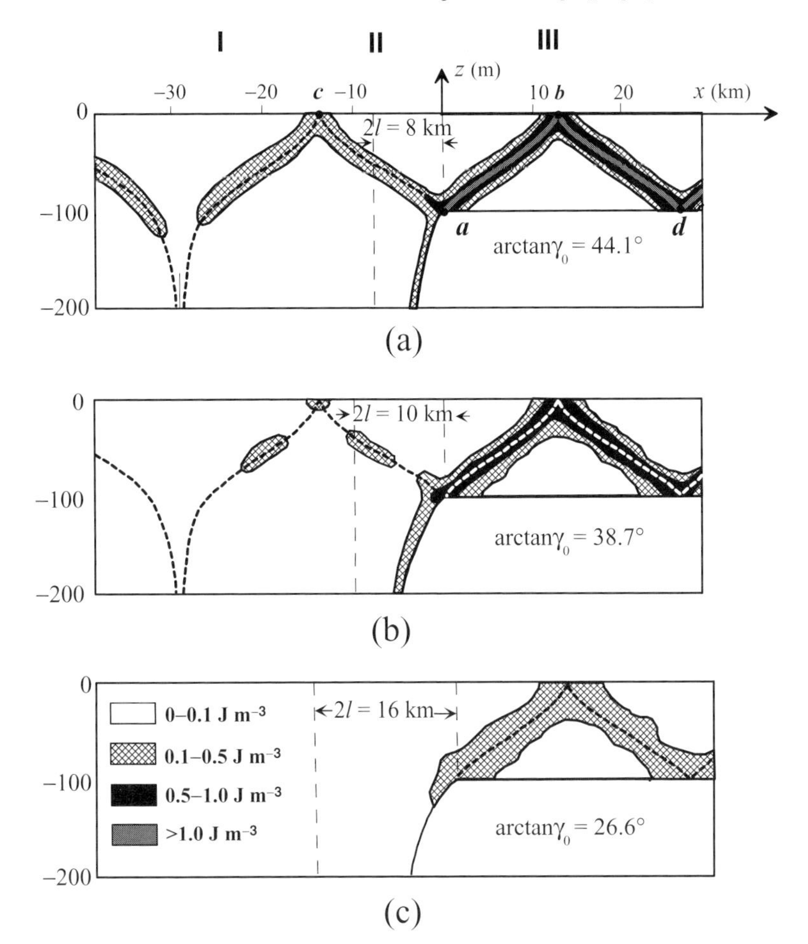

Figure 2.19. Fields of the averaged density of the wave energy (in J m^{-3}) for a continental slope linear in the characteristic variables (x, z_1) at different values of the bottom steepness. Dashed lines are characteristic lines.

for such an intensification is the bottom steepness. With its growth the number of generated baroclinic modes increases and their amplitudes also become larger (see Table 2.2). So one can conclude that the specific superposition of a large number of baroclinic modes leads to the formation of these zones where the basic part of the wave energy is concentrated.

The analysis of the spatial structure of these energy saturated spots shows a high correlation of their shape and position in space with the bottom topographic features and with the characteristic lines. It is seen that the axial lines of these zones (see the dashed lines in Figure 2.19) coincide with the characteristic lines $\mathscr{X} = x + \alpha^{-1}z$ and $\mathscr{Z} = x - \alpha^{-1}z$ of the wave equation (1.42). Thus, the wave beams emanating from the shelf break (from point ***a*** in Figure 2.19(a)) are formed along

Table 2.3. *Amplitudes* $a_\xi^{(j)}$ *(in meters) of the first ten internal modes in zone I at* arctan $\gamma_0 = 44.1°$ *for different values of the shelf depth* H_1.

	Number of baroclinic mode, j									
H_1 (m)	1	2	3	4	5	6	7	8	9	10
300	2.1	2.5	3.6	4.7	5.7	6.6	7.2	8.0	7.9	6.8
150	4.7	6.0	7.9	8.6	7.8	12.4	19.0	21.0	12.0	12.3
75	9.8	4.0	12.5	11.3	18.0	22.7	21.7	17.2	3.4	15.8

the characteristic lines (***a–b*** and ***a–c***) in both directions (shoreward and seaward) where the basic part of the baroclinic energy is concentrated.

The interesting feature of the wave beam is its ability to reflect from boundaries (from the free surface and the bottom). In Figure 2.19 such "turning" points are marked by the letters ***c***, ***b***, and ***d***, from which the "reflected" beam propagates along the characteristic line of the other family. With a decrease of the bottom steepness this effect becomes weaker (compare Figures 2.19 (a), (b), and (c)).

The decrease in efficiency of the generation of high modes (and consequent weakening of the wave beam) accompanies an increase of the depth of the shallow water – in other words, with the diminishing of the jump in depth between zones I and III – even though the bottom steepness remains constant. Table 2.3 illustrates this fact rather clearly.

If the bottom topography is supercritical in a certain sector and, consequently, if the characteristic lines intersect the bottom surface more than once, then the numerical procedure presented in Section 2.2 is no longer valid for this sector, and one must introduce some modification to the method so that it again becomes valid. Let us consider this problem.

The formulation of the problem of generation of baroclinic tides over supercritical bottom features is the same as that considered above. We consider the BVP (2.1) for which the solution at any point $A(x, z_1)$ (see Figure 2.20) can be expressed via (2.47) in terms of the horizontal velocity at the free surface between the points $B\,(x - z_1, 0)$ and $C\,(x + z_1, 0)$ and the Riemann function $\mathcal{R}(x, z_1; \zeta, 0)$ defined in (2.45). This yields

$$\Psi(x, z_1) = \frac{\Psi_0}{\alpha^{1/2}(0)} \times \left[1 - \frac{1}{2}\int_{x-z_1}^{x+z_1} \left.\frac{\partial \mathcal{R}(x, z_1; \zeta, \eta)}{\partial \eta}\right|_{\eta=0} d\zeta\right] + \frac{1}{2}\int_{x-z_1}^{x+z_1} \mathcal{R}(x, z_1; \zeta, 0) \left.\frac{\partial \Psi(\zeta, \eta)}{\partial \eta}\right|_{\eta=0} d\zeta. \tag{2.74}$$

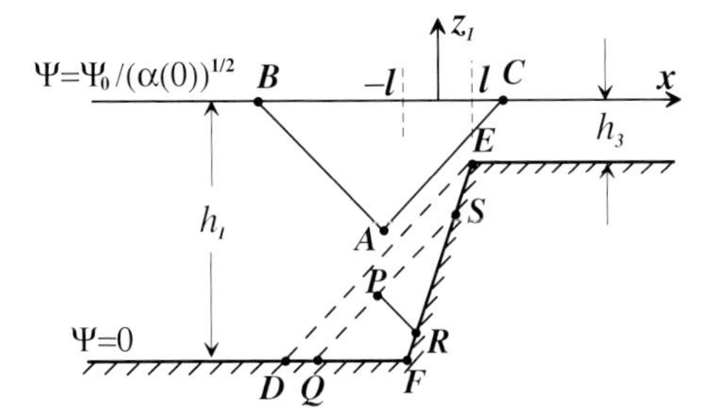

Figure 2.20. Schematic representation of the computational domain for steep bottom topography.

Beyond the sector *DFE* of Figure 2.20 the application of the bottom boundary condition leads to the functional equation

$$\int\limits_{x-h_2(x)}^{x+h_2(x)} \mathcal{R}(x, z; \zeta, 0) \frac{\partial \Psi(\zeta, \eta)}{\partial \eta}\bigg|_{\eta=0} d\zeta$$

$$= \frac{\Psi_0}{\alpha^{1/2}(0)} \left[2 - \int\limits_{x-h_2(x)}^{x+h_2(x)} \frac{\partial \mathcal{R}(x, z_1; \zeta, \eta)}{\partial \eta}\bigg|_{\eta=0} d\zeta \right], \tag{2.75}$$

which is similar to (2.49) if we account for the modification of the boundary conditions (we consider problem (2.1) instead of (2.37)). Within the "shadow zone" inside the triangle *DEF* in Figure 2.20, formula (2.75) is not valid and we should find some additional functional relations which are valid there and which permit us to describe the wave field.

If we consider any point $P(x, z_1)$ inside the triangle *DEF*, we can apply the Riemann formula (2.47) twice. For the region *QFRP* we have

$$\Psi(x, z_1) = \frac{1}{2} \int\limits_{QFR} \mathcal{R}_1(x, z_1; \zeta, h_2(\zeta))[\Psi_\zeta \, d\eta + \Psi_\eta \, d\zeta]. \tag{2.76}$$

Here $\mathcal{R}_1(x, z_1; \zeta, h_2(\zeta))$ is the Riemann function defined in region *QFRP*.

Similarly, for the point $P(x, z_1)$ we can write

$$\Psi(x, z_1) = -\frac{1}{2} \int\limits_{RS} \mathcal{R}_2(x, z_1; \zeta, h_2(\zeta))[\Psi_\zeta \, d\eta + \Psi_\eta \, d\zeta], \tag{2.77}$$

where $\mathcal{R}_2(x, z_1; \zeta, h_2(\zeta))$ is the Riemann function defined within the triangle *RSP*.

Equating the right-hand sides of (2.76) and (2.77) we find

$$\int_{QFR} \mathcal{R}_1(x, z_1; \zeta, h_2(\zeta))[\Psi_\zeta \, d\eta + \Psi_\eta \, d\zeta]$$

$$+ \int_{RS} \mathcal{R}_2(x, z_1; \zeta, h_2(\zeta))[\Psi_\zeta \, d\eta + \Psi_\eta \, d\zeta] = 0, \tag{2.78}$$

which can be rewritten as

$$\int_{x(Q)}^{x(S)} \mathcal{R}(x, z_1; \zeta, h(\zeta)) \frac{1 - h_{2\zeta}^2(\zeta)}{(1 + h_{2\zeta}^2(\zeta))^{1/2}} \Psi_n(\zeta) \, d\zeta = 0. \tag{2.79}$$

Here we have substituted the integration along the contour *QFRS* by performing the integration along x from the horizontal coordinate $x(Q)$ of the point Q to the coordinate $x(S)$ of the point S. Note that $\Psi_n(\zeta)$ is the derivative of the stream function normal to the bottom surface (the tangent velocity), $\mathcal{R} = \mathcal{R}_1$ within the region *QFRP*, and $\mathcal{R} = \mathcal{R}_2$ within the triangle *RSP*. Thus, (2.79) is an additional relation which must be satisfied in the "shadow" zones of supercritical bottom features.

2.6 Internal waves near the critical latitude

One of the basic parameters controlling the characteristics of the internal waves is the Coriolis parameter. In this section we consider the influence of the rotation of the Earth in the dynamics of internal gravity waves in a horizontally nonuniform ocean (according to ref. [259]), but we do not consider here the inertial-gravitational frequency band.

The kinematic characterization of internal waves, such as wavelength λ (wavenumber k), phase speed c_{p} (group speed c_{g}) and the vertical structure $q(z)$ can be estimated with the solution of the standard BVP (1.67). For a realistic ocean stratification it can only be solved numerically. It is obvious that the dependence of such a solution on all input parameters is not as evident as it is for the analytical solution. Fortunately, for the three-parameter family (1.70) which describes typical oceanic stratifications relatively well (see Figure 1.7), such an analytical solution was found; for the analysis, (1.74)–(1.78) can be used.

The phase speed and the group speed can be readily found from (1.78) provided that $N_{\mathrm{p}}^2 \gg \sigma^2$ and $N_{\mathrm{p}}^2 \gg f^2$, conditions that are commonplace for the

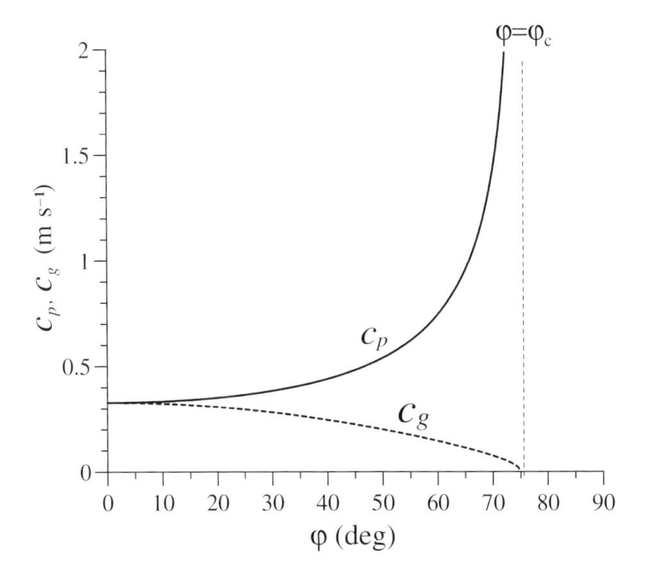

Figure 2.21. Group, c_g, and phase, c_p, velocities versus latitude ($N_p = 0.015$ s^{-1}, $\Delta H_p = 30$ m, $H_p = 35$ m, $H_0 = 300$ m).

ocean. So,

$$\left.\begin{aligned} c_p &= \frac{\sigma}{k} \cong \frac{\Delta H_p N_p}{2} \frac{\sigma}{\sqrt{\sigma^2 - f^2}} \frac{1}{\sqrt{p_0}}, \\ c_g &= \frac{d\sigma}{dk} \cong \frac{\Delta H_p N_p}{2} \frac{\sqrt{\sigma^2 - f^2}}{\sigma} \frac{1}{\sqrt{p_0}}, \end{aligned}\right\} \tag{2.80}$$

where

$$p_0 = \frac{(j\pi)^2}{\left\{\arctan\left[\frac{2(H_p - H)}{\Delta H_p}\right] - \arctan\left(\frac{2H_p}{\Delta H_p}\right)\right\}^2 - 1}.$$

Scrutiny of these formulas shows that near the equator ($\sigma \gg f$) the group and phase speeds coincide. Strong rotation violates this rule, and, when the Coriolis parameter tends to the tidal frequency, the wavelength and the phase velocity tend to infinity, whilst the group velocity tends to zero. The graphical representation of these dependences is shown in Figure 2.21. The above-mentioned peculiarities of the internal tides near the critical latitude will be useful for the subsequent analysis.

As was mentioned above, the efficiency of the generation of internal waves is strongly exposed by the resonance properties of the system. The multiplier $\sin(\pm k_j l)$ in (2.22) indicates this fact. The amplitude of the jth generated mode is equal to zero when $l = n\pi/k_j$, and has a maximum for the half length of the obstacle $l \approx n\pi/2k_j$. For a given water stratification and with a constant value of the Coriolis

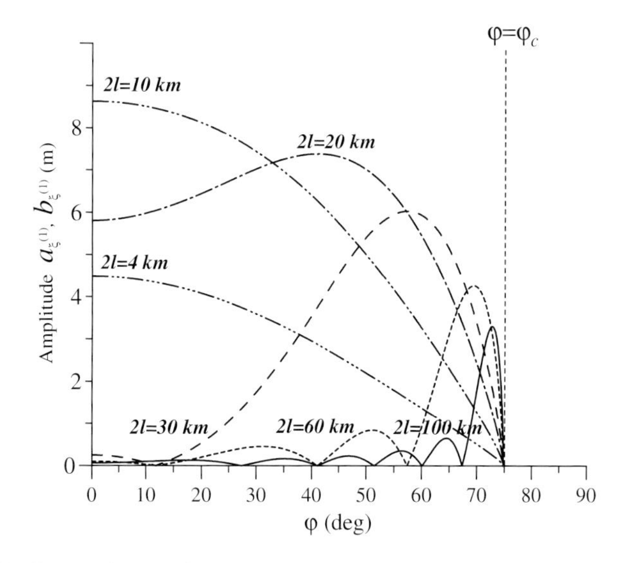

Figure 2.22. Dependence of the amplitude of the generated internal semidiurnal wave (the first mode) upon the latitude, calculated for different values of the width $2l$ of the underwater ridge (2.18).

parameter, the change in efficiency goes together with the change of the ridge width. Alternatively, the amplitudes of the generated waves for the oceanic ridge with fixed width also strongly depend on the latitude because the wavenumber k_j is a function of the Coriolis parameter; see (1.78).

This dependence can be seen more clearly in Figure 2.22, where the amplitude of the vertical isopycnal displacement $a_\xi^{(1)}$ for the first mode is shown as a function of the latitude φ for six different values of the ridge width ($2l = (4; 10; 20; 30; 60; 100)$ km) when $\Psi_0 = 50$ m^2 s^{-1}. The parameters for the stratification (1.70) were chosen as for Figure 2.21. The total water depth, H_0, and the ridge height, $H_{\max}$, were 300 m and 175 m, respectively.

The curves $a_\xi^{(1)}(\varphi)$ and $b_\xi^{(1)}(\varphi)$ in Figure 2.22 reveal the above-mentioned resonance character of the internal wave generation over the ridge. This fact is more remarkable for wide obstacles (see the solid and dotted lines in Figure 2.22). An increase of the latitude φ decreases the wavenumber k_j (according to (2.80), $k_j \sim \sqrt{\sigma^2 - f^2}/j\sigma$) and, in turn, leads to the periodicity of the multiplier $\sin(k_j l)$ in (2.22).

The next important conclusion is that all curves in Figure 2.22 approach zero when the latitude tends to the critical value, $\varphi \to \varphi_c$. Thus, within the context of a linear model, the rotation of the Earth suppresses the generation of internal waves at high latitudes; in fact, the closer the local latitude is to the critical one, the weaker will be the generated waves. Of course, this dependence arises in the immediate proximity of the critical latitude.

Note that, above the critical latitude ($\sigma < f$), (2.1) changes its type from hyperbolic to elliptic; no free propagating waves are allowed in this case. This problem is discussed in detail in, for instance, refs. [123] and [132]. The change of sign in front of the second term in the BVP (2.8) leads to an exponential (instead of a trigonometric) representation of the Green function and to an exponential attenuation of the disturbances from the point of generation (the ridge).

The above-mentioned arguments concern the problem of the local wave generation above the critical latitude. This does not mean, however, that internal waves approaching the critical latitude cannot penetrate above it. From WKB theory it is known that propagating waves are reflected from caustics. For the northward propagating internal waves the critical latitude acts as a wall. However, this wall is not "rigid" but provides a "soft" reflection. The amplitude of the incident wave approaching the critical latitude behaves as an Airy function: it is almost constant below the level $\sigma = f$ and decreases exponentially above it.

3

Combined effect of horizontal density gradient and bottom topography on the dynamics of linear baroclinic tides

One type of interesting phenomenon which makes the dynamics of the World Ocean rather complicated and surprisingly manifold is a "hydrological front." The term *front* characterizes regions in which the hydrodynamic fields possess large horizontal gradients. These manifest themselves in the thermohaline characteristics. Fronts in the ocean or in the atmosphere can be defined as regions where background long-term averaged properties of the medium change substantially over a relatively short distance. Such sharp boundaries usually separate two adjacent water masses with different properties. In the World Ocean, however, hydrological fronts can also reach more than 100 km width [233].

We emphasize that a strict standardized definition of the term "oceanic front" does not exist. As a quantitative measure for the systematization of oceanic fronts, the change of any hydrological property across a localized area – say temperature, or salinity, or both – can be used. If such gradients are an order of magnitude larger than similar changes adjacent to this area, then the presence of a front can be identified.

From the viewpoint of water dynamics, the hydrological fronts in the World Ocean can be classified by their influence on the distribution of the density field. Since the sea water density is a function of temperature and salinity, any horizontal gradients associated with a change of one or both components across a localized area can affect the density distribution (Figure. 3.1) and lead to the formation of a strong density gradient, the *density front*. Examples of such fronts can be found in shelf-slope areas as a result of a near coastal upwelling, or in the region of interaction of two water masses with different characteristics (i.e. in the regions of river runoff, estuary, or straight zones).

Sometimes the horizontal gradients of salinity and temperature are in opposite directions, so that their effects on the density compensate each other; this happens, for instance, in the Sargasso Sea. Consequently, in these cases, the boundary

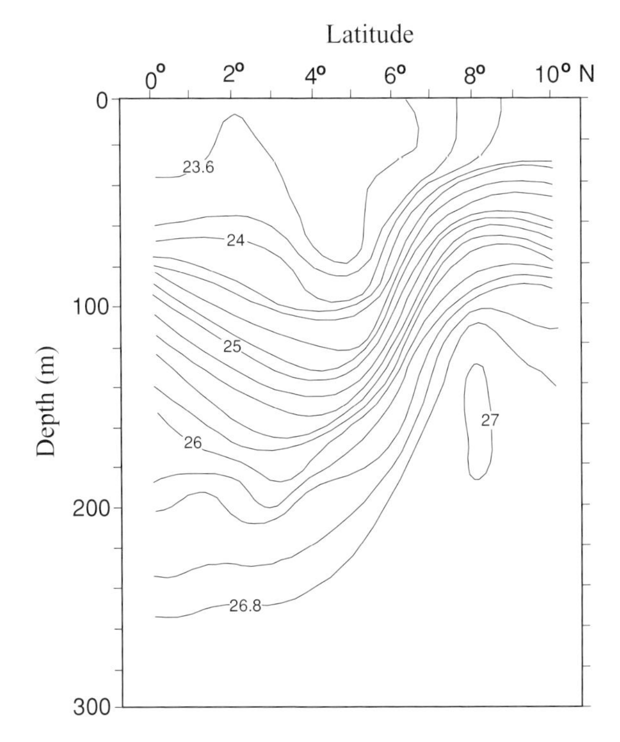

Figure 3.1. Zonal vertical section of the conventional density distribution along the latitude $\varphi = 45^\circ$ E measured in the area of the frontal zone in the northern part of the Tropical Atlantic. The plot shows isolines of the density (isopycnals).

between two water masses exists only as a *temperature front* or as a *salinity front* but it cannot be identified in the density field.

Interest in investigations of frontal zones can be explained by high level dynamic activity of all oceanic processes taking place near the frontal boundaries and their strong impact on many oceanic processes. For instance, strong oceanic currents such as the Gulf Stream, Kuroshio, Oyasio, and other currents, are the principal source of mass, heat, and passive admixture transport in the World Ocean; they are areas of sharp density gradients across their main flow direction.

An example of such a density front is illustrated in Figure 3.2. This front is located in the tropical part of the Western Atlantic near the Brazilian coast in the area of the North Equatorial Counter Current. Figure 3.2(a) shows positions of CTD (conductivity, temperature, depth) stations, where measurements were carried out during Cruise 23 of Research Vessel (R/V) *Akademik Vernadskiy*. The density profiles across the front are shown in Figure 3.2(b). It is clearly seen that the vertical fluid stratification changes remarkably from station to station. The indicated isopycnals (dashed lines) rise more than 100 m to the free surface over a distance of

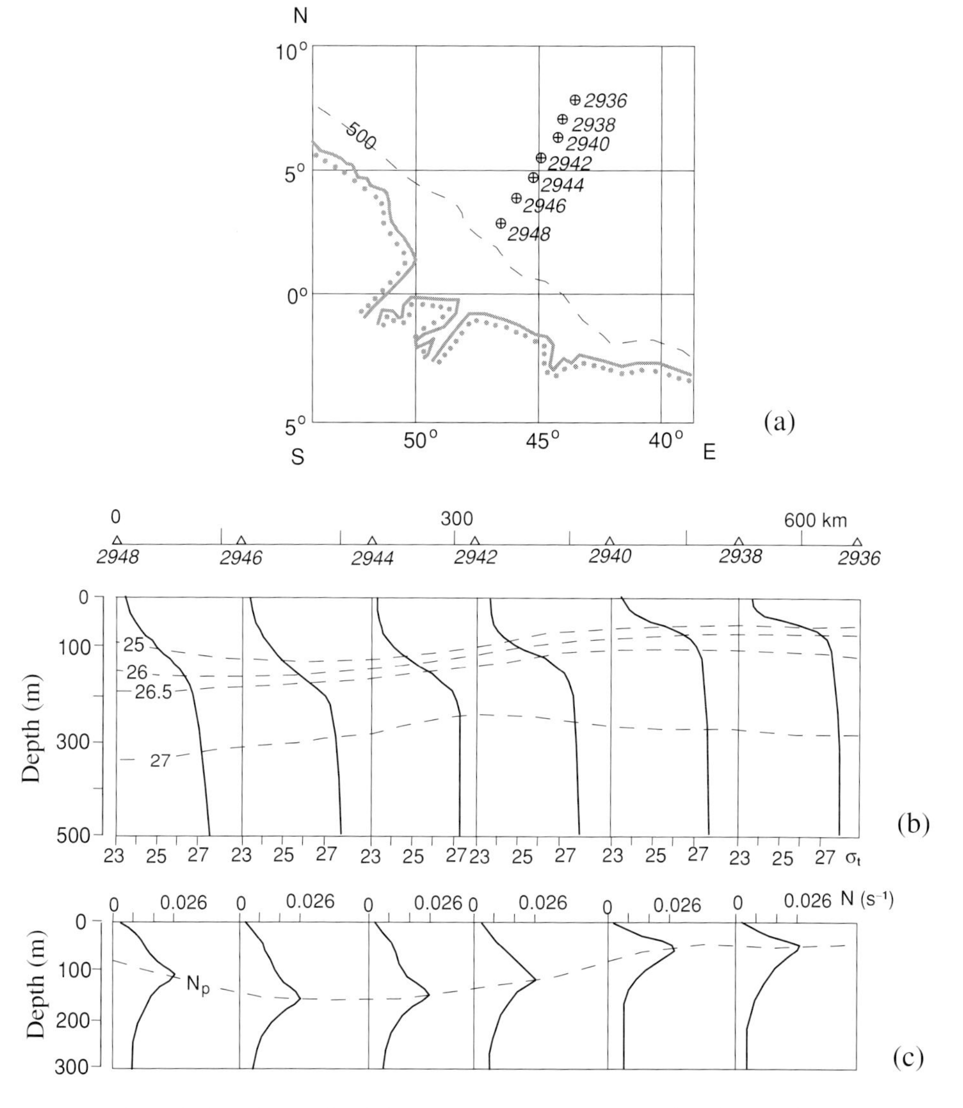

Figure 3.2. (a) Position of the CTD stations in Cruise 23 of R/V *Akademik Vernadskiy*. (b) Measurements of conventional density at stations 2948–2938. (c) Vertical distribution of the buoyancy frequency calculated from the measurements at the CTD stations. The position of the buoyancy frequency maximum across the front is shown by a dashed line.

300 km from station 2946 to station 2940. The position of the density front is even more clearly seen in the variation of the pycnocline depicted by the dashed line in Figure 3.2(c), in which the vertical profiles of the buoyancy frequency are displayed.

Note that the above-considered frontal zone near the Brazilian coast is known as a region of high baroclinic tidal activity [128]: both low-frequency semidiurnal

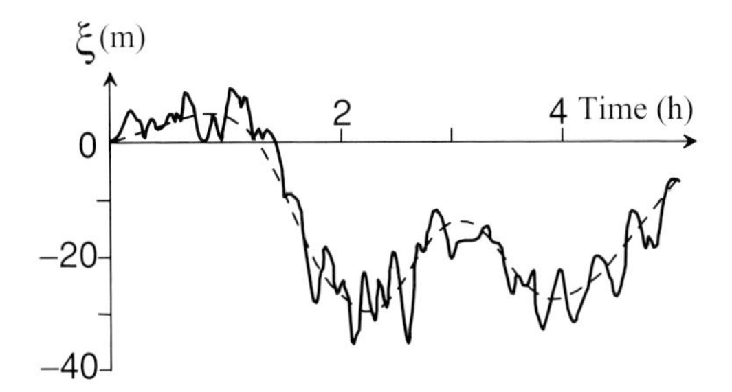

Figure 3.3. Typical record of thermocline oscillations measured during the 37th cruise of the R/V *Akademik Vernadskiy* during the passage of the frontal face of the wave train. High-frequency oscillations are clearly seen on the background of a low-frequency wave.

baroclinic tides and tidally generated short solitary internal waves and wave trains are regularly observed there. An example of such an observation in the region of the North Equatorial Counter Current is presented in Figure. 3.3. Here, long baroclinic tidal waves together with packets of short but strong internal waves having amplitudes of several tens of meters were registered. The purpose of this chapter is to answer the following questions. (i) What is the specific input of the horizontal density gradients to the dynamics of internal waves? (ii) Can these waves be substantially distorted or changed by frontal zones?

In Chapter 2 we showed that, in a vertically stratified ocean, an oscillating barotropic tidal flux interacting with the inclined bottom topography can be responsible for the generation of baroclinic tides. As just shown above, the ocean can also be stratified horizontally. Thus, since isopycnals in frontal zones are also tilted, it is important to understand how strong the effects of horizontal stratification on the dynamics of internal waves can be: in particular, whether or not the hydrological front can be a generating mechanism of internal waves as effective as the bottom topography.

Unfortunately, little attention has been paid to the investigation of the effects introduced into the dynamics of baroclinic tides by frontal zones. In this connection we can mention the first fundamental theoretical work by Mooers [160], who recognized the importance of the problem and formulated a linear wave equation which took into account both vertical and horizontal density gradients. On this basis, and with the help of the Riemann method for partial differential equations, he described the peculiarity of baroclinic tides inside a frontal zone. We will discuss this equation in detail in the following.

The next step was taken by Chuang and Wang [40], who developed a finite difference model based on the linear wave equation which not only reproduces standing internal waves inside a density front but also calculates amplitudes of several baroclinic modes radiated from frontal zones. It was found that a shelf-slope front exerts strong effects on the generation and propagation of internal tides and that these effects are highly sensitive to the orientation of the front. Note that the applicability of this model was restricted by application of a monotonic profile for the buoyancy frequency ((1.70) with $c_3 = 0$).

Another study of relevance here is the theoretical work of Ou and Maas [184]. They examined the influence of a density front upon the dynamics of internal tides by using a two-layer analytical model in which the interface intersects both the bottom and the surface. According to their solution, the presence of the density front can lead to a remarkable intensification of the semidiurnal and diurnal internal tides. In particular, the semidiurnal tide should be preferentially amplified due to the presence of the superinertial eigenmodes. Of additional interest are Chen and Beardsley's numerical studies using the Princenton Ocean Model (POM) [32], [33]. They examined the tidal rectification and mixing over Georges Bank and were able to produce not only the internal tides, but also a tidal front at the edge of the bank. However, the relation between the two was not investigated because the baroclinic tidal currents could not clearly be separated from those of the barotropic tide.

Substantial progress was achieved in a more recent paper [34], in which internal tides near a mid-latitude shelf-slope front were studied using a numerical model incorporating the turbulence closure for vertical mixing. With this model it was found that the properties of internal tides are highly dependent on the configuration of the front and the tidal frequency. In fact, the presence of the frontal zone may substantially enhance the generation of internal waves. The baroclinic tide can be either arrested in the frontal zone in the form of an internal circulation cell or radiated from the place of generation, depending on the stratification.

To answer this and many other related questions, we must first understand the conditions of stability of the density front in isolation, without the action of any external perturbation such as internal waves. In other words, we must find the balance conditions between all forces which maintain the dynamic equilibrium of a system with density front. To this end, let us estimate the different terms of the governing equation (1.6).

Oceanic hydrological fronts are large-scale geophysical objects. Horizontal density gradients are usually observed in the upper stratified layer, the thickness of which lies in the range 10^2–10^3 m. So, these values can be taken as the vertical scale $\mathbb{H}$ of the considered phenomenon. At the same time its horizontal scale $\mathbb{L}$ is usually about 10^4–10^5 m. This means that the aspect ratio $\delta = \mathbb{H}/\mathbb{L}$ is a small parameter and that the trajectories of fluid particles within the density front lie very close together in horizontal planes. This is why the vertical motions in the dynamics of the density front can be neglected and the corresponding terms in (1.6) set to zero.

In other words, we assume that for oceanic fronts the hydrostatic approximation for the pressure is valid.

To estimate the remaining terms in the momentum balance equation (1.6), let the symbols $\mathbb{U}$ and $\mathbb{T}$ be scales for the horizontal velocity and time, respectively. It is known that oceanic fronts can be characterized as quasistationary phenomena, which exist for a long time and whose structure can be subjected to a synoptic variation with a characteristic time scale of several days. Thus, far from the equator comparison of the dynamical and the Coriolis acceleration is given by

$$\frac{|d\mathbf{v}/dt|}{2|\boldsymbol{\Omega} \times \mathbf{v}|} = \frac{|\partial \mathbf{v}/\partial t|}{2|\boldsymbol{\Omega} \times \mathbf{v}|} + \frac{|(\mathbf{v} \cdot \nabla)\mathbf{v}|}{2|\boldsymbol{\Omega} \times \mathbf{v}|} = O\left[(2\Omega\mathbb{T})^{-1}, \frac{\mathbb{U}}{2\Omega\mathbb{L}}\right],$$

which, in view of the values of $\mathbb{U} \sim 1$ m s $^{-1}$, $\mathbb{T} \sim 10$ days and $\mathbb{L} \sim 100$ km, is evidently small for oceanic fronts. This indicates that the acceleration terms in (1.6) can be neglected. The dissipative term $\mathbf{F}(\mathbf{v})$ is usually important in strong shear currents, for instance in surface and bottom boundary layers, which, with sufficient accuracy, can be omitted here.

Summarizing, the equations for the description of the dominant force balance within oceanic density fronts are

$$-fV = -\frac{1}{\bar{\rho}_0}\frac{\partial P_0}{\partial x}, \qquad 0 = -\frac{1}{\bar{\rho}_0}\frac{\partial P_0}{\partial z} - g\frac{\rho_0}{\bar{\rho}_0}, \tag{3.1}$$

where V is the velocity of the current and x is perpendicular to it. This is the so-called "geostrophic" approximation. It is not difficult to reduce (3.1) to a relation for the "thermal wind," a term originating in meteorology,

$$fV_z = -g\frac{\rho_{0x}}{\bar{\rho}_0}. \tag{3.2}$$

It is seen that *the vertical geostrophic shear is directly related to the horizontal density gradient (and vice versa)*. Equation (3.2) also implies that, within the geostrophic approximation, V must have a z dependence if ρ_0 has a nonvanishing horizontal gradient. This equation also suggests the possible introduction of a horizontal buoyancy frequency, namely that given by

$$M^2 = g\frac{\rho_{0x}}{\bar{\rho}_0} \tag{3.3}$$

via the analogy with the vertical buoyancy frequency N [160].

Our study of the generation and propagation of baroclinic tides in the presence of horizontal density gradients is restricted by the consideration of the following two models. The first is the two-layer model and will be developed in Section 3.1. For this model we will study the formation and scattering of internal waves by a frontal zone in Section 3.2. Then, in Section 3.3, the range of applicability of the two-layer models in the generation of baroclinic tides will be discussed. The second method is concerned with the construction of the solution for a fluid that is continuously stratified in both the horizontal and vertical direction, and the Riemann method for solving the hyperbolic differential equations, already encountered earlier, is applied. The numerical procedure is described in Section 3.4, and in Section 3.5 some effects on the dynamics of internal waves due to horizontal density gradients are explained.

3.1 Semianalytical two-layer model

This section is devoted to the description of a mathematical model for internal tides when the fluid stratification can be approximated by two homogeneous layers (with a sharp density jump) [222]. The shallow-water theory will be applied. It is valid for linear baroclinic tides when the vertical component of the accelerations of the fluid particles can be neglected in comparison with the horizontal ones. So, after omitting the vertical acceleration terms, the momentum balance equations for a homogeneous inviscid fluid take the simplified form

$$\left.\begin{aligned} u_t + uu_x - fv &= -P_x/\bar{\rho}_0, \\ v_t + uv_x + fu &= 0, \\ P_z &= -g\bar{\rho}_0, \end{aligned}\right\} \tag{3.4}$$

where the density $\rho_0(z)$ has been replaced by the constant averaged value $\bar{\rho}_0$. The last equation defines the hydrostatic law for the pressure. Designating the deviation of the free surface from its rest position by ξ and integrating the last equation in (3.4) over z, we find

$$P = g\bar{\rho}_0(\xi - z) + P_{\mathrm{a}}, \tag{3.5}$$

where P_{a} is the atmospheric pressure at the free surface. We should add an additional equation to complete system (3.4). It can be obtained by integrating the mass balance equation (1.21) over z from $z = -H$ to $z = \xi$. With the use of the boundary conditions (1.32), we find

$$\xi_t + [u(\xi + H)]_x = 0. \tag{3.6}$$

Equations (3.4)–(3.6) are valid for a homogeneous fluid. However, they can also be used as a basis for the construction of tidal models in a stratified fluid, if stratification can be approximated by several homogeneous layers with different densities.

The advantage of layered models is in their relative simplicity. In addition, they provide us with the opportunity to simulate relatively complex hydrological situations by combining a large number of layers. Indeed, theoretical studies of internal waves in oceanography started with the development of layered models. The first work in this regard was a laboratory investigation by Franklin [65], which was devoted to the study of the eigenfluctuations of the interface between oil and water. The increased interest in investigations on internal waves, both theoretically and experimentally, arose after Nansen's expedition to the Barents Sea [174], during which the phenomenon of "dead water," which is closely connected to the two-layer structure of sea water, was discovered. Subsequently, Ekman [56] described the first observation of internal waves. He confirmed Nansen's conclusion that the fluctuations of the interface in a two-layer fluid are responsible for the abnormal resistance felt by ships in the stratified sea. Much later, *in situ* measurements showed that the density of a fluid may suffer considerable changes, not only in the vertical, but also in the horizontal direction [62], [237].

For a two-layer ocean, the most elementary model of a frontal zone can be represented by a tilted interface, which separates two water masses with different densities; see Figure 3.4. In the area of the frontal zone the horizontal density gradient, $\partial\rho_+^{(2)}(x)/\partial x$, can also be taken into account [222], as we shall soon see. The complex density field is accompanied by a geostrophic flow directed perpendicular to the front.

The problem formulation here is similar to the formulations considered in Chapter 2: internal waves are studied which are generated by (1) baroclinic or (2) barotropic tidal waves interacting with the frontal zone II that is located in $|x| < l$, where the basin depth may also vary.

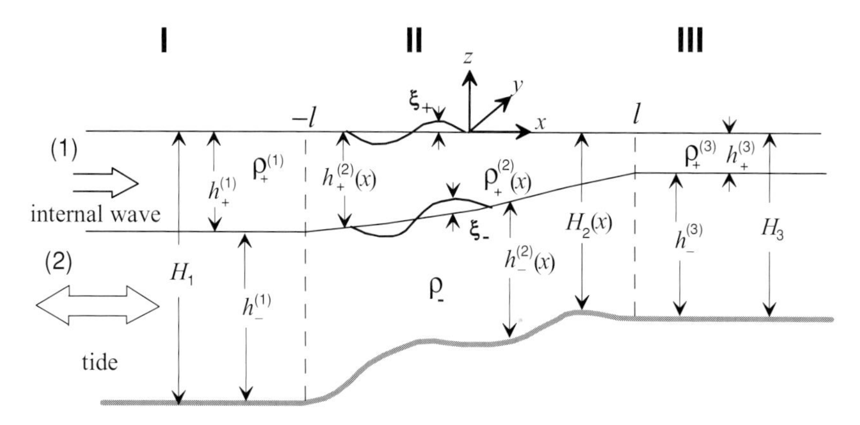

Figure 3.4. Sketch of the two-layer model of an ocean with a frontal zone.

We consider a two-layer fluid stratification. In areas I and III the densities and thicknesses of the upper layer are constant and are designated by $\rho_+^{(1)}$, $\rho_+^{(3)}$ and $h_+^{(1)}, h_+^{(3)}$, respectively. In area II the horizontally dependent density, $\rho_+^{(2)}(x)$ and the thickness of the upper layer, $h_+^{(2)}(x)$, are given by

$$\rho_+^{(2)}(x) = \rho_+^{(1)} + \varrho(x + l), \quad h_+^{(2)}(x) = h_+^{(1)} + \varsigma(x + l), \tag{3.7}$$

where

$$\varrho = (\rho_+^{(3)} - \rho_+^{(1)})/2l, \quad \varsigma = (h_+^{(3)} - h_+^{(1)})/2l. \tag{3.8}$$

The water density in the lower layer is assumed to be constant through the entire basin; its value, ρ_-, is the same for zones I, II, and III. The basin depths, H_1 for $x < -l$ and H_3 for $x > l$, are constant, but in region II for $|x| < l$ the variable bottom topography is defined by the function $H_2(x)$.

The dynamical "equilibrium" or stability of the horizontal density gradient and tilted interface in area II is maintained by a geostrophic current along the isobaths (in the y-direction perpendicular to the plane of Figure 3.4). We define its velocity in the upper and lower layers by $V_+^{(2)}(x, z)$ and $V_-^{(2)}(x, z)$, respectively. To find the explicit expressions for these functions, equation (3.2) should be used; it yields

$$\left.\begin{aligned} V_+^{(2)}(x, z) &= -\frac{g}{f\bar{\rho}_0}\frac{\partial[\rho_+^{(2)}(x)z]}{\partial x}, \\ V_-^{(2)}(x, z) &= \frac{g}{f\bar{\rho}_0}\frac{\partial[\rho_+^{(2)}(x)h_+^{(2)}(x) - \rho_-(h_+^{(2)}(x) + z)]}{\partial x}. \end{aligned}\right\} \tag{3.9}$$

The terms on the right-hand sides denote the pressure in the upper and lower layers, respectively. It is clear that outside the region with horizontal density gradient, i.e. in areas I and III, where the fluid is not horizontally stratified, the geostrophic current vanishes.

Now we assume that the specific contributions of the nonlinear terms in (3.4) to the wave dynamics are small. So, performing a linearization, and treating the stable geostrophic current $(0, V, 0)$ as being of lowest order and all other terms of first order, the equations of motion for infinitesimal waves take the following forms. For the upper layer,

$$\left.\begin{aligned} \frac{\partial u_+^{(i)}}{\partial t} - f v_+^{(i)} &= -\frac{g}{\bar{\rho}_0}\frac{\partial}{\partial x}(\rho_+^{(i)}\xi_+^{(i)}), \\ \frac{\partial v_+^{(i)}}{\partial t} + \left(\frac{\partial V_+^{(i)}}{\partial x} + f\right) u_+^{(i)} &= 0; \end{aligned}\right\} \tag{3.10}$$

and for the lower layer,

$$\left.\begin{aligned} \frac{\partial u_-^{(i)}}{\partial t} - f v_-^{(i)} &= -\frac{g}{\bar{\rho}_0}\frac{\partial}{\partial x}[\rho_+^{(i)}(\xi_+^{(i)} - \xi_-^{(i)}) + \rho_- \xi_-^{(i)}], \\ \frac{\partial v_-^{(i)}}{\partial t} + \left(\frac{\partial V_-^{(i)}}{\partial x} + f\right) u_-^{(i)} &= 0. \end{aligned}\right\} \tag{3.11}$$

Here, $\xi_+^{(i)}$ and $\xi_-^{(i)}$ are the displacements of the free surface and interface, respectively (see Figure 3.4), and the value of the superscript "(i)" of the variables is equal to 1, 2, or 3, dependent upon to which area I, II, or III the equations are applied. As can be seen from (3.10) and (3.11), the background geostrophic current appears in the governing system not via the velocity itself but through its derivative with respect to x. These derivatives for every layer can be found from (3.9):

$$\left.\begin{aligned} \frac{\partial V_+^{(2)}}{\partial x} &= -\frac{g}{f\bar{\rho}_0}\frac{\partial^2[\rho_+^{(2)}(x)]}{\partial x^2} z, \\ \frac{\partial V_-^{(2)}}{\partial x} &= \frac{g}{f\bar{\rho}_0}\frac{\partial^2[(\rho_+^{(2)}(x) - \rho_-)h_+^{(2)}]}{\partial x^2}. \end{aligned}\right\} \tag{3.12}$$

For the linear law of the density gradient (see the first equation in (3.7)), it is evident that $[V_+^{(2)}]_x = 0$. The mass balance equation (1.21) for the two layers reads

$$\frac{\partial(\xi_+^{(i)} - \xi_-^{(i)})}{\partial t} + \frac{\partial(u_+^{(i)} h_+^{(i)})}{\partial x} = 0, \tag{3.13}$$

$$\frac{\partial \xi_-^{(i)}}{\partial t} + \frac{\partial[u_-^{(i)}(H_i - h_+^{(i)})]}{\partial x} = 0. \tag{3.14}$$

To solve the homogeneous equations (3.10)–(3.14), we assume further harmonic periodicity of all unknown functions with frequency σ, i.e.

$$\left\{u_\pm^{(i)}, v_\pm^{(i)}, \xi_\pm^{(i)}\right\} = \left\{\overset{*(i)}{u}_\pm, \overset{*(i)}{v}_\pm, \overset{*(i)}{\xi}_\pm\right\}\exp(\iota\sigma t).$$

Substituting these into (3.10) and (3.11) and eliminating $\overset{*(i)}{v}_\pm$ yields the expressions

$$\overset{*(i)}{u}_+ = \left(\frac{\iota\sigma g}{\bar{\rho}_0}\right)\frac{1}{\sigma^2 - f^2}\frac{\partial}{\partial x}\left(\rho_+^{(i)}\overset{*(i)}{\xi}_+\right) \tag{3.15}$$

and

$$\overset{*(i)}{u}_- = \left(\frac{\iota\sigma g}{\bar{\rho}_0}\right)\frac{1}{\sigma^2 - f^2 - f[V_-^{(i)}]_x}\frac{\partial}{\partial x}\left[\rho_+^{(i)}\left(\overset{*(i)}{\xi}_+ - \overset{*(i)}{\xi}_-\right) + \rho_- \overset{*(i)}{\xi}_-\right] \tag{3.16}$$

for the layer velocities in the various regions. Finally, substituting (3.15) and (3.16) into (3.13) and (3.14), we find the following system of differential equations for $\overset{*}{\xi}{}^{(i)}_{\pm}$:

$$\left.\begin{aligned}\frac{\partial}{\partial x}\left[\frac{gh_+^{(i)}}{\bar{\rho}_0(\sigma^2-f^2)}\frac{\partial}{\partial x}\left(\rho_+^{(i)}\overset{*}{\xi}{}^{(i)}_{+}\right)\right]+\left(\overset{*}{\xi}{}^{(i)}_{+}-\overset{*}{\xi}{}^{(i)}_{-}\right)=0,\\ \frac{\partial}{\partial x}\left\{\frac{g(H_i-h_+^{(i)})}{\bar{\rho}_0(\sigma^2-f^2-f[V_-^{(i)}]_x)}\frac{\partial}{\partial x}\left[\rho_+^{(i)}\left(\overset{*}{\xi}{}^{(i)}_{+}-\overset{*}{\xi}{}^{(i)}_{-}\right)+\rho_-\overset{*}{\xi}{}^{(i)}_{-}\right]\right\}+\overset{*}{\xi}{}^{(i)}_{-}=0.\end{aligned}\right\} \tag{3.17}$$

These are a pair of second-order ODEs.

In areas I and III, where there is no horizontal density gradient and the bottom is flat, system (3.17) takes the simpler form ($i = 1, 3$)

$$\frac{\partial^4\overset{*}{\xi}{}^{(i)}_{+}}{\partial x^4}+\frac{\bar{\rho}_0(\sigma^2-f^2)\left[\left(H_i-h_+^{(i)}\right)\rho_-+h_+^{(i)}\rho_+^{(i)}\right]}{g\rho_+^{(i)}(H_i-h_+^{(i)})h_+^{(i)}(\rho_--\rho_+^{(i)})}\frac{\partial^2\overset{*}{\xi}{}^{(i)}_{+}}{\partial x^2}$$
$$+\frac{(\sigma^2-f^2)^2\bar{\rho}_0^2}{g^2\rho_+^{(i)}(H_i-h_+^{(i)})h_+^{(i)}(\rho_--\rho_+^{(i)})}\overset{*}{\xi}{}^{(i)}_{+}=0, \tag{3.18}$$

$$\overset{*}{\xi}{}^{(i)}_{-}=\overset{*}{\xi}{}^{(i)}_{+}+\frac{g\rho_+^{(i)}h_+^{(i)}}{\bar{\rho}_0(\sigma^2-f^2)}\frac{\partial^2\overset{*}{\xi}{}^{(i)}_{+}}{\partial x^2}. \tag{3.19}$$

Equation (3.18) is a fourth-order ODE for $\overset{*}{\xi}{}^{(i)}_{+}$ ($i = 1, 3$) with constant coefficients (∂ could be replaced by d). Once (3.18) is solved, $\overset{*}{\xi}{}^{(i)}_{-}$ then follows from (3.19).

It is natural to assume that the solution outside the region of the frontal zone and bottom topography consists of reflected and transmitted waves. So, the solution of (3.18) can be expressed as a periodic function $\overset{*}{\xi}{}^{(i)}_{+} \sim \exp(\imath k_i x)$, where the wavenumber k_i is defined by the dispersion relation

$$k_i^2=\frac{\bar{\rho}_0(\sigma^2-f^2)\left[\left(H_i-h_+^{(i)}\right)\rho_-+h_+^{(i)}\rho_+^{(i)}\right]}{2g\rho_+^{(i)}(H_i-h_+^{(i)})h_+^{(i)}(\rho_--\rho_+^{(i)})}$$
$$\times\left\{1\pm\left[1-\frac{4(H_i-h_+^{(i)})h_+^{(i)}(\rho_--\rho_+^{(i)})\rho_+^{(i)}}{\left[\left(H_i-h_+^{(i)}\right)\rho_-+h_+^{(i)}\rho_+^{(i)}\right]^2}\right]^{1/2}\right\}, \tag{3.20}$$

obtained from (3.18) by solving the bi-quadratic equation for k_i^2.

Let us analyze (3.20). Depending upon the $\pm$ signs in (3.20), two different waves, with wavenumbers $k_i^{(0)}$ and $k_i^{(1)}$, emerge. One of them corresponds to the barotropic, the other one to the baroclinic mode. To identify this more explicitly, let us make some simplifications. For a typical oceanic situation, the upper layer is usually much thinner than the lower layer, $(H_i - h_+^{(i)})h_+^{(i)}/H_i^2 < 1$, and $(\rho_- - \rho_+^{(i)})/\rho_- \sim 10^{-3}$. Thus it is evident that $\left[4(H_i - h_+^{(i)})h_+^{(i)}(\rho_- - \rho_+^{(i)})\rho_+^{(i)}\right] \Big/ \left[\left(H_i - h_+^{(i)}\right)\rho_- + h_+^{(i)}\rho_+^{(i)}\right]^2 \ll 1$. Therefore using a Taylor series expansion, (3.20) can be rewritten as

$$k_i^2 \cong \frac{\bar{\rho}_0(\sigma^2 - f^2)\left[\left(H_i - h_+^{(i)}\right)\rho_- + h_+^{(i)}\rho_+^{(i)}\right]}{2g\rho_+^{(i)}(H_i - h_+^{(i)})h_+^{(i)}(\rho_- - \rho_+^{(i)})} \times \left\{1 \pm \left[1 - \frac{2(H_i - h_+^{(i)})h_+^{(i)}(\rho_- - \rho_+^{(i)})\rho_+^{(i)}}{\left[\left(H_i - h_+^{(i)}\right)\rho_- + h_+^{(i)}\rho_+^{(i)}\right]^2}\right]\right\}. \quad (3.21)$$

It follows that $(k_i^{(0)})^2 = (\sigma^2 - f^2)/gH_i$. This wavenumber corresponds to the barotropic wave. The other wavenumber, $k_i^{(1)}$, which is not simplified by (3.21), must then correspond to the baroclinic wave. Thus, the two-layer model allows for two modes and no more: one barotropic mode and one baroclinic mode.

We study the interaction of the barotropic or baroclinic tidal wave with the frontal zone and bottom topography. In other words, the incident wave propagates from area I to area III through area II. This wave generates two reflected (barotropic and baroclinic) waves in area I and two transmitted barotropic and baroclinic waves in area III. Thus, the solutions ahead of and behind the transition zone II can be written as

$$\left.\begin{aligned} \overset{*(1)}{\xi}_{\pm} &= a_0^{\mathrm{inc}} D_{(0)\pm}^{(1)} \exp(-\imath k_0^{(1)} x) + b_0^{\mathrm{refl}} D_{(0)\pm}^{(1)} \exp(\imath k_0^{(1)} x + \chi_0^{(1)}) \\ &\quad + a_1^{\mathrm{inc}} D_{(1)\pm}^{(1)} \exp(-\imath k_1^{(1)} x) + b_1^{\mathrm{refl}} D_{(1)\pm}^{(1)} \exp(\imath k_1^{(1)} x + \chi_1^{(1)}), \\ \overset{*(3)}{\xi}_{\pm} &= a_0^{\mathrm{tr}} D_{(0)\pm}^{(3)} \exp(-\imath k_0^{(3)} x + \phi_0^{(3)}) + a_1^{\mathrm{tr}} D_{(1)\pm}^{(3)} \exp(-\imath k_1^{(3)} x + \phi_1^{(3)}), \end{aligned}\right\} \quad (3.22)$$

where a_0^{inc} and b_0^{refl} and $\chi_0^{(1)}$ or a_1^{inc} and b_1^{refl} and $\chi_1^{(1)}$ are the amplitudes and phases of the incident and reflected barotropic or baroclinic waves; a_0^{tr} and $\phi_0^{(3)}$, a_1^{tr} and $\phi_1^{(3)}$ are the amplitudes and phases of the transmitted barotropic and baroclinic waves. The subscript index "+" identifies the free surface, whilst "−" refers to the interface. We consider situations when only one wave, either barotropic or baroclinic, interacts with an obstacle. This means that, if we define a_0^{inc} as nonzero, then we should take $a_1^{\mathrm{inc}} = 0$, and vice versa.

The coefficient $D^{(i)}_{(j)\pm}$ is found from (3.19) by substituting (3.22). It has the following representation:

$$D^{(i)}_{(j)\pm} = \begin{cases} 1, & \text{free surface} \quad (+), \\ 1 - \dfrac{g h^{(i)}_{+} (k^{(i)}_{j})^2 \rho^{(i)}_{+}}{(\sigma^2 - f^2)\bar{\rho}_0}, & \text{interface} \quad (-). \end{cases}$$

Thus, it is clearly seen that the value of the amplitude of the wave at the interface differs from the value at the free surface by the coefficient $D^{(i)}_{(j)-}$.

The solution of equation (3.17) in area II cannot be separated into barotropic and baroclinic parts and is found numerically as a sum of four unknown linearly independent functions $F_i(x)$ and $R_i(x)$, i.e.

$$\overset{*(2)}{\xi}_{+} = \sum_{j=1}^{4} B_j F_j(x), \quad \overset{*(2)}{\xi}_{-} = \sum_{j=1}^{4} B_j R_j(x), \tag{3.23}$$

in which, according to (3.17), $R_j(x)$ is connected to $F_j(x)$ by the following relation:

$$R_j(x) = F_j(x) + \frac{g}{\bar{\rho}_0(\sigma^2 - f^2)} \left\{ \frac{h^{(i)}_{+}}{\left[\rho^{(i)}_{+} F_j(x)\right]_x} \right\}_x .$$

This result justifies the selection of the same amplitudes B_j $(j = 1, \ldots, 4)$ in the second of the representations of (3.23).

The displacements of the free surface and interface and velocities in areas I: $\overset{*(1)}{\xi}_{\pm}$, $\overset{*(1)}{u}_{\pm}$, II: $\overset{*(2)}{\xi}_{\pm}$, $\overset{*(2)}{u}_{\pm}$, and III: $\overset{*(3)}{\xi}_{\pm}$, $\overset{*(3)}{u}_{\pm}$, satisfy continuity conditions at $x = \pm l$ as follows:

$$\left.\begin{aligned} \overset{*(1)}{\xi}_{\pm}(-l) &= \overset{*(2)}{\xi}_{\pm}(-l), \quad \overset{*(2)}{\xi}_{\pm}(l) = \overset{*(3)}{\xi}_{\pm}(l), \\ \overset{*(1)}{u}_{\pm}(-l) &= \overset{*(2)}{u}_{\pm}(-l), \quad \overset{*(2)}{u}_{\pm}(l) = \overset{*(3)}{u}_{\pm}(l). \end{aligned}\right\} \tag{3.24}$$

This yields a system of linear equations with unknown values of b_0, $b^{(1)}_{\xi}$, a_0, $a^{(1)}_{\xi}$, and B_1, B_2, B_3, and B_4. They can be found by Gaussian elimination, if the functions $F_j(x)$ and $R_j(x)$ $(j = 1, \ldots, 4)$ are known.

When substituting (3.23) into (3.17), the resulting right second-order ODEs for F_j and R_j can be expressed in the form

$$\frac{d^2 F_j}{dx^2} = \Lambda\left(R_j, F_j, \frac{dF_j}{dx}\right),$$

$$\frac{d^2 R_j}{dx^2} = \Upsilon\left(R_j, \frac{dR_j}{dx}, F_j, \frac{dF_j}{dx}\right),$$

with known functions Λ and Υ. Introducing the auxiliary variables $X_j = dF_j/dx$ and $Z_J = dR_j/dx$ translates the above problem into a set of first-order ODEs,

$$\frac{dF_j}{dx} = X_j = \Upsilon_j^{(1)}(R_j, Z_j, F_j, X_j, x),$$
$$\frac{dR_j}{dx} = Z_j = \Upsilon_j^{(2)}(R_j, Z_j, F_j, X_j, x),$$
$$\frac{dX_j}{dx} = \Upsilon_j^{(3)}(R_j, Z_j, F_j, X_j, x),$$
$$\frac{dZ_j}{dx} = \Upsilon_j^{(4)}(R_j, Z_j, F_j, X_j, x),$$

with the formal solution at some point x_1

$$F_j(x_1) = F_j(x_0) + \int_{x_0}^{x_1} \Upsilon_j^{(1)}(R_j, Z_j, F_j, X_j, x)\, dx,$$
$$R_j(x_1) = R_j(x_0) + \int_{x_0}^{x_1} \Upsilon_j^{(2)}(R_j, Z_j, F_j, X_j, x)\, dx,$$
$$X_j(x_1) = X_j(x_0) + \int_{x_0}^{x_1} \Upsilon_j^{(3)}(R_j, Z_j, F_j, X_j, x)\, dx,$$
$$Z_j(x_1) = Z_j(x_0) + \int_{x_0}^{x_1} \Upsilon_j^{(4)}(R_j, Z_j, F_j, X_j, x)\, dx,$$

of which $F_j(x_0)$, $R_j(x_0)$, $X_j(x_0) = F_{jx}(x_0)$ are the initial values at the starting point x_0, which must be determined. Taking into account that the functions F_j, R_j, F_{jx}, and R_{xj} must all be linearly independent, by choosing $x_0 = -l$ and selecting

$$\left.\begin{array}{llll} F_1(-l) = 1, & F_2(-l) = 0, & F_3(-l) = 0, & F_4(-l) = 0, \\ R_1(-l) = 0, & R_2(-l) = 1, & R_3(-l) = 0, & R_4(-l) = 0, \\ F_{1x}(-l) = 0, & F_{2x}(-l) = 0, & F_{3x}(-l) = 1, & F_{4x}(-l) = 0, \\ R_{1x}(-l) = 0, & R_{2x}(-l) = 0, & R_{3x}(-l) = 0, & R_{4x}(-l) = 1, \end{array}\right\} \tag{3.25}$$

eight linearly independent functions $F_j(x)$, $R_j(x)$, $j = 1, \ldots, 4$, are found that are used in (3.23) and (3.24) to determine the unknown b_0^{ref}, b_1^{ref}, a_0^{tr}, b_1^{tr}, B_1, B_2, B_3, and B_4 with the help of the Runge–Kutta method.

3.2 Wave characteristics derived from the two-layer model

In this section we describe the results obtained by the previously formulated linear two-layer BVP. We will study both the generation and the scattering of internal tidal waves by frontal zones. Two specific cases are considered. First, we will estimate

Table 3.1. *Parameters used in the numerical calculations of the wave and frontal zone interaction.*

H_1 (km)	$h_+^{(1)}$ (km)	$h_+^{(3)}$ (km)	φ (°)	T (h)	ρ_- (kg m^{-3})	$\rho_+^{(1)}$ (kg m^{-3})	$\rho_+^{(3)}$ (kg m^{-3})	$2l$ (km)
4	0.2	0.1	30	12.4	1027	1024	1025	50–500

the effect of a front in which the transition zone is characterized only by horizontal density gradients. Such a situation is typical for the deep parts of the World Ocean where the effects connected with the bottom topography can be neglected. Then, we will study the combined effect of a density front together with variations of the bottom topography – a situation which is typical for areas of oceanic ridges or continental slopes. Calculations are performed with the set of input parameters listed in Table 3.1. The underwater ridge, given by the formula

$$H_2(x) = \begin{cases} H_0, & x < -l, \\ H_0 - H_{\max}\cos^2(\pi x/2l), & -l \le x \le l, \\ H_0, & x > l, \end{cases} \tag{3.26}$$

is used as our test bottom topography.

The results will be presented as dependences of the amplitudes of the interface displacements, normalized to the amplitude of the incident, baroclinic or barotropic, wave. We define the wave amplitude in area I as $\bar{b}_1 = b_1^{\mathrm{refl}} D_-^{(1)}/a_0^{\mathrm{inc}}$ (reflected wave) and as $\bar{a}_1 = a_1^{\mathrm{tr}} D_-^{(3)}/a_0^{\mathrm{inc}}$ in area III (transmitted wave) if the incident barotropic wave with amplitude a_0^{inc} is considered; for the scattering problem we choose $\bar{b}_1 = b_1^{\mathrm{refl}}/a_1^{\mathrm{inc}}$ and $\bar{a}_1 = a_1^{\mathrm{refl}}/a_1^{\mathrm{inc}}$.

3.2.1 Generation of internal waves

We start by considering the generation of internal waves by the barotropic tide when it interacts with the isolated frontal zone in an ocean of constant depth. As mentioned above, the result of such interactions manifests itself as two internal waves propagating in opposite directions from the source of generation (the frontal zone). The dependence of their amplitudes $\bar{a}_1(l)$ (at $x < -l$) and $\bar{b}_1(l)$ (at $x > l$) upon the width of the frontal zone, $2l$, is shown in Figure 3.5. The minimum value of the frontal zone width, $2l_{\min}$, considered here was defined by simple oceanographic reasoning: the maximum value of the geostrophic current $V_{\max}$ observed in the World Ocean in frontal zones of strong currents like the Gulf Stream or Kuroshio is of the order of 1 m s^{-1}. These currents basically occupy the upper layer with thickness h_+. So, this maximum value of the geostrophic current (which, in fact,

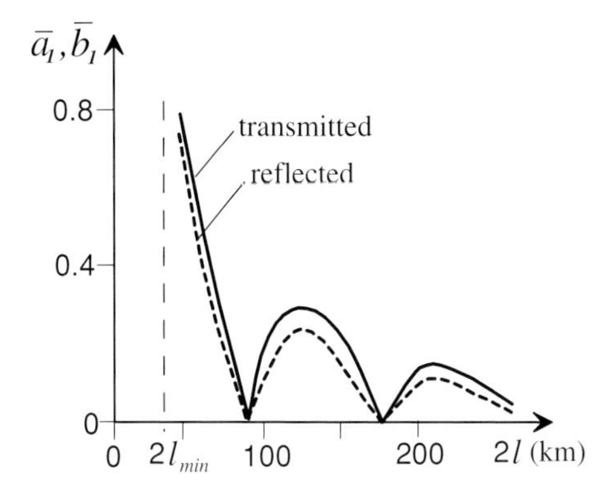

Figure 3.5. Amplitudes $\bar{b}_1$ (transmitted) and $\bar{a}_1$ (reflected) of internal waves generated in area I (short-dashed line) and area III (solid line), respectively, versus the width of the frontal zone $2l$.

defines the magnitude of $2l_{\text{min}} \cong -g(\rho_+^{(3)} - \rho_+^{(1)})h_+/\bar{\rho}_0 f V_{\text{max}}$, by formula (3.2)) was taken as a natural bound for our calculations.

Figure 3.5 shows that the amplitude $\bar{b}_1$ of the transmitted wave (solid line) is larger than or equal to the amplitude of the reflected wave, $\bar{a}_1$ (short-dashed line). For variable bottom topography, the mechanism of interaction of the barotropic tidal wave with a local density front also enjoys a resonance character: both dependences, $\bar{b}_1(l)$ and $\bar{a}_1(l)$, possess local minima and maxima which occur when certain resonance conditions are satisfied. For instance, the first minimum in Figure 3.5 arises when $2l = 92$ km. On the other hand, the "mean wavelength" of the internal wave, λ_{mean}, generated in zones I and III and defined as $(\lambda_1 + \lambda_3)/2$ ($\lambda_1 = 116$ km is the wavelength in area I and $\lambda_3 = 68$ km is the wavelength in area III), is also equal to 92 km. In addition, the value $T_1(2l)$ of the integral (2.63) which defines the propagation time of the wave between the boundaries $x = -l$ and $x = l$ of the obstacle is equal to the tidal period.

The next finding is that the positions of the local maxima of the functions $\bar{b}_1(2l)$ and $\bar{a}_1(2l)$ are divisible by $\lambda_{\text{mean}}/2$ (as was found for bottom obstacles), and the maximum values decrease asymptotically with an increase of $2l$ (see also Sections 2.1 and 2.3). For example, at $2l = 130$ km, the amplitude of the transmitted wave is $\bar{a}_1 = 0.3$ and at $2l = 220$ km we have $\bar{a}_1 = 0.16$. The decrease of the efficiency of the generation of internal waves with the growth of the width of the frontal zone is connected with a decrease in the density gradient inside it. Since the value of the density difference between regions I and III is kept constant, the increase in the width of the frontal zone reduces the density gradient.

Above, the amplitudes of the generated waves outside the frontal zone, for $|x| > l$, were considered. Now, we investigate the wave field inside the frontal zone, for

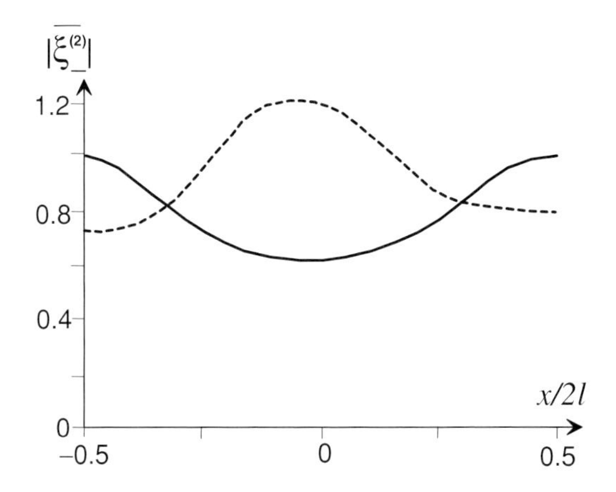

Figure 3.6. Maximum deviations of the interface inside the frontal zone plotted against the local normalized coordinate $x/2l$. The solid line corresponds to the first zero values of the amplitudes of generated waves $\bar{a}_1$ and $\bar{b}_1$ in Figure 3.5 when the width of the frontal zone $2l = \lambda_{\text{mean}}/2 = 92$ km, and the dotted line corresponds to the maximum values of the amplitudes when $2l = \lambda_{\text{mean}} = 130$ km.

$|x| < l$. We consider two extreme situations: the maximum and minimum of the wave amplitudes outside the frontal zone, i.e. when the value of its width is $2l = \lambda_{\text{mean}}/2$ and $2l = \lambda_{\text{mean}}$, respectively. Because inside the region with nonvanishing density gradient the total wave field cannot be decomposed into baroclinic and barotropic waves – here modes are not separated – the total normalized interface-displacement amplitude $|\bar{\xi}_{-}^{(2)}(x/2l)| = |\xi_{-}^{(2)}(x/2l)|/(a_1^{\text{inc}} D_{-}^1)$ in area II is produced by both waves. This is shown in Figure 3.6. The dashed line depicts $\bar{\xi}_{-}^{(2)}$ for the condition of maximum amplitude generation in zones I and III (occurring at $2l = \lambda_{\text{mean}}$) and the solid line displays it for minimum efficiency ($\bar{a}_1 = \bar{b}_1 = 0$).

The two curves in Figure 3.6 exhibit quite different behavior. The solid line has a minimum in the center of the frontal zone at $x = 0$, whereas the dashed line $|\bar{\xi}_{-}^{(2)}(x/2l)|$ has its maximum there. This can be easily explained. In fact, inside the frontal zone standing internal waves exist which are produced by two progressive baroclinic waves propagating in opposite directions. An input of barotropic waves into the total wave field can be excluded because of their large wavelengths, which are $\sim 10^5$ km; this is much more than those of the internal waves which have a spatial scale of $\sim 10^2$ km). Thus, the graph of the function $|\bar{\xi}_{-}^{(2)}(x/2l)|$ depends strongly on the conditions of superposition of these two internal waves. These conditions are defined by the relation between the wavelength λ_{mean} and the width $2l$: co-phase wave displacements at liquid boundaries lead to wave intensification, whilst counter-phase relations yield wave depression.

A further inference from Figure 3.6 is that weak wave activity (small or zero values of the amplitudes of generated waves) outside the density front (at $x > l$)

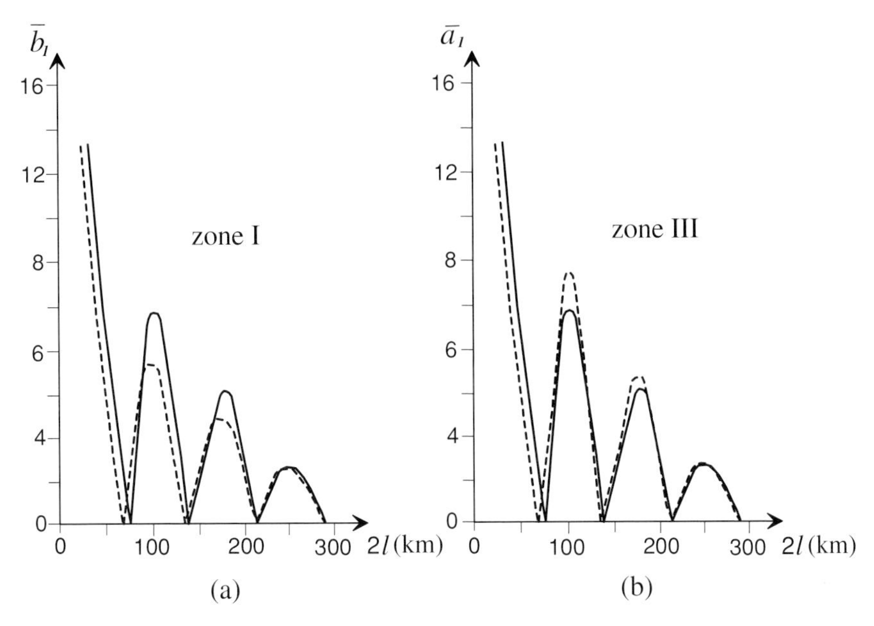

Figure 3.7. Normalized amplitudes of waves generated in (a) zone I and (b) zone II by the interaction of a barotropic tidal wave with a density front located above the oceanic ridge (3.26) with $H_{max} = 3.4$ km versus the width of the transition zone $2l$. Solid lines correspond to the case of zero horizontal density gradient (oceanic ridge only); dashed lines were calculated for parameters of the density front as listed in Table 3.1.

does not mean that the wave motions inside it are also negligible. On the contrary, they can be quite strong when $2l$ is close to the value $n\lambda_{mean}$ ($n = 1, 2, 3, \ldots$).

In conclusion, we may state that similar resonance peculiarities of the generation of internal waves, formulated in Section 2.1.1 for barotropic tides interacting with bottom topography, arise here because the generating areas are density fronts. Next, let us study the generation process when the barotropic tidal wave propagates through a frontal zone that is located above an underwater ridge, e.g. (3.26), and compare these results with those obtained for the ridge without a frontal zone. The comparison, illustrating the specific input of the density gradients into the generated wave field, is shown in Figure 3.7.

The solid lines in Figures 3.7(a) and (b) were obtained for a vanishing horizontal density gradient. Obviously, the results obtained for the two-layer model must be in accordance with the results of Chapter 2, where a continuously stratified fluid was considered. We note that these curves agree well with one another, which is evidence for the model symmetry. This symmetry was shown above in Section 2.1.1 by the analytical solution (2.21), (2.22). Next, both curves also exhibit resonance character of the generation of internal waves: the dependences $\bar{a}_1(2l), \bar{b}_1(2l)$

have local maxima at points $2l = n\lambda_{mean}/2$ (n is an integer number) and minima when $2l = (2n - 1)\lambda_{mean}$. This is in agreement with results described earlier (see Figures 2.3, 2.6, and 2.7).

The presence of the frontal zone above the bottom topographic ridge modifies the internal wave fields. However, these modifications are not crucial as they do not change the qualitative behavior. Obviously, the density front changes slightly the local wavenumber and local phase speed of the first baroclinic mode. Consequently, the propagation time of the wave between the boundaries of the obstacles changes (decreases in our case); see (2.63). This, in turn, leads to a shift of the positions of the extrema to the left; however, this shift is appreciable only for small values of the width $2l$ of the transition zone. This is not surprising, because the density gradient decreases with the growth of $2l$ and fixed density difference between zones I and III.

The last remark concerns the magnitude of the peaks of the functions $\bar{a}_1(2l)$ and $\bar{b}_1(2l)$. It is clear that the density front also introduces an asymmetry to the model: the amplitudes of the waves in zone I are usually smaller than those for the waves in zone III. This difference can, however, be regarded as a secondary effect.

3.2.2 Internal wave scattering

The interaction of internal tidal waves with a bottom obstacle in a continuously stratified ocean was considered in Chapter 2. It was shown that the incident wave of the nth baroclinic mode transfers part of its energy to the neighboring baroclinic modes (with numbers $n - 1, n + 1, n - 2, n + 2, \ldots$). In accordance with the resonance properties of the scattering mechanism, the amplitudes of the generated modes depend upon the width of the bottom obstacle. Reflected waves were negligible in comparison with those transmitted. This means that the overwhelming part of the energy of the incident wave passes through the transition zone and is redistributed behind the obstacle between several neighboring modes.

In a two-layer representation of the ocean, only the first baroclinic mode can be resolved. Thus, we cannot expect an energy transfer from the incident wave to the other baroclinic modes. We can only estimate how strong the effect of the density front will be on the propagating internal wave. Generally, a certain part of the energy of the incident baroclinic wave must also be transferred to a long barotropic tidal wave. However, the results of direct numerical calculations show that this part of the energy is vanishingly small, and the resulting amplitudes of the barotropic waves generated in areas I and III are so minute that this effect can safely be ignored. This is not surprising if one tries to compare the energy of the baroclinic and barotropic tidal waves of the same amplitude.

Thus, the wave field consists, in our case, of the incident, reflected, and transmitted first-mode baroclinic waves. Note that the amplitude of the wave that is

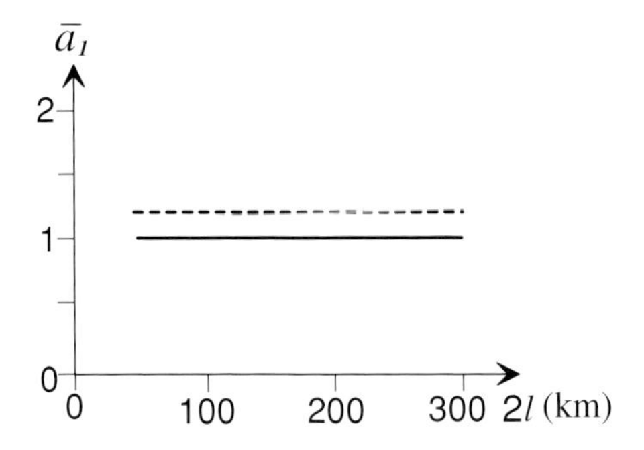

Figure 3.8. Normalized amplitudes of the first baroclinic mode penetrating behind the transition zone which contains only the effect of the density front (dashed line) or only that of the oceanic ridge (3.26) (solid line) versus the width of the transition zone, $2l$. The ridge height equals 3.4 km, and all other input parameters are as in Table 3.1.

reflected from the density front is also at least two orders of magnitude smaller than that transmitted. This is in agreement with results obtained in Section 2.4 concerning the weak reflection of internal wave energy from a bottom topographic disturbance.

The next result is a consequence of the first two conclusions. If only one baroclinic mode is resolved by the model, and if, in addition, the reflected wave is infinitely small, the total wave energy must penetrate through the density front. It follows that in such a case the amplitude of the transmitted wave cannot depend on the width of the frontal zone $2l$ (see Figure 3.8). It will be related sensitively only to the hydrological constraints in zone III (i.e. the thickness of the upper layer and the density difference between the layers). The amplitude becomes larger if the thickness of the upper layer, h_+, decreases or the horizontal density gradient increases. For the parameters given in Table 3.1, the value of the amplitude of the wave in area III, $\bar{a}_1$, is about 1.2.

The independence of the amplitude of the transmitted wave of the width of area II remains valid when not only the density front but also the bottom obstacle is placed in the transition area II. In the two-layer model they depend neither on the width of the area II nor on the height of the bottom obstacle. Of course, this is only so when the amplitude of the wave reflected from the bottom obstacle is small. However, as was found in Section 2.4, the result of a weak reflection is valid in a wide range of oceanic conditions (see also Figure 2.18).

3.3 Applicability of layer models

The results of Section 3.2 show that a two-layer model of fluid stratification cannot be a good approximation for the density if one wishes to study the interaction of internal waves with density fronts or underwater obstacles. However, it is not clear whether this conclusion is valid also for the generation of internal waves when the barotropic tidal wave interacts with a bottom protuberance. The purpose of this section is to define the ranges of applicability of layered models in the generation of baroclinic tides by topographic variations. This purpose is achieved by performing a comparative analysis of the results obtained for both two-layer and continuously stratified models [255].

The method of solution for a continuously stratified fluid is described in Chapter 2 and that for the two-layer stratification is described in Section 3.1. One of the basic questions of the coordination of two models is the correct choice of the input parameters for the two-layer model: the interface depth, $z = -h_+$, and the values of the density, ρ_+ and ρ_-, in the upper and lower layers, respectively. These parameters must be determined on the basis of the employed law of the buoyancy frequency, $N(z)$, in the model for the continuously stratified fluid. We shall use the smooth pycnocline profile (1.70) for qualitative and quantitative estimations. The values of the input parameters N_p, H_p, and ΔH_p for the continuously stratified model are defined in (1.71).

For the two-layer model, the parameters h_+, ρ_+, and ρ_- will be defined as follows: the interface between the two layers at $z = -h_+$ will be chosen to agree with the depth of the pycnocline, $z = -H_p$, and the quantities ρ_+ and ρ_- will be defined as averaged densities of the fluid above and below the pycnocline:

$$\rho_+ = \frac{1}{H_p} \int\limits_{-H_p}^{0} \rho_0(z)\, dz; \quad \rho_- = \frac{1}{H_0 - H_p} \int\limits_{-H_0}^{-H_p} \rho_0(z)\, dz.$$

This choice of parameters is the most natural one, because, in a limiting transition when $\Delta H_p \to 0$, the law (1.70) reduces to the two-layer stratification.

The main distinction between the two techniques to account for the wave fields consists in the use of different laws of stratification. In addition, in the model with a continuous density distribution, the "rigid lid" condition is applied, whereas in the two-layer model the complete kinematic conditions are taken into account at $z = 0$; the two-layer model uses the hydrostatic approximation for the pressure, whilst the alternative model also takes into account the dynamic part of the pressure.

These distinctions imply corresponding discrepancies between the values of the internal wave amplitudes provided by the two models. We will make an attempt to find out on which parameters these discrepancies depend. To this end, we will compare the amplitude of the interface deviation, a_0^{inc}, and the velocities, u_1, u_2, v_1, v_2, in the two-layer model, with the amplitude of the first baroclinic mode, $a_\xi^{(1)}$, at the depth $z = -H_p$, and the values of the mean (in terms of the averages (1.70)) horizontal velocities obtained for the continuously stratified model. The analysis will only be performed when the first baroclinic mode predominates. Otherwise, the comparison suggests that the first and higher modes are important, and it is impossible to approximate the processes by a two-layer model.

Calculations for the ridge (3.26) with $H_0 = 4000$ m, $0 \leq H_{\max} \leq 3800$ m, $20 \leq 2l \leq 600$ km were carried out for the following values of input parameters of stratification, (1.70):

$$N_{\mathrm{p}} = 4 \times 10^{-2}\mathrm{s}^{-1},\ H_{\mathrm{p}} = 100\ \mathrm{m},\ \Delta H_p = (10; 40; 75; 220; 345)\ \mathrm{m}.$$

A tidal flux with value $\Psi_0 = 40\ \mathrm{m}^2\,\mathrm{s}^{-1}$ is considered, which corresponds to the amplitude of a propagating barotropic surface displacement equal to 0.2 m. It is clear that, in the linear theory, the results depend linearly on the intensity of the baroclinic tidal flow. So, the deduced relative estimations will also be valid for all other values of the amplitude for which the linear theory is valid.

Figure 3.9 shows the wave amplitudes of the isopycnals at the depth of the density jump, at $z = -H_{\mathrm{p}}$, derived by the two models outside the ridge area for $\Delta H_{\mathrm{p}} = 75$ m. Evidently, in the range 20 km $\leq 2l \leq$ 300 km the $a_\xi^{(1)}(2l)$ diagrams are characterized by two maxima and one minimum in accordance with the resonance conditions; the case $2l < 20$ km will be examined separately. The amplitude values for the continuously stratified fluid are always slightly larger than those of the two-layer fluid. However, for the given ridge height, the relative error $\delta = [(a_\xi^{(1)} - a_1^{\mathrm{tr}})/a_\xi^{(1)}]\cdot 100\%$ between the solutions varies between 6 and 8%. Similar results were obtained when the horizontal wave velocities were compared.

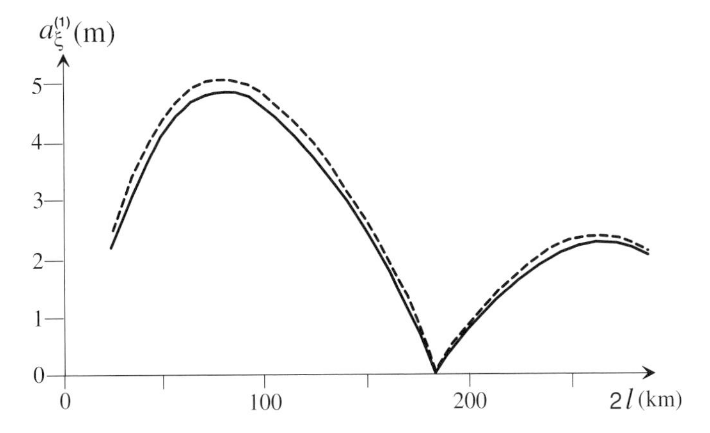

Figure 3.9. Amplitude of the first baroclinic mode generated by the barotropic tide over the oceanic ridge (3.26) with height $H_{\max} = 3.5$ km in an ocean of 4 km depth plotted against the width $2l$, obtained for the two-layer (solid line) and continuously stratified (dashed line) models.

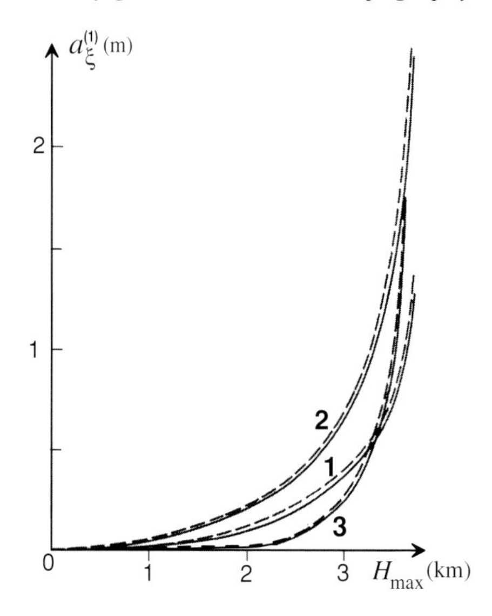

Figure 3.10. Amplitude of the first baroclinic mode obtained for the two-layer (solid lines) and continuously stratified (dashed lines) models versus the height of the ridge, $H_{\max}$. Curves 1–3 are for the ridge widths $2l = 60$, 180, 400 km, respectively. All other parameters are the same as for Figure 3.9.

The situation is not so simple for narrow oceanic ridges ($2l < 20$ km). The decrease of the parameter $2l$ with fixed height $H_{\max}$ leads to a steepening of the slopes, the bottom inclination γ being close to the inclination α of the characteristic lines of the wave equation. As was shown in Section 2.5, in such circumstances the specific input of high baroclinic modes increases, and the first baroclinic mode ceases to be predominant. Thus, when $\gamma \sim \alpha$, the application of the two-layer model is not justified, because it does not permit us to determine the amplitudes of the higher modes.

The dependence of the first mode amplitude on the ridge height $H_{\max}$ computed for $H_{\text{p}} = 75$ m and $2l = 60, 180, 400$ km, is presented in Figure 3.10 (curves 1–3, respectively). With these, one can estimate how much the considered models differ from one another as functions of the bottom height. The specific feature, common to all curves, is the monotonic rise with growing $H_{\max}$. The solid and dashed lines have a spacing between them, but it is small and implies a satisfactory coincidence of the results obtained by the use of the two different models. The error, δ, does not exceed 10%.

Let us turn to the question of how the value of δ is affected by the degree of fluid stratification, i.e. the pycnocline width ΔH_{p}. The maximum errors, $\delta_{\max}$, at various values of $2l$ and $H_{\max}$ are listed in Table 3.2 for five values of ΔH_{p}. It supports the above inference which implies that the error, δ, depends weakly on

Table 3.2. *Maximum relative discrepancy δ_{max} for different values of width 2l and height H_{max} of the ridge depending on the pycnocline width ΔH_p.*

Pycnocline width, ΔH_p (m)	δ_{max} (%) for different H_{max} (km) ($20 \leq 2l \leq 600$ km)					δ_{max} (%) for different $2l$ (km) ($0 \leq H_{max} \leq 3.8$ km)				
	1.0	2.0	3.0	3.5	3.8	50	100	200	400	600
10	5	3	2	3	3	3	3	4	3	4
40	8	8	7	4	5	6	7	5	6	6
75	9	7	8	6	7	8	8	7	7	9
220	13	13	12	10	10	16	14	13	15	16
345	21	19	18	17	18	23	20	20	19	20

the geometric dimensions of the ridge. In fact, with constant ΔH_p, the H_{max}, $2l$-dependent deviations of δ_{max} account for 2–4%. The value of δ_{max} is most sensitive to the width of the pycnocline, ΔH_p. For example, we have $\delta_{max} \sim$ 3–4% when $\Delta H_p = 10$ m, and it attains roughly 20% when $\Delta H_p = 345$ m, Thus, with a large pycnocline width, the error δ yielded by the two models increases.

3.4 Riemann method for a continuously stratified fluid

The two-layer model of baroclinic tides is relatively simple and effective and therefore very attractive for efficient estimations of the combined effects of variable bottom topography and density fronts on tidal dynamics (see Sections 3.1 and 3.2.1); it possesses, however, imperfections. As was mentioned above, such models are inappropriate for the description of the generation of complex, multimodal baroclinic responses which usually occur near steep bottom features. More significantly, in Section 3.2.2 it was shown that the scattering of internal waves by bottom obstacles or frontal zones cannot be described correctly by such models, because they cannot reproduce the energy leakage from the incident wave to the neighboring baroclinic modes. Yet this process even arises over a slightly inclined bottom. The mentioned restrictions can be eliminated by mathematical models such as those in refs. [40] and [223], which are able to account correctly for a continuous fluid stratification. This section is devoted to the description of such models.

The simplest model of a frontal zone in area II is a linear interpolation of the density difference for $-l \leq x \leq l$ in an ocean with linear vertical fluid stratification (Figure 3.11). The transition zone II can also include bottom variations. It is assumed that a barotropic or baroclinic tidal wave propagates from area I through area II into zone III. While interacting with the bottom and the density front it generates internal waves, possibly with several baroclinic modes, which propagate in both directions

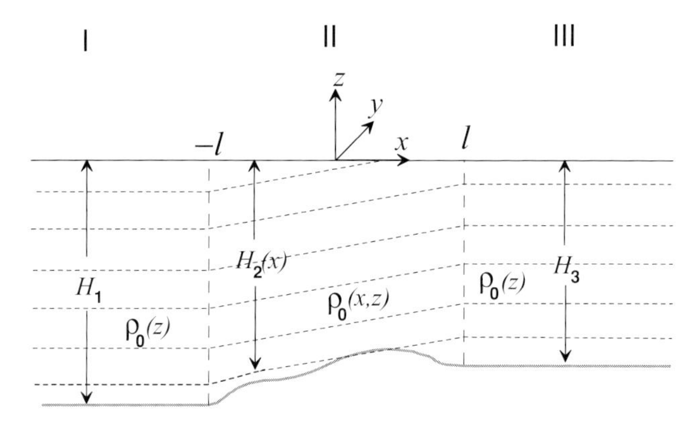

Figure 3.11. Schematic representation of the model domain with a linear stratified frontal zone. Dashed lines represent isopycnals.

from the source of generation. The problem consists in defining the amplitudes of these waves.

To derive the equations describing the wave motions in a horizontally and vertically continuously stratified fluid, we should first linearize the momentum and mass balance equations (1.6) and (1.21) with respect to the geostrophic current $(0, V, 0)$, which is nontrivial due to the presence of the frontal zone. Subsequently, the governing equations are linearized. This yields the following system of partial differential equations:

$$\left.\begin{aligned} u_t - fv &= -\tilde{P}_x/\bar{\rho_0}, \\ v_t + uV_x + wV_z + fu &= -0, \\ w_t &= -\tilde{P}_z/\bar{\rho_0} - g\tilde{\rho}/\bar{\rho_0}, \\ u_x + w_z &= 0, \\ \tilde{\rho}_t + u\rho_{0x} + w\rho_{0z} &= 0, \end{aligned}\right\} \tag{3.27}$$

in which $\tilde{P}$ and $\tilde{\rho}$ are the pressure and density fields caused by wave perturbations. To simplify these equations even further, let us eliminate the transverse velocity component, v. Inroducing the stream function ψ from (1.40) and eliminating $\tilde{\rho}$ then yields

$$\left(f^2 + \frac{\partial^2}{\partial t^2}\right)\psi_z + (fV_z)\psi_x = -\frac{\tilde{P}_{xt}}{\bar{\rho}_0},$$
$$\left(N^2 + \frac{\partial^2}{\partial t^2}\right)\psi_x + M^2\psi_z = -\frac{\tilde{P}_{zt}}{\bar{\rho}_0}.$$

Taking into account the harmonic periodicity of the tidal forcing $\sim\exp(\imath\sigma t)$ for the generated waves in the linear case, we also can assume a harmonic dependence of the remaining variables (1.41). This allows us to reduce the preceding two equations to the final form of a wave equation, which is valid in a region with nonuniform depth [160],

$$(N^2 - \sigma^2)\overset{*}{\psi}_{xx} + 2M^2\overset{*}{\psi}_{xz} - (\sigma^2 - f^2)\overset{*}{\psi}_{zz} = 0. \tag{3.28}$$

This equation is written in terms of the periodic stream function $\overset{*}{\psi}$ defined in (1.41). For the derivation of (3.28), we eliminated the pressure $\tilde{P}$ and used the definitions of the "thermal wind," (3.2), and of the horizontal buoyancy frequency, (3.3).

The dynamics of internal waves in a horizontally stratified ocean of variable depth is described by the following BVP:

$$\left.\begin{array}{c}(N^2 - \sigma^2)\overset{*}{\psi}_{xx} + 2M^2\overset{*}{\psi}_{xz} - (\sigma^2 - f^2)\overset{*}{\psi}_{zz} = 0,\\ \overset{*}{\psi} = 0, \quad z = 0,\\ \text{barotropic case} \quad \overset{*}{\psi} = \Psi_0, \quad z = -H(x),\\ \text{baroclinic case} \quad \overset{*}{\psi} = 0, \quad z = -H(x).\end{array}\right\} \tag{3.29}$$

At the free surface, as before, we use the "rigid lid" approximation, so the boundary condition at $z = 0$ has the form (1.59). At the bottom we must satisfy conditions (1.61) if the problem of generation of internal waves is considered. For the problem of the scattering of internal waves, the boundary conditions should be taken in the form (1.60). All this is noted in (3.29).

In Section 1.5 it was shown that for an ocean of constant depth and a stratification without a horizontal density gradient the solution of the wave equation (3.28) can be found by the method of separation of variables in the form (1.66). Even though somewhat restrictive, the fluid in the vertical direction is linearly stratified (i.e. $N(z) = N_0 =$ const.), and the vertical structure of progressive internal waves and the dispersion relation can be found analytically, as given in (1.68).

Bearing this reasoning in mind, the wave fields $\overset{*}{\psi}_1$ and $\overset{*}{\psi}_3$ outside the transition zone II (i.e. in regions of constant depth, I and III) can be represented as the superposition of the incident internal wave or barotropic tidal flow Φ, and the generated internal waves radiated from the source of generation:

$$\overset{*}{\psi}_1(x, z) = \Phi_1(x, z) + \sum_{j=1}^{\infty} B_j \sin\left(\frac{j\pi z}{H_1}\right) \exp(\imath k_{1j} x), \qquad x < -l, \tag{3.30}$$

$$\overset{*}{\psi}_3(x, z) = \Phi_3(x, z) + \sum_{j=1}^{\infty} A_j \sin\left(\frac{j\pi z}{H_3}\right) \exp(-\imath k_{3j} x), \qquad x > l. \tag{3.31}$$

Here, $B_j = b_j \exp(\imath \chi_j)$ and $A_j = a_j \exp(\imath \phi_j)$, where b_j and a_j are the amplitudes of the generated modes, χ_j and ϕ_j are the phases of these waves, and

$k_{ij} = (j\pi/H_i)[(\sigma^2 - f^2)/(N_0^2 - \sigma^2)]^{1/2}$ $(i = 1, 3)$ are the wavenumbers (see (1.68)). If the incoming wave is barotropic, then, as shown previously, the functions $\Phi_i(x, z)$ can be represented by the barotropic tidal flux as follows:

$$\Phi_1(x, z) = -\frac{\Psi_0 z}{H_1}, \quad \Phi_3(x, z) = -\frac{\Psi_0 z}{H_3}. \tag{3.32}$$

When the baroclinic mode with number m $(m = 1, 2, \ldots)$ interacts with the transition zone, the functions $\Phi_i(x, z)$ can be expressed as

$$\Phi_1(x, z) = a_m^{\text{inc}} \sin\left(\frac{m\pi z}{H_1}\right) \exp(-\imath k_{1m} x), \quad \Phi_3(x, z) = 0. \tag{3.33}$$

In the frequency band $f < \sigma < N$, (3.28) is hyperbolic. So it can be transformed to the canonical form which, for constant N and constant M, takes the form

$$\overset{*}{\psi}_{\mathscr{Z}\mathscr{X}} = 0, \tag{3.34}$$

with the characteristic variables

$$\mathscr{X} = z - \alpha_- x, \quad \mathscr{Z} = z - \alpha_+ x, \tag{3.35}$$

where

$$\alpha_\pm = \frac{M^2}{(N^2 - \sigma^2)} \pm \left[\frac{M^4}{(N^2 - \sigma^2)^2} + \frac{(\sigma^2 - f^2)}{(N^2 - \sigma^2)}\right]^{1/2}, \tag{3.36}$$

with $\alpha_+ > 0$ and $\alpha_- < 0$.

Let us apply the Riemann integration method [42] to find the solution of (3.34). Note that, due to the presence of the horizontal density gradient in zone II, characteristic lines of the different families (3.35) have different tangents of inclination (3.36). To this end, consider Figure 3.12.

Beginning at $x = -l$, two characteristic lines,

$$z - \alpha_+ x = z_B + \alpha_+ l, \quad z - \alpha_- x = z_C + \alpha_- l,$$

are drawn from the points $C(-l, z_C)$ and $B(-l, z_B)$ of the vertical boundary between areas I and II. They intersect at the point $A(x, z)$.

Taking into account that the Riemann function $\mathcal{R}$ is equal to unity, the solution $\overset{*}{\psi}_2(x, z)$ of (3.34) at point $A(x, z)$ within the transition zone II can be found from formula (2.46), which, in the present case, simplifies to

$$\overset{*}{\psi}_2(x, z) = \frac{1}{\alpha_+ - \alpha_-}\left[\alpha_+ \overset{*}{\psi}_2(-l, z_C) - \alpha_- \overset{*}{\psi}_2(-l, z_B) + \int_{z_C}^{z_B} \overset{*}{\psi}_{2\zeta}\, d\eta\right]. \tag{3.37}$$

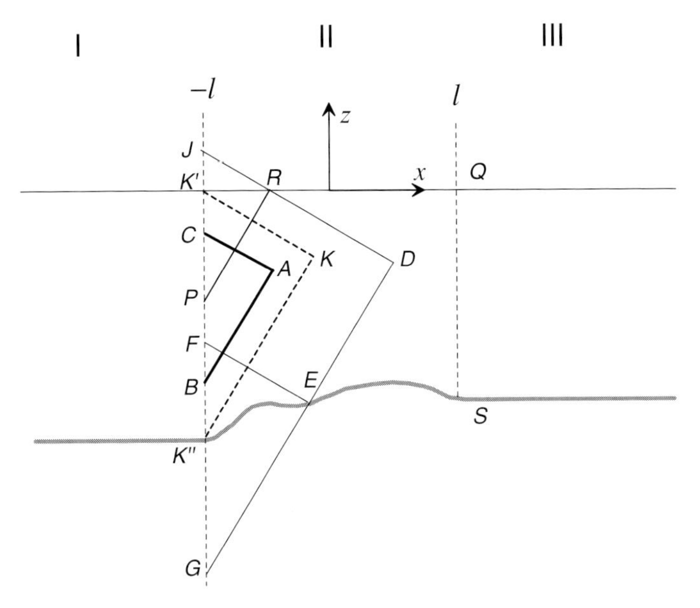

Figure 3.12. Schematic diagram for finding the solution within the continuously stratified frontal zone.

Thus, the stream function $\overset{*}{\psi}_2(x, z)$ inside area II can be found if the function $\overset{*}{\psi}_2(x, z)$ and its derivative $\overset{*}{\psi}_{2x}(x, z)$ are known at the left boundary of the frontal zone, $x = -l$. The connection with the solution in region I can be established by demanding continuity of the horizontal, u, and vertical, w, velocities at the liquid vertical boundary at $x = -l$ (and $x = l$ for region III) of the transition zone II:

$$\left.\begin{array}{ll} \overset{*}{\psi}_1(-l, z) = \overset{*}{\psi}_2(-l, z), & \overset{*}{\psi}_{1x}(-l, z) = \overset{*}{\psi}_{2x}(-l, z), \\ \overset{*}{\psi}_2(l, z) = \overset{*}{\psi}_3(l, z), & \overset{*}{\psi}_{2x}(l, z) = \overset{*}{\psi}_{3x}(l, z). \end{array}\right\} \tag{3.38}$$

Using this result, (3.37) can be rewritten as

$$\overset{*}{\psi}_2(x, z) = \frac{1}{\alpha_+ - \alpha_-}\left[\alpha_+ \overset{*}{\psi}_1(-l, z_C) - \alpha_- \overset{*}{\psi}_1(-l, z_B) + \int_{z_C}^{z_B} \overset{*}{\psi}_{1\zeta}\, d\eta\right]. \tag{3.39}$$

If (3.30) is now substituted into (3.39), one finds that the stream function $\overset{*}{\psi}_2(x, z)$ at any point A in area II can be expressed by the trigonometric series (3.30) with unknown amplitudes B_j ($j = 1, 2, 3, \ldots$).

Of course, some difficulties can arise when characteristic lines cross the upper or bottom surface. For instance, two characteristics emanating from point $D(x_D, z_D)$

(see Figure 3.12), cross the free surface at point $R(x_R, 0)$, and the bottom topography at point $E(x_E, -H_E)$. Obviously, the stream function and its x-derivative are unknown at the sections JK' (above the upper surface) and GK'' (below the bottom). If this happens, an additional relation must be found in order to find the solution of (3.34) at point $D(x_D, z_D)$. One possible way is to use the representations of the stream functions at the points $D(x_D, z_D)$, $R(x_R, 0)$, and $E(x_E, -H_E)$ as follows:

$$\overset{*}{\psi}_2(x_D, z_D) = \frac{1}{\alpha_+ - \alpha_-}\left[\alpha_+\overset{*}{\psi}_1(-l, z_J) - \alpha_-\overset{*}{\psi}_1(-l, z_G) + \int_{z_G}^{z_J} \overset{*}{\psi}_{1\zeta}\, d\eta\right],$$

$$\overset{*}{\psi}_2(x_R, 0) = \frac{1}{\alpha_+ - \alpha_-}\left[\alpha_+\overset{*}{\psi}_1(-l, z_J) - \alpha_-\overset{*}{\psi}_1(-l, z_P) + \int_{z_P}^{z_J} \overset{*}{\psi}_{1\zeta}\, d\eta\right],$$

$$\overset{*}{\psi}_2(x_E, -H_E) = \frac{1}{\alpha_+ - \alpha_-}\left[\alpha_+\overset{*}{\psi}_1(-l, z_F) - \alpha_-\overset{*}{\psi}_1(-l, z_G) + \int_{z_G}^{z_F} \overset{*}{\psi}_{1\zeta}\, d\eta\right].$$

Then, by subtracting the last two equations from the first one, and by taking into account the boundary conditions from (3.29), $\overset{*}{\psi}_2(x_R, z_R) = 0$ and $\overset{*}{\psi}_2(x_E, z_E) = 0$ (for the barotropic wave we take $\overset{*}{\psi}_2(x_E, z_E) = \Psi_0$), the following formula can be derived:

$$\overset{*}{\psi}_2(x_D, z_D) = \frac{1}{\alpha_+ - \alpha_-}\left[\alpha_+\overset{*}{\psi}_1(-l, z_P) - \alpha_-\overset{*}{\psi}_1(-l, z_F) + \int_{z_F}^{z_P} \overset{*}{\psi}_{1\zeta}\, d\eta\right].$$

Now $\overset{*}{\psi}_2(x_D, z_D)$ can be found from the known values of the stream functions at the points $P(-l, z_P)$ and $F(-l, z_E)$, and $\overset{*}{\psi}_{1x}$ at section PF (Figure 3.12).

Thus, using (3.39) and the boundary conditions of the BVP (3.29), we can express the stream function at the vertical section QS at the right liquid boundary $x = l$, denoted by $\overset{*}{\psi}_2(l, B_j, z)$, in terms of a trigonometric expansion with unknown coefficients B_j $(j = 1, 2, \ldots)$. On the other hand, the stream function at the vertical $x = l$ is defined by formula (3.31). That is why, according to (3.38), we can write

$$\overset{*}{\psi}_2(l, B_j, z) = \overset{*}{\Phi}_3(x, l) + \sum_{j=1}^{\infty} A_j \sin\left(\frac{j\pi z}{H_3}\right)\exp(-\imath k_{3j} l). \tag{3.40}$$

The next step is the determination of the still unknown coefficients A_j and B_j $(j = 1, 2, \ldots)$. To this end, the following orthogonalization procedure is used. We multiply (3.40) by $\sin(n\pi z/H_3)$, $n = 1, 2, 3, \ldots$, and integrate the resulting equation with respect to z from 0 to $-H_3$ (similarly to Step 4 in Section 2.2). The result of this operation is an infinite system of equations for the unknown amplitudes B_j and A_j. This system is truncated at a finite number $n = \mathcal{J}_0$ (similarly to Step 6 in Section 2.2). As a result of this truncation, the following system of $\mathcal{J}_0$ linear inhomogeneous equations with $2\mathcal{J}_0$ unknown amplitudes B_j and A_j is found:

$$\int_{-H_3}^{0} \overset{*}{\psi}(l, B_j, z) \sin\left(\frac{n\pi z}{H_3}\right) dz = \frac{1}{2} A_n H_3 \exp(-\imath k_{3n} l), \tag{3.41}$$

where $n = 1, 2, \ldots, \mathcal{J}_0$. This system is not complete. To close the problem, we repeat the analogous procedure in the downstream direction and build the stream function $\overset{*}{\psi}_2(-l, A_j)$ at the left boundary, $x = -l$, using the data at the vertical section $x = l$. Taking into account (3.39), we thus find

$$\overset{*}{\psi}_2(-l, A_j, z) = \overset{*}{\Phi}_1(-l, z) + \sum_{j=1}^{\infty} B_j \sin\left(\frac{n\pi z}{H_1}\right) \exp(-\imath k_{1j} l). \tag{3.42}$$

Multiplying (3.42) by $\sin(n\pi z/H_3)$ $(n = 1, 2, \ldots)$ and integrating the resulting equation with respect to z from 0 to $-H_1$, we obtain, after truncation of the number of equations to $\mathcal{J}_0$ terms, the additional $\mathcal{J}_0$ equations:

$$\int_{-H_1}^{0} \overset{*}{\psi}(-l, A_j, z) \sin\left(\frac{n\pi z}{H_1}\right) dz = \frac{1}{2} B_n H_1 \exp(-\imath k_{1n} l), \tag{3.43}$$

where $n = 1, 2, \ldots, \mathcal{J}_0$. The solution of systems (3.41) and (3.43) for the amplitudes A_j and B_j can be found by standard Gaussian elimination. This integration technique for solving the wave equation (3.34) in a continuously stratified ocean through a stratified frontal zone has been proposed in ref. [223], and was numerically implemented and applied to realistic oceanic frontal zones.

3.5 Propagation of internal waves through a frontal zone

We now discuss some results obtained with the method developed for a continuously stratified fluid on the interaction of internal waves with horizontal density gradients. Such a study cannot be carried out in the context of a two-layer model because the energy distribution between several baroclinic modes resulting from a

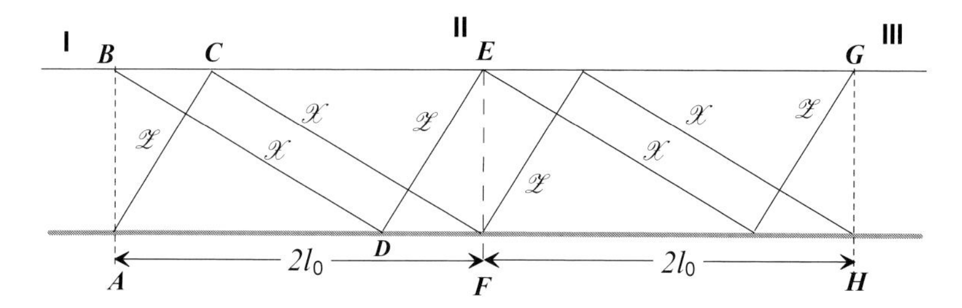

Figure 3.13. Sketch explaining the geometrical meaning of the term "effective width" in an ocean of constant depth. Inclined solid lines $\mathscr{Z}$ and $\mathscr{X}$ are characteristics of different families.

wave–front interaction cannot be properly taken into account with an idealized layer model. However, as will be shown below, this process plays an important role in the mechanism of internal wave scattering. First, we consider the interaction of internal waves with a density front located in a basin of constant depth. This study will highlight the effects connected with horizontal density gradients but without the influence of a variable bottom topography.

One of the basic parameters controlling the process of interaction is the width of the frontal zone $2l$. The second one is the *effective width* in the terminology of Mooers [160], defined as $2l_0 = H_1(1/\alpha_+ - 1/\alpha_-)$, see Figure 3.13. At this distance, two characteristic lines $\mathscr{X}$ and $\mathscr{Z}$ emanating from points A and B after the reflection from the free surface and the bottom at points C and D, respectively, return as characteristic lines of the other family to the fluid boundaries at points E and F, which are again located on a vertical line. From the Riemann method, it follows that the stream function at section FE is the same as at section AB. This means that, under conditions of weak wave reflection from the density front, the internal wave at section FE coincides with the incident wave. This is also valid for all other sections (GH and so on) which are $2l_0n$ units away from AB, where n is an integer. Needless to say, this holds true only if M in equation (3.28) is constant, i.e. for constant ρ_x.

The oceanographic consequence of this mathematical reasoning is the following. If the width of the frontal zone $2l$ coincides with $2l_0n$, the incident wave passes through the frontal zone without any changes. Alternatively, we would expect some effects of decomposition of the incident wave into other, neighboring baroclinic modes in zone III. How strong this energy transfer is depends on the relation between $2l$ and $2l_0$, and will be considered below.

Figure 3.14 displays the dependences of the normalized amplitudes of the internal waves $\bar{a}_j(2l)$ transmitted into area III when the first (Figure 3.14(a)) or the second (Figure 3.14(b)) baroclinic mode interacts with the density front. Here, $j = 1, 2, 3, 4, 5$ are the identifiers of the first five generated modes. Calculations

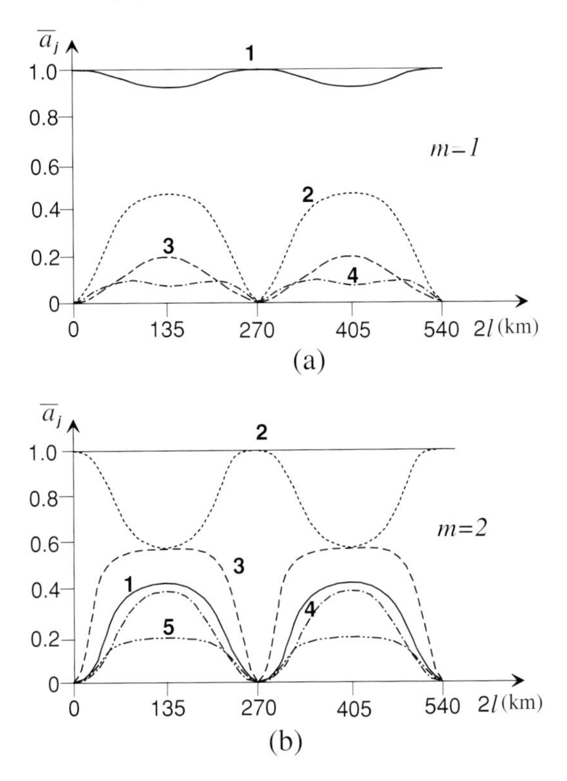

Figure 3.14. Normalized amplitudes of generated internal waves in zone III plotted against the width of the density front, $2l$, when (a) the first or (b) the second incoming baroclinic mode is considered. The numbers labeling the curves correspond to the numbers of the excited baroclinic modes.

were performed with the following values of input parameters:

$$\left.\begin{aligned} N = 4 \times 10^{-3}\,\mathrm{s}^{-1}, \quad M = 3.1 \times 10^{-4}\,\mathrm{s}^{-1}, \\ \varphi = 30^\circ, \quad T = 12.4\,\mathrm{h}, \quad H_1 = 4\,\mathrm{km}. \end{aligned}\right\} \tag{3.44}$$

Figure 3.14 suggests that the values of the amplitudes $\bar{a}_j$ depend strongly upon the value of the width of the density front, $2l$. First, it is evident that in the present case the frontal zone has an effective width $2l_0$ equal to 270 km. When $2l = 270$ or 540 km, the incident wave passes the density front without any changes. In the interval between these values a multimodal regime of the solution is realized in zone III. The maximum of the energy transfer of the incoming wave to the neighboring baroclinic modes occurs when $2l$ is divisible by $2l_0$ (at $2l = 135$ and 405 km in our case). Moreover, the effect of the mode splitting is quite pronounced. For instance, at the maximum splitting of the second mode, its amplitude coincides with the amplitude of the third mode. The other waves also have comparable amplitudes.

Note that the curves $\bar{a}_j(2l)$ in Figure 3.14 are symmetric relative to the verticals at $2l = l_0$ and $1.5l_0$. Their periodicity expressed by $\bar{a}_j(2l) = \bar{a}_j(2l + 2l_0)$ is clearly seen. Figure 3.14 shows also a dependence of the mechanism of the wave–front interaction on the number of the incident baroclinic mode. The larger this number is, the greater is the energy transformation from the incident to the other generated modes. For instance, at $2l = l_0$ the first and second baroclinic modes have the largest amplitudes in Figure 3.14(a) ($m = 1$), whereas in Figure 3.14(b) ($m = 2$) the first, second, third, and fourth modes have comparable amplitudes.

Consider next the joint effect of the density front and variable bottom topography: in our realization, the incident internal wave propagates through the frontal zone with parameters (3.44) located above the ridge (3.26) with height $H_{\max} = 2$ km and width $2l = 260$ km. This last value does not coincide with the effective width ($2l_0 = 270$ km), and this implies that energy splitting will occur. To illustrate the results, we choose the field of the vertical velocities w, which represents the baroclinic wave motion most strikingly. Figure 3.15(a) shows the amplitudes of the vertical velocities for the first baroclinic mode propagating through the density front located above the oceanic ridge. For comparison, a similar field was also constructed without horizontal density gradient in Figure 3.15(b). The intensity of the wave motion increases as the shading becomes darker in color.

The fact that the incident wave is the first baroclinic mode is clearly seen from the wave field in area I: the maximum values of w in Figure 3.15(a) and (b) are located in the middle of the water column (see also Figure 1.4(a)). The local distortions of the shaded zone boundaries and the gaps can be explained by the influence of the reflected waves.

Above the ridge and behind it, in areas II and III, the wave fields reveal a well developed ray structure: baroclinic wave motions are concentrated in narrow stretched zones oriented along the characteristic lines. They are "reflected" from the surface and bottom as characteristic lines of the other family. Such behavior of the wave fields was already discussed in Section 2.5 (see also Figure 2.19), where steep bottom topography was considered. The beam-like structure of the wave field results from the superposition of a large number of baroclinic modes which are generated during internal wave scattering.

As was shown above, the effective splitting of the incident wave into several baroclinic modes can be produced by both bottom obstacles and density fronts. However, in the real ocean the effect of bottom topography most likely has a much stronger influence on the internal wave scattering than the density fronts. This can be illustrated by comparing Figures 3.15(a) and (b), where the amplitude values of the vertical velocity-wave field are presented for the cases with and without a horizontal density gradient. In the two cases, similar ray structures of the wave field are produced in areas II and III. The presence of the frontal zone is capable of

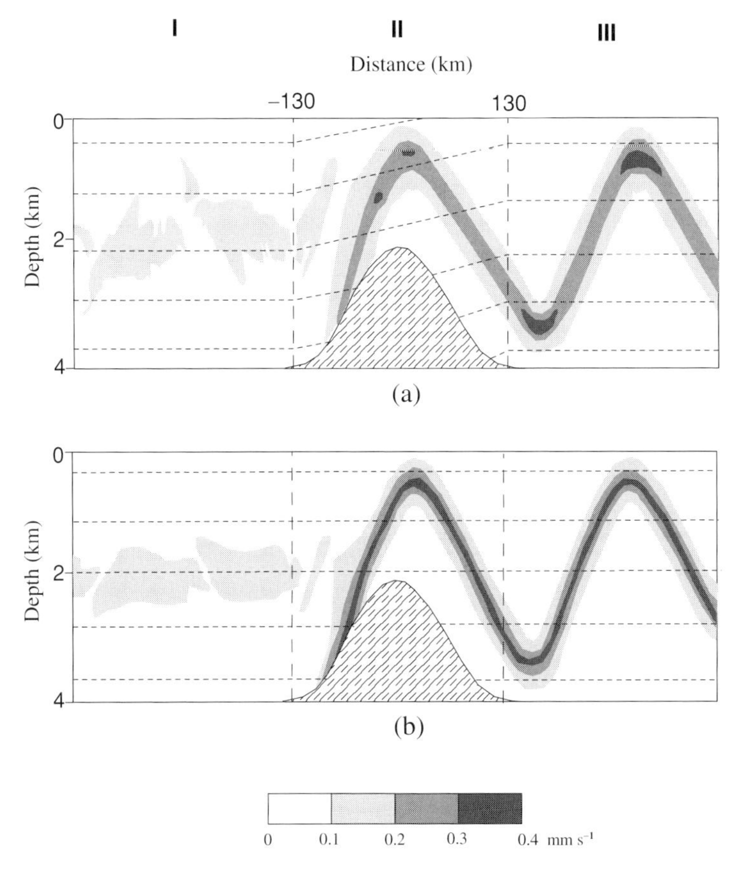

Figure 3.15. Fields of the vertical velocity amplitude produced by the first baroclinic incident mode interacting with the underwater ridge. (a) The density front is located in the ridge zone; (b) the background fluid stratification is horizontally homogeneous. Dashed lines are isopycnals.

modifying the scattering mechanism (by either increasing or decreasing the energy splitting). For instance, the presence of the frontal zone in the considered case makes the ray structure less pronounced, since the intensity of the wave motions within the ray decreases. The effect can also be a substantial intensification of the wave beam if the density gradient is directed in the opposite direction.

We considered above the situation when the width of the transition area II (density front together with underwater ridge) equals 260 km whereas the effective width $2l_0$ is 270 km. So, such a ratio between two basic parameters does not favor the energy transfer from the incident wave to a group of neighboring modes if such a process would take place in the presence of a frontal zone only (without bottom obstacle). From Figure 3.14(a) it is seen that, besides the first baroclinic mode, only the second one would have a perceptible amplitude in this case. At the same time,

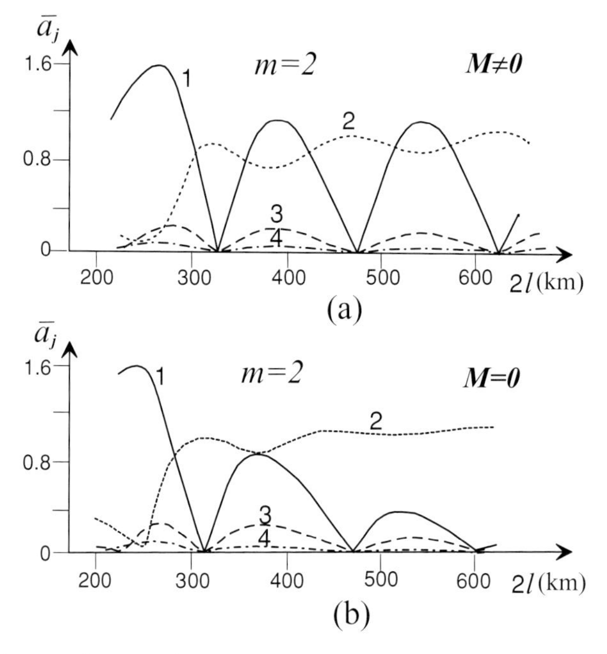

Figure 3.16. Amplitudes of the generated baroclinic modes in zone III plotted against the width $2l$ of the underwater ridge (3.26) (a) with and (b) without a frontal zone above it. Scattering of the second baroclinic mode is considered. The height of the ridge was equal to 2 km. All other input parameters are taken from (3.44). Numbers on the curves designate the generated modes.

the presence of the underwater ridge changes the situation dramatically: the bottom topography leads to the generation of a large number of baroclinic modes, which together result in the pattern of the wave beam shown in Figure 3.15. The next question is, what will be the wave transformation for different values of the width $2l$ of area II when both effects are present? To clarify this point, we investigate the dependences $\bar{a}_j(2l)$ shown in Figure 3.16 for the second incident mode.

Several inferences can be drawn on inspection of Figure 3.16. Of fundamental importance is that the dependences $\bar{a}_j(2l)$ in both cases, with and without the density front, have a quasiperiodic (garlandic) character with alternations of maxima and zeros. This is in accordance with the resonance mechanism of the wave–bottom interaction formulated in Chapter 2 and does not contradict the idea of "effective width" for the frontal zones introduced above.

Some unusual features of these curves are, however, evident which have to be explained. The second mode is only generated in zone III (amplitudes of the other modes are equal to zero) when the width of area II with density front, $2l$, equals 330 km (Figure 3.16(a)). The similar value for zone II without density front equals 310 km (Figure 3.16(b)). This 20 km discrepancy evidently arises because of the

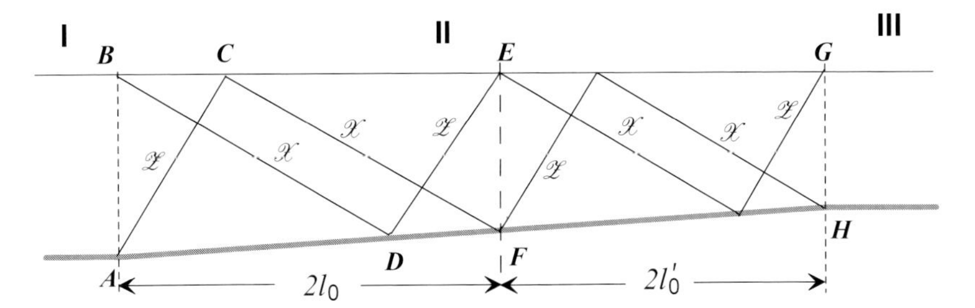

Figure 3.17. Schematic diagram explaining the geometrical meaning of the term "effective width" in an ocean of variable depth. Inclined lines $\mathscr{Y}$ and $\mathscr{X}$ are characteristics of the wave equation (3.28).

presence of the density gradients. Next, we should note that the functions $\bar{a}_j(2l)$ have a period $2l_p = 150$ km, whereas the effective width of the frontal zone without ridge found above was 270 km. Finally, the maxima of the function $\bar{a}_j(2l)$ attenuate with an increase of the width $2l$ when $M = 0$, whereas for $M \neq 0$ they are not damped.

All these facts can be explained in terms of the "effective width" of the transition zone in a basin of variable depth. In other words, the "effective width" can be interpreted as simply the local wavelength $\lambda(x)$ in area II introduced in Chapter 2. To corroborate this, consider a simplified model situation in which an analytical solution of the problem can be found. We assume that the bottom topography is linear (see Figure 3.17):

$$H_2(x) = H_1 - \gamma(x + l). \tag{3.45}$$

Simple geometric reasoning allows us to find the expression for the effective width: the characteristic lines emanating from the points A (bottom) and B (free surface) cross the bottom at D and the free surface at C. Two other characteristics emanating from these points cross the bottom at F and the free surface at E. The line EF is vertical and parallel to AB. The distance $2l_0$ between AB and EF is the effective width,

$$2l_0 = H_1(1 - \alpha_-\alpha_+^{-1})(\gamma - \alpha_-)^{-1}. \tag{3.46}$$

This is the minimum value of the width of area II for undisturbed propagation of internal waves. Similar properties are also exhibited by the frontal zones with width $2l_0 + 2l_0'$, $2l_0 + 2l_0' + 2l_0''$, etc., where $2l_0' = H_2(2l_0)2l_0/H_1$, $2l_0'' = H_2(2l_0 + 2l_0')2l_0'/H_1$; see Figure 3.17. It is evident that $2l_0 > 2l_0' > 2l_0''$.

Thus, in a basin of variable depth, the effective width $2l_0$ depends not only on the value of the horizontal density gradient, but also on the bottom profile. This is also clearly seen in Figure 3.18, where the dependences $\bar{a}_j(2l)$ of the amplitude of

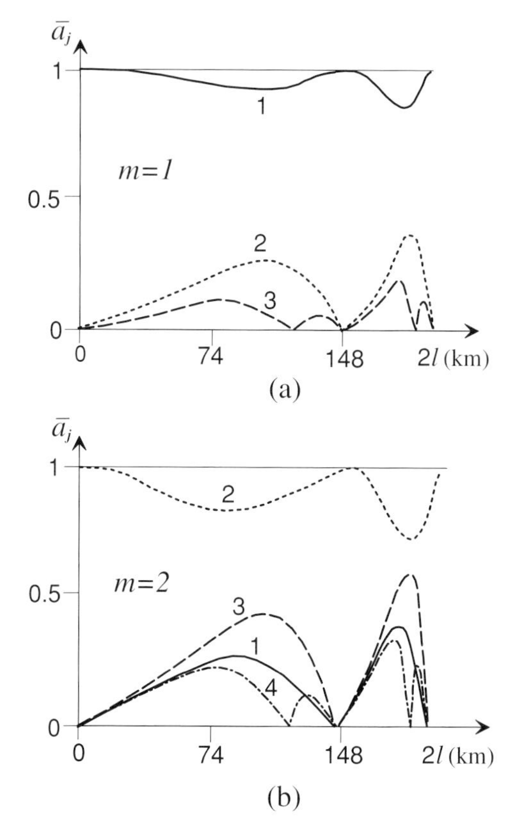

Figure 3.18. Amplitudes of internal waves generated behind an obstacle plotted against the width of the transition zone II when (a) the first and (b) the second mode interact with the density front located over the linear bottom topography (3.45). The input model parameters were $H_1 = 4$ km, $\gamma = 0.02$, $N = 4 \times 10^{-3}$ s^{-1}, and $M = 3.1 \times 10^{-4}$ s^{-1}.

the waves generated in zone III are presented as a result of the interaction of the internal wave with a density front located over the inclined bottom topography. The maximum width of the transition zone II in this series of runs was restricted to two lengths of the effective width.

Only those baroclinic modes for which the number coincides with the mode number of the incident wave are generated in zone III when the width of the frontal zone is equal to $2l_0 = 148$ km, $2l_0 + 2l_0^{'} = 197$ km. It is seen also from Figure 3.18 that the regular periodic structure of the function $\bar{a}_j(2l)$, which was obtained above for a basin of constant depth (see Figure 3.14), is violated for a variable bottom topography: functions $\bar{a}_j(2l)$ are compressed in the band $2l_0 < 2l < 2l_0 + 2l_0^{'}$ as compared with the band $0 < 2l < 2l_0$, and their maxima increase. A similar behavior was already seen for the function $\bar{a}_j(2l)$ shown in Figure 2.12.

Table 3.3. *Normalized amplitudes $\bar{a}_j$ of the baroclinic modes generated in zone III ($2l = 100$ km) as a result of wave–front interaction in a basin of constant depth ($H_0 = 4$ km).*

The value of the horizontal density gradient (horizontal buoyancy frequency) and the mode number of the incident wave, m, are given in the first two columns, respectively.

		j							
M (s^{-1})	m	1	2	3	4	5	6	7	8
1.1×10^{-4}	1	[1.0]							
	2	0.022	[0.998]	0.055					
	3	0	0.052	[0.995]	0.073				
	4	0	0	0.073	[0.992]	0.093			
2.1×10^{-4}	1	[0.997]	0.110						
	2	0.110	[0.996]	0.245	0	0.050			
	3	0	0.245	[0.905]	0.337	0.041	0.077	0.048	
	4	0.035	0.032	0.339	[0.829]	0.418	0.077	0.102	0.063
3.1×10^{-4}	1	[0.982]	0.257	0.055	0.063				
	2	0.258	[0.828]	0.495	0.141	0.123	0.095	0.059	
	3	0.066	0.0485	[0.572]	0.575	0.220	0.167	0.163	0.103
	4	0.063	0.147	0.585	[0.290]	0.646	0.290	0.200	0.179

Thus, the density fronts in a real ocean can be an important factor influencing the redistribution of the wave energy between the baroclinic modes. This conclusion is formulated on the basis of results of model runs which were performed for a single value of the horizontal density gradient ($M = 3.1 \times 10^{-4}$ s^{-1}) that is typical for the ocean. Some additional runs were carried out to study how sensitive the results are to the value of the horizontal density gradient and to estimate how strong the effects of energy splitting can be in various oceanographic situations.

Table 3.3 shows the results of the wave–front interaction for three different values of M. The model parameters in this series of runs were $N = 4 \times 10^{-3}$ s^{-1}, $2l = 100$ km, $H_0 = 4$ km, $\varphi = 30°$. To make the results more representative, the amplitudes of the generated waves are given in boxes when their numbers j coincide with the number of the incident waves, m. It is clearly seen that the amplitudes of these modes usually predominate in zone III. Moreover, the higher baroclinic modes transfer their energy to the neighboring modes more efficiently than the lower ones: the larger the number of the incident baroclinic mode, the smaller the amplitude of the generated mode with the same number. Finally, Table 3.3 shows that sharper density fronts produce a wider spectrum of the generated modes. In other words, the number of modes with remarkable amplitudes becomes larger with an increase of the horizontal density gradient.

3.6 Generation of baroclinic tides in the presence of a frontal zone

The generation mechanism of linear internal waves by barotropic tides interacting with variable bottom topography was considered in Chapter 2 in great detail. Here we dwell on the effects which introduce to the model the presence of a stationary horizontal density gradient located at the place of wave generation. As was shown above, these effects can be rather remarkable for propagating internal waves. How strong they can be for the generation of waves will now be discussed.

We consider here the joint effect of a variable bottom topography and a density front. As already shown in Section 3.2.1, the direct generation of internal waves by an isolated frontal zone is relatively inefficient in comparison with that due to topographic obstructions; compare Figures 3.5 and 3.7. So, our model here includes both the density front and a bottom topographic bump specified by the formula

$$H_2(x) = H_1 - \frac{(H_1 - H_3)}{2}\left(1 + \sin\frac{\pi x}{2l}\right), \quad -l \le x \le l; \tag{3.47}$$

it represents the transition zone between two regions of the ocean with depths H_1 and H_3.

We will focus attention on the "multimodal" case, i.e. upon the situation of near-critical bottom features (in the sense of Section 2.5) when several baroclinic modes are generated. The role of the density front in such a case is clearer and can be better understood than an analysis of both in isolation. Wave motions in a two-layer model are not so interesting from this point of view, because in this case internal waves are visible only via the oscillating interface, which mimics only the first baroclinic mode.

An example of the excitement of a baroclinic tidal beam near a steep bottom topography without horizontal density gradients is shown in Figure 3.19(b). The beam pattern obtained here is created by the superposition of a large number of baroclinic modes. The spatial configuration of the ray repeats the behavior of the characteristic lines if one takes into account their "reflection" from the free surface and the bottom. The presence of the density front above the bottom obstacle modifies the wave field. These modifications depend considerably on the direction of the density gradient (whether it is positive or negative). Comparison of Figures 3.19(a), (b), and (c) shows that a positive density gradient in Figure 3.19(a) partly compensates for the narrowing of the wave guide. The wave beam in this case becomes

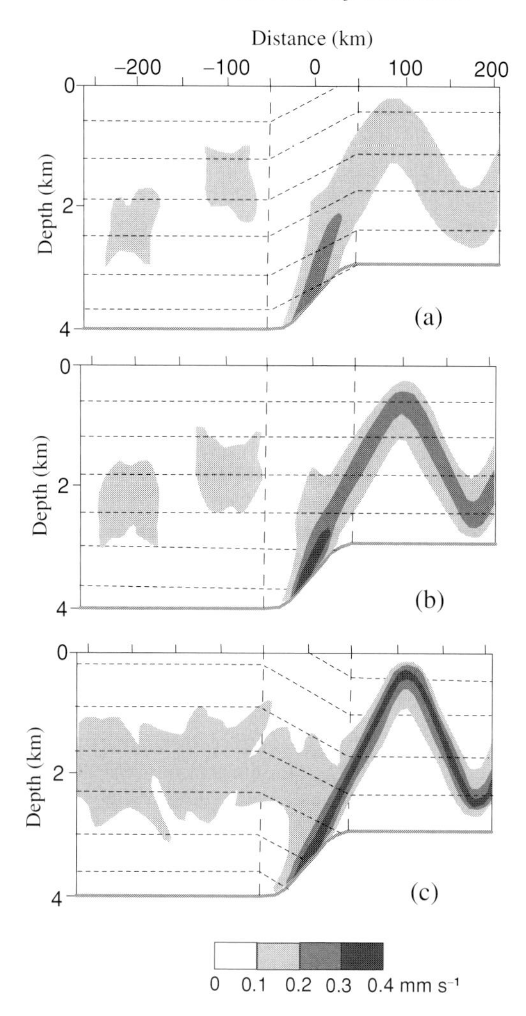

Figure 3.19. Model-predicted fields of the vertical velocity maximum produced by a semidiurnal barotropic tidal flux ($\Psi_0 = 40\ \mathrm{m^2\ s^{-1}}$) interacting with the transition zone (3.47) ($2l = 100$ km, $H_1 = 4$ km, $H_3 = 3$ km) in an ocean with $N = 4 \times 10^{-3}\ \mathrm{s^{-1}}$ and $\varphi = 30°$. The fluid is horizontally stratified in (a) ($M = 3.1 \times 10^{-4}\ \mathrm{s^{-1}}$, $M^2 > 0$) and (c) ($|M| = 3.1 \times 10^{-4}\ \mathrm{s^{-1}}$, $M^2 < 0$), and in (b) horizontally homogeneous ($M = 0$). Dashed lines are isopycnals.

weaker and wider. On the other hand, when the bottom surface and the isopycnals are converging, which occurs at negative density gradients, a contrary behavior is observed, i.e. the wave motion intensifies within the ray (Figure 3.19(c)). As was shown in Section 3.5, this goes in parallel with a more efficient energy transfer from the barotropic to the baroclinic motions, especially with a more efficient excitement of the high baroclinic modes. How large the difference in all three cases is quantitatively can be estimated from Figure 3.20, where the dependences $a_\xi^{(1)}(2l)$ and

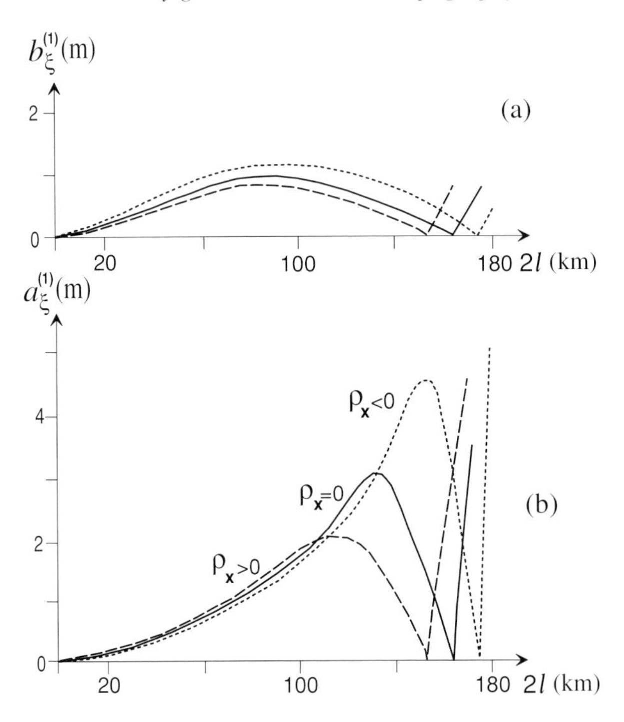

Figure 3.20. Amplitudes of the first baroclinic mode generated in (a) the deep and (b) the shallow zone of the ocean by the barotropic tidal flux interacting with the slope (3.47) plotted against the width of the transition zone $2l$. All input parameters are the same as for Figure 3.19. Solid lines correspond to the horizontally homogeneous ocean (Figure 3.19(b)), dotted lines hold for positive density gradient (Figure 3.19(a)), while dashed lines are for negative horizontal density gradients (Figure 3.19(c)).

$b_\xi^{(1)}(2l)$ of the first baroclinic mode are presented for positive and negative density gradients, as well as for a horizontally homogeneous fluid.

It is seen that the resonance character of the wave excitement remains valid in the presence of the density gradient. However, the density front shifts the positions of the zeros of the curves $a_\xi^{(1)}(2l)$ and $b_\xi^{(1)}(2l)$ in accordance with the change of the effective width. Note also that the internal waves generated in the deep part of the basin, to the left of the slope in Figure 3.19, are usually weaker than those in the shallow-water zone to the right of the slope.

Of course, it is not only the first but also higher baroclinic modes that are generated near steep bottom features. For the case presented in Figure 3.19(a), the dependences of the amplitudes of the first three baroclinic modes on the width of the slope $2l$ are shown in Figure 3.21. It is seen that the amplitudes of the second and third modes are usually smaller than those for the first mode. However, their

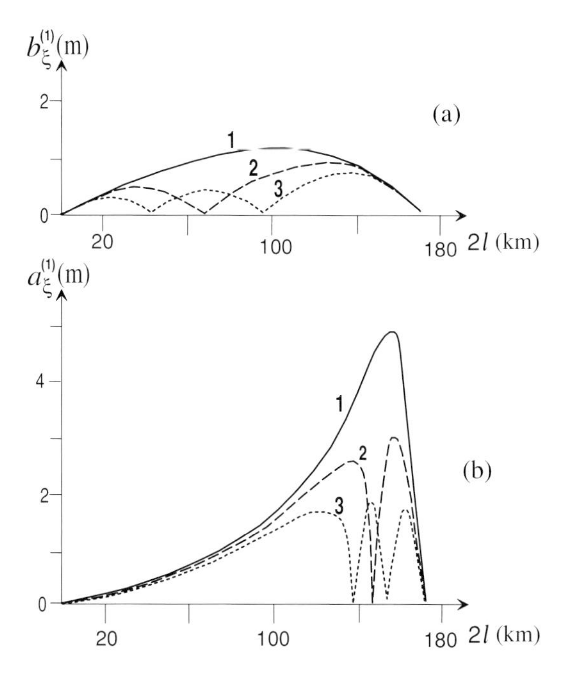

Figure 3.21. Amplitudes of the first three baroclinic modes generated by the barotropic tidal flux in deep (a) and shallow (b) zones of the ocean interacting with the slope (3.47), plotted against the width of the transition zone $2l$. All input parameters are the same as for Figure 3.19(a) (negative density gradient). The numbers on the curves identify the orders of the generated baroclinic modes.

magnitudes are comparable, and this is a condition of the generation of the wave beam.

It is also clear that the values of the term "effective width" introduced above for high baroclinic modes are smaller than for the first one, due to their shorter length scales. As a consequence, the dependences $a_\xi^{(j)}(2l)$ and $b_\xi^{(j)}(2l)$, $j = 2, 3$, have additional extrema in comparison with $a_\xi^{(1)}(2l)$ and $b_\xi^{(1)}(2l)$.

4

Topographic generation of nonlinear baroclinic tides

The basic relations of the linear theory presented in Chapters 2 and 3, and the results obtained with them, are valid within the framework of the linear approximation; i.e. the assumption that the amplitudes of the internal waves are infinitely small is assumed to be satisfied with sufficient accuracy. In such circumstances, the role of the nonlinear terms of the governing equations describing the dynamics of waves could be shown to be negligible, and the terms marked by boxes in system (1.39) could be omitted. This assumption is commonly used at low intensity of the external tidal forcing. However, such conditions are not valid for the whole area of the World Ocean. In many places they are not satisfied, because of either the strength of the forcing or the steepness of the topography, and numerous *in situ* data reveal a nonlinear wave response.

Obviously, a mathematical model which adequately describes nonlinear wave dynamics must be based on the complete system of nonlinear equations; it includes both the full advective nonlinearities and the nonhydrostatic law for the pressure. The model must also incorporate all those factors which control the structure of the wave fields: variable bottom topography, fluid stratification, external forcing, parameterization of background mixing, etc. The parameter

$$\varepsilon_1 = \frac{\text{tidal excursion amplitude}}{\text{topographic length scale}} \times \frac{\text{topographic height}}{\text{fluid depth}} \tag{4.1}$$

can be used as a measure for the estimation of the specific input, which the nonlinear terms introduce into wave dynamics during the generation of internal waves by topographic features [73], [75]. If $\varepsilon_1 \ll 1$, it is very likely that the linear theory is a good approximation for the generation of internal waves; otherwise, the nonlinear terms of the governing system must be taken into account.

Nonlinear numerical models of baroclinic tides have been actively developed only during the late 1980s; this is the reason why fewer investigations were performed using the nonlinear than the linear approach. In the first nonlinear models,

[90], [150], [268], the hydrostatic approximation for the pressure was used; it excludes the important (and stabilizing) nonhydrostatic wave dispersion that emerges from the inclusion of the vertical acceleration terms in the vertical momentum balance. In these models the effect of nonlinear (destabilizing) steepening of the internal wave is compensated for by dissipation. Therefore, the process of nonlinear disintegration and transformation of baroclinic tidal waves into a sequence of solitary internal waves and wave trains, which is frequently observed in the ocean, was neither obtained nor described. For the formation of nonlinear periodic (cnoidal) or solitary (solitons) waves, the nonhydrostatic dispersion absolutely must be accounted for. The existence of such stationary solutions is achieved within the framework of the so-called weakly nonlinear theories and arises because of the balance between nonlinear steepening and nonhydrostatic dispersion [264].

The weakly nonlinear models of baroclinic tides which take into account the nonhydrostatic dispersion are presented in refs. [23], [73], and [75]. In these models, the two-layered law of fluid stratification was used. The more general case of a continuously stratified fluid was implemented in the nonlinear nonhydrostatic numerical models described in refs. [95], [129], [148], [242], and [243].

Various aspects of the generation of baroclinic tides are considered in the references cited above. Among these are the influence of the rotation of the Earth, stratification, bottom profiles, intensity of tidal forcing, etc. In particular, the process of nonlinear disintegration of tidal waves to a train of short waves was described. The most complete study, analyzing the influence of the Coriolis dispersion (due to the rotation of the Earth) on the formation of a nonlinear wave train, is presented in refs. [73] and [75]. There, as well as in ref. [129], it was shown that the rotation of the Earth suppresses the disintegration of a baroclinic tidal bore into secondary nonlinear internal waves. On the basis of the modified Boussinesq equations, valid for weakly nonlinear waves in a two-layer fluid of nonuniform depth, a critical value of the Coriolis parameter was found, above which the effects of nonlinearity are completely compensated for by Coriolis dispersion, and, consequently, wave trains are not formed. This conclusion, however, is not corroborated by results of field observations in high-latitude seas [121], [196], where short nonlinear wave trains having a tidal nature are conspicuously evident in satellite images of the sea surface.

There are further unsolved problems concerning nonlinear baroclinic tidal dynamics in the horizontally nonuniform ocean. The role of the steepness of the bottom topography (subcritical, critical, or supercritical inclination) is neither recognized nor completely understood in the nonlinear case. As mentioned previously, in linear models a large number of baroclinic modes are generated above supercritical underwater obstacles. In turn, with amplification of the tidal forcing, this situation should also have an effect on the structure of the fields of nonlinear tides both in the area of generation and beyond it. The first attempts to describe the multimodal effects, which may occur during a nonlinear generation of baroclinic tides, were undertaken in refs. [74] and [246]; however, this problem is still far from its final solution. Within the framework of two-layer models of fluid stratification, [23], [73], [75], a satisfactory result obviously cannot be obtained.

Three-dimensional nonlinear effects of baroclinic tides (i.e. the influence of nonlinearities on the local wave generation, divergence and convergence effects, refraction, etc.) are also relatively poorly understood. Several attempts have been undertaken to calculate the global baroclinic tidal energy distribution; see, e.g., ref. [177], as well as the local three-dimensional characteristics of baroclinic tides [48], [101], [152]. These calculations were carried out with the use of the primitive equation models, e.g. the POM.[1] They reproduce the basic three-dimensional features of the generated fields near seamounts, ridges, and islands quite successfully, and can provide an estimation of the global sink of the barotropic tidal energy into the baroclinic component. However, due to the hydrostatic approximation for pressure, their application is restricted to the consideration of the weakly nonlinear cases only, when nonhydrostatic dispersion can be neglected.

This chapter and the remaining part of the book are devoted to the analysis of nonlinear effects in the generation of baroclinic tides. In this chapter we will construct the numerical model, valid under conditions when the value of the parameter ε_1 is not small, and then, on the basis of this model, we will analyze the influence of some typical nonlinear effects which frequently occur in the study of tidal wave dynamics of the World Ocean. However, first we give some examples which set in evidence the existence of nonlinear effects in the generated wave fields as observed in the sea.

4.1 Experimental evidence for nonlinear baroclinic tides

In this section we will describe some of the typical features inherent in nonlinear baroclinic tides which were inferred from sea measurements. For this purpose, we consider the region of the New York Bight in the North Atlantic (Figure 4.1), from where some representative data were collected and described in refs. [260] and [261].

The Joint United States–Russian Experiment JUSREX-92 was carried out in July–August 1992 in the rectangular domain 39°–40°30′ N, 71°–73° W to study the dynamics of baroclinic tides. The field campaign, performed during Cruise 7

[1] There are many other models available which serve the same purpose as the POM. We mention the POM as one of the choices, to illustrate the numerical scale involved.

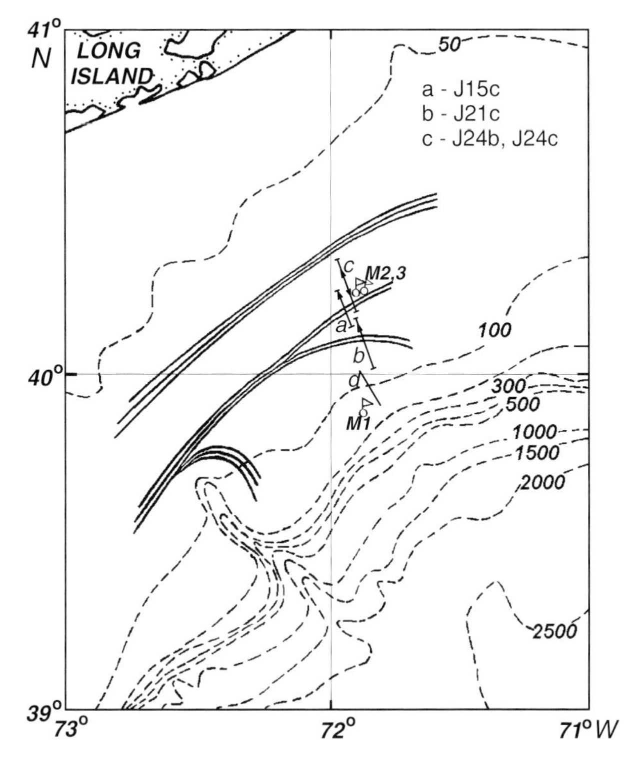

Figure 4.1. Chart of the New York Bight. The bathymetric level lines are presented by dashed lines. Circles labeled M1, M2, and M3 show the mooring positions. Lines with arrows, labeled *a*, *b*, *c*, and *d*, are the ship tracks. Solid lines are the surface signatures of internal waves from SAR images.

of the R/V *Academik Ioffe*, was accompanied by remote sensing measurements of the sea surface with the images from ERS-1 synthetic aperture radar (SAR). In New York Harbor, the amplitude of the M_2-tide in July 1992 was $\cong 2.0$ m for spring tides and $\cong 0.9$ m for neap tides. Taking into account the parameters of the New York Bight topography, the value of ε_1 is estimated to be about 0.05.

Three moorings, M1, M2, and M3, with current and temperature sensors were deployed on the shelf at the points 39°52.92′ N, 71°53.03′ W, 40°10.55′ N, 71°56.18′ W, and 40°11.05′ N, 71°54.08′ W; see Figure 4.1. The current meters at these moorings were located both above and below the thermocline. The conductivity, temperature, depth (CTD) measurements and towings of a tow-yo CTD profiler were carried out along the sections perpendicular to the isobaths along the ship tracks *a*, *b*, *c*, and *d* shown in Figure 4.1.

A summer-type water stratification with the seasonal thermocline positioned at a depth of 20 m and strong temperature gradients between 10 and 30 m is inherent in the hydrological structure of the tested area. There were no significant horizontal density gradients associated with frontal zones inside the studied area. The buoyancy frequency in the layer of the density jump reached a value of 5.5×10^{-2} s^{-1}, whereas beneath this layer its value did not exceed 10^{-2} s^{-1}.

A preliminary analysis of the remote sensing measurements reported in refs. [4], [70], and [140] showed that, usually, the internal waves in the New York Bight are registered as packets of short waves moving shoreward with tidal periodicity. The signature of the waves, seen from the SAR images, is sketched in Figure 4.1. The strong correlation of the periodicities of their registrations with the tidal period and the dependence of their amplitudes on the moon phase (spring or neap tide) corroborates the tidal nature of their origin.

It ought to be mentioned that the continental slope of the New York Bight is preferentially two-dimensional – the basic changes of the bottom relief occur across the coast – but it is crossed by a narrow canyon. Such a topography of the slope results in a three-dimensional structure of the internal wave fields, with the secondary waves reflected from the canyon in close proximity to it (see Figure 4.1). Far away from the canyon the wave fronts were basically two-dimensional.

The data obtained at the M3 mooring (Figure 4.2) show that the water dynamics in the studied area is basically driven by tides. The semidiurnal tidal periodicity of the horizontal velocity with a clockwise rotation of the velocity vector, Figures 4.2(a) and (b), is clearly seen. The time series of the vertical velocity, w, and temperature, T, also reveal the presence of semidiurnal oscillations. A typical value of the amplitude of the horizontal speed is about 10–15 cm s^{-1}, and its maximum value is estimated as 30 cm s^{-1}. Fluctuations of the vertical velocity related to the internal waves reach values of 1–2 cm s^{-1}, although the tidal periodicity of the $w(t)$-series is less pronounced. The four isolated spikes attain values $\cong$5 cm s^{-1} and are said to be connected with the passage of intensive internal waves. The strongest peaks are marked in Figure 4.2(c) by P18, P21, and P24. Packets of short internal waves are also clearly seen in the measurements of the temperature in Figure 4.2(d) as separate fluctuations up to 6–7 °C.

A signal that is related to the internal wave packet P18 from the same mooring M3, but taken at a depth level of $z = -17$ m is shown in Figure 4.3 in greater detail. Analyses of the w and T time series in particular clarify that the internal waves in the packet have the form of separate depressions: a downward displacement of the pycnocline (Figure 4.3(c)) is accompanied by an increase of the temperature (Figure 4.3(d)). After the passage of the internal wave, the thermocline returns to its initial position, and the temperature is restored to its initial background value.

The value of the horizontal velocity at the depth $z = -17$ m (Figure 4.3(b)) increases during the passage of the internal wave from a background level $\cong$ 20 cm s^{-1} to a maximum value of about 40 cm s^{-1}. Figure 4.3 also clearly illustrates the rotation of the vector of the horizontal velocity. In this case (Figure 4.3(a)), the

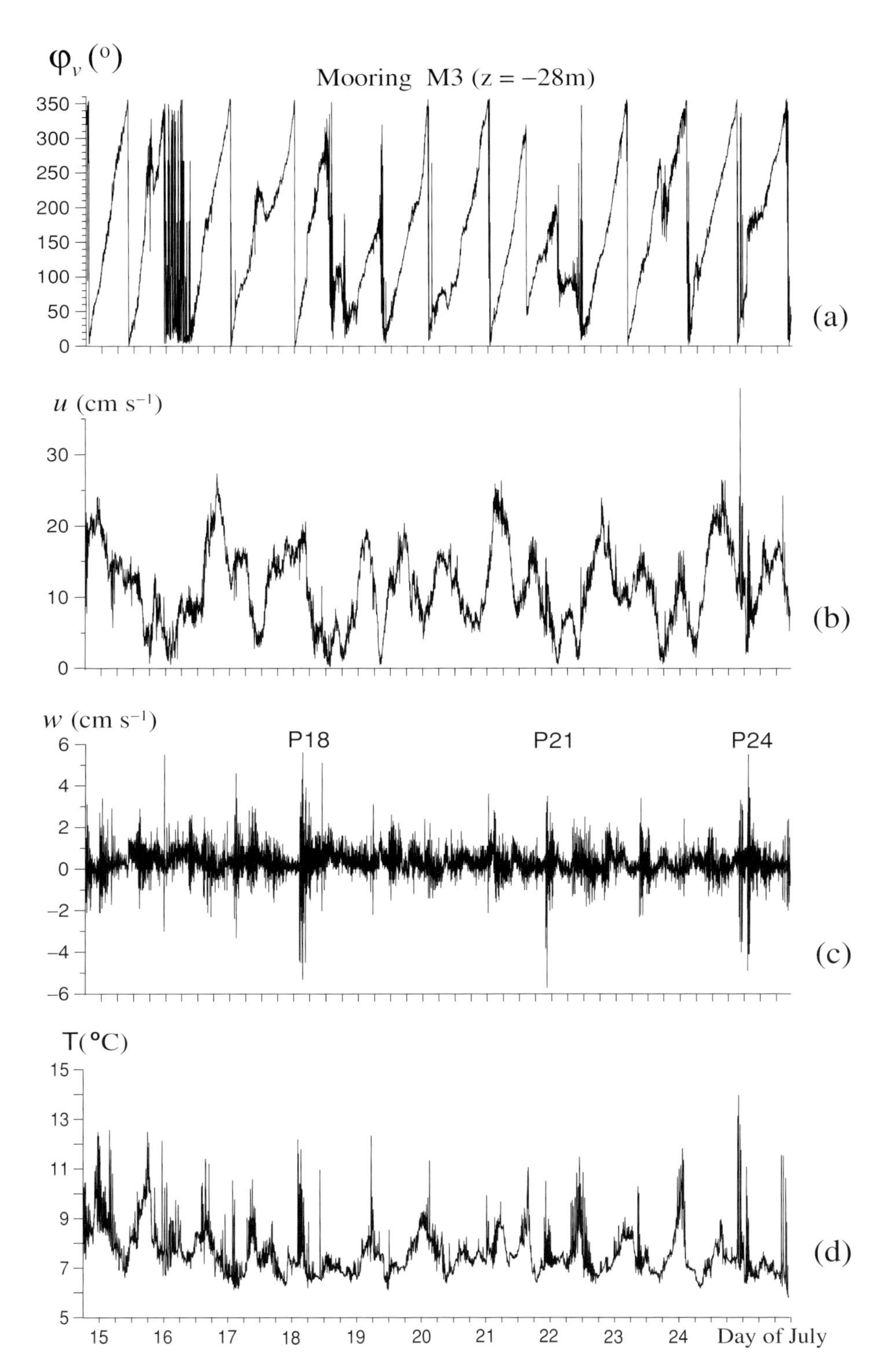

Figure 4.2. Time series at mooring M3 of (a) the angle of rotation of the horizontal velocity vector φ_v, (b) the absolute value of the horizontal velocity u, (c) the vertical velocity w, and (d) the temperature T at the level $z = -28$ m.

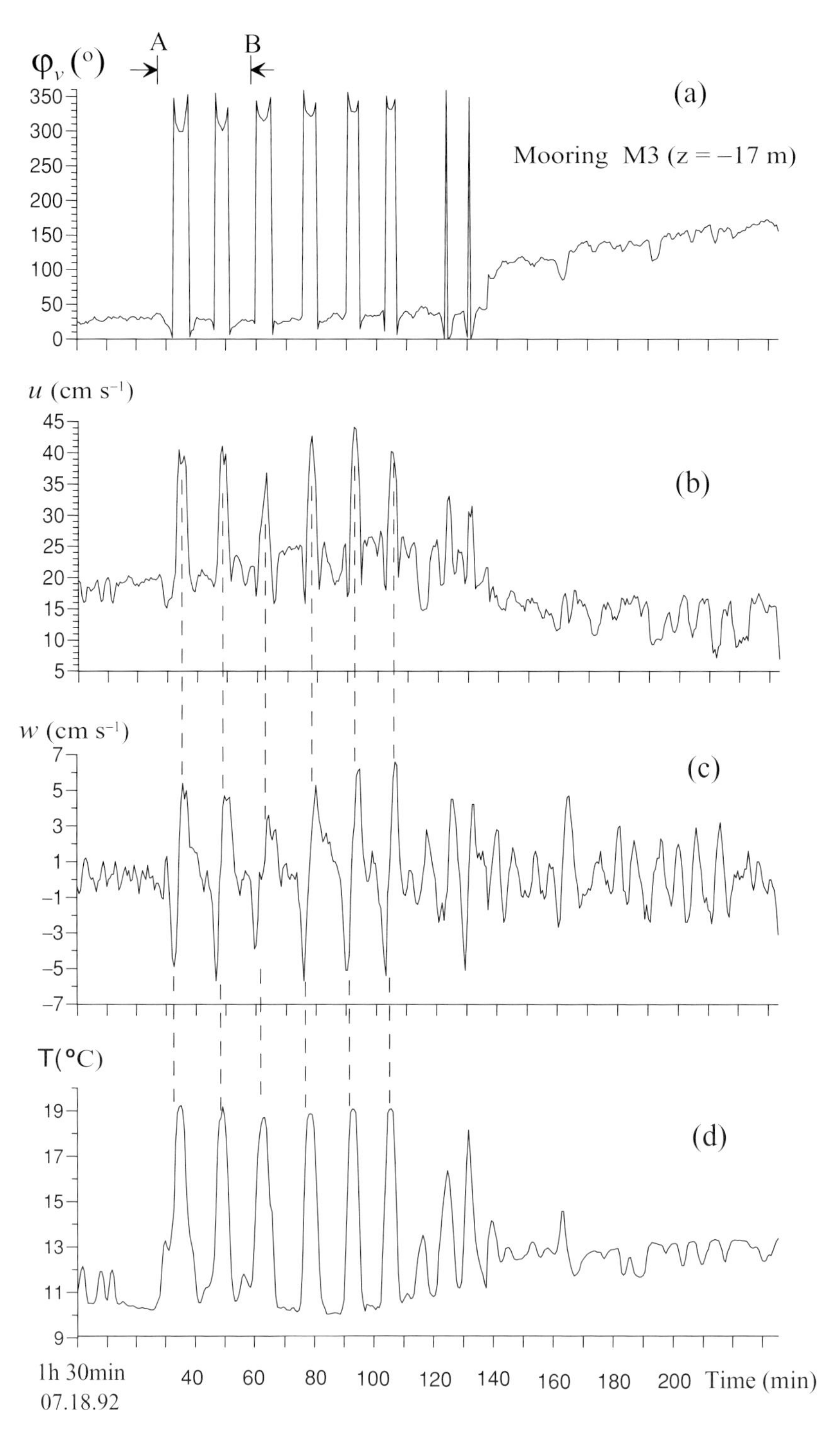

Figure 4.3. Detailed time series of mooring M3 at 17 m depth of the fragment marked in Figure 4.2(c) by P18. Designations are the same as in Figure 4.2.

background tidal flux before the arrival of the wave is directed towards north-northeast ($\cong 30°$). However, during the passage of the internal wave it turned toward northwest, i.e. to the coast ($\cong 300$–$330°$).

This example shows the wave characteristics measured above the thermocline level, at depth $z = -17$ m. Current meters located below this level at the depth of the density jump, $z = -28$ m, did not show any essential signals of the horizontal velocity during the passage of packet P18 (see Figure 4.2(b)); the perturbations of the horizontal velocity u are less than 5 cm s^{-1}. Consequently, the deployed current meters measured no perturbations in the value of the angle of rotation, φ_V, of the horizontal velocity of the background tidal currents. The theory is not in contradiction with these measurements, as one might infer, because, in accordance with the behavior of the eigenfunction $q_j(z)$ of the first baroclinic mode ($j = 1$; see Figure 1.7), the horizontal velocity changes its sign at the depth of the density jump ($u \sim dq_j/dz$). At the same time, the vertical velocity assumes its maximum value. The course of the time series for w displayed in Figure 4.2(c), where the maximum intensity reaches 5–6 cm s^{-1}, confirms this fact.

Similar results of the wave behavior have also been identified by analysis of the time series for the wave packets P21 and P24 in Figure 4.2(c). They are qualitatively very similar to those shown in Figure 4.3 for wave train P18. But, despite the significant temporal separation (six days) of the registration of packets P18, P21, and P24, they have identical characteristics. Such a qualitative agreement and the strict correlation with the tidal frequency are manifestations of the existence of an identical mechanism of their generation. They equally demonstrate the persistence of stability of this mechanism through space and time.

Additional information about the dynamics of the internal tides in the area of the New York Bight can also be obtained if one takes into account the data obtained at mooring M1, which is located at the upper edge of the shelf break (isobath 120 m, see Figure 4.1). Here, the measurements revealed strong fluctuations (up to 10 cm s^{-1}) in both the vertical and horizontal velocity fields. However, these fluctuations did not have a deterministic but rather a turbulent "white noise" character and reflected the presence of fully developed background turbulence.

The variations of the temperature–time series at mooring M1 did not exceed 2 °C. No packets of short internal waves were detected here, by either the *in situ* measurements or the remote sensing data. The absence of the internal wave packets above the shelf break and their presence on the shelf allow us to assume that these waves were generated in the area somewhere between moorings M1 and M3. Obviously, it is difficult to study the process of their excitation and evolution only on the basis of fixed mooring data. Such an interpretation, however, may become successful if the data of the thermistor-chain towings a, b, and c (Figure 4.1) across the wave fronts are also considered. Examples of the density field obtained by the

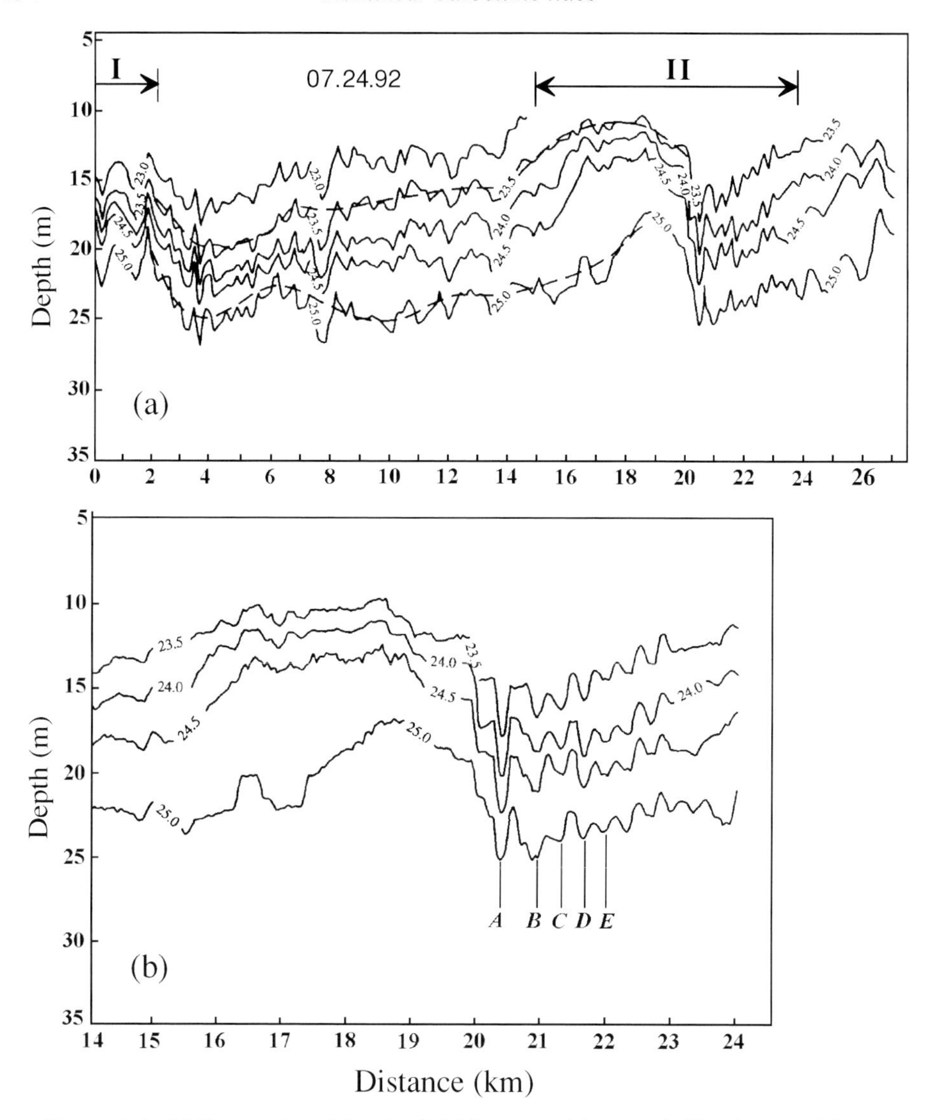

Figure 4.4. (a) Conventional density field (isopycnals), recorded by the towed CTD profiler along section *b* of Figure 4.1. (b) Detailed presentation of the fragment marked by II in part (a). The separate internal waves in the wave packet are designated by *A*, *B*, *C*, *D*, and *E*.

tow-yo CTD profiler along ship tracks *b* and *c* are presented in Figures 4.4 and 4.5. These measurements were performed in order to trace the packet of the internal waves during its propagation across the shelf. This packet was also measured at the M3 mooring and marked by P24 in Figure 4.2(c). Now the spatial structure of the internal waves and their evolution in time and space can be estimated more precisely.

Figure 4.4 displays the density field measured between the 80 and 95 m isobaths during towing along the ship track *b*, Figure 4.1. Two internal wave packets, marked

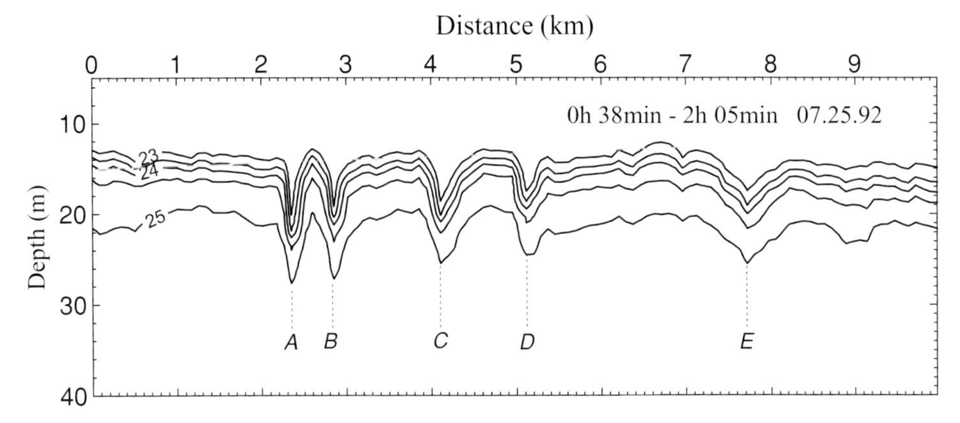

Figure 4.5. Conventional density field recorded by the towed CTD profiler along section c; see Figure 4.1. Designations of wave peaks by capital letters A, B, C, D, and E correspond to similar designations in Figure 4.4(b).

in Figure 4.4 by I and II, with a vertical extent of the isopycnal displacements of about 2–4 m, were crossed during this towing. The measured waves are characterized by co-phase vertical isopycnal displacements, which correspond to the first baroclinic mode, with a horizontal length scale of several hundred meters. Between the two packets, the structure of the long internal tidal wave with slightly deformed horizontal profile, marked by a dashed line, is clearly seen: isopycnals gently rise by approximately 10 m over a distance of 15 km. The distance between the internal wave packets I and II can be estimated as $\cong 20$ km.

The structure of wave packet II is shown in Figure 4.4(b) in more detail. This fragment of the baroclinic tidal wave can be interpreted as the initial stage of the formation of short-periodic internal waves when the trailing edge of the long progressive internal tidal wave transforms into a baroclinic bore. The amplitudes of the vertical oscillations in the packet are larger in the frontal part of it than in the tail. These waves are identified in Figure 4.4(b) by A, B, C, D, and E. Similar wave structures were qualitatively obtained during the towings carried out during the days on which further measurements were taken.

Let us consider now the further evolution of wave packet II. It was measured again several hours later on the isobath $\sim$70 m, 25 km closer to the coast. The depressions marked with letters A, B, C, D, and E in Figure 4.5 identify the same first five short waves as shown in Figure 4.4.

The changes of the wave packet II after 11 hours of its evolution across the shelf are prominent. It is clearly seen that the amplitudes of the separate waves became larger. For instance, the amplitudes of the first two waves, A, B, now have values of approximately 8 and 7.5 m, respectively. Moreover, the wave packet is stretched in space; in Figure 4.5 it must be considered as a series of solitary internal waves of depression, not as a compact wave train (which is the case for fragment II in

Figure 4.4). In fact, the distance between the first and fourth wave troughs (A and D) almost tripled (from $\cong 1.3$ km to $\cong 3.6$ km).

Thus, summarizing this analysis of the *in situ* and remote sensing data collected during JUSREX-92, we conclude that in the New York Bight several specific features inherent in the nonlinear baroclinic tides were observationally identified: one feature is the steepening of a long internal tidal wave and its disintegration into a packet of short internal waves, and the second is the evolution of these waves during propagation in the ocean of variable depth. These cannot be explained on the basis of linear theories such as those discussed in Chapters 2 and 3. A correct theoretical explanation is possible only in the framework of the fully nonlinear system of equations. Such a model is presented in Section 4.2.

4.2 Numerical model for the description of nonlinear waves

This section is devoted to the construction of a numerical algorithm for the solution of the generation and evolution of baroclinic tides in an ocean of variable depth. To this end, the fundamental equations (1.79), (1.80), or (1.81) and the initial and boundary conditions (1.82) to either (1.85a) or (1.85b) are used. The numerical implementation of the governing equations will be carried out by means of a finite difference approach, as in refs. [243] and [260]. For the construction of an optimal grid attention should be paid to the following points. First, the real ocean is nonuniformly stratified in the vertical direction. The dominant density changes occur in the seasonal pycnocline layer, which is located near the free surface and usually occupies several tens of meters. As shown in Section 1.5, this implies equally dominant vertical changes of the wave parameters in this layer, as seen already in Figure 1.7. So, to obtain more reliable results, it is reasonable to introduce in this region a finer vertical spatial resolution than elsewhere. This procedure of exaggerating the grid will speed up the calculations. The second point concerns the geometry of the basin. The water depth in the shallow-water zones is usually more than ten times smaller than in regions off the shelf. Consequently, the vertical gradients of the wave characteristics here are considerably larger than beyond the shelf. Both facts dictate an increase of the resolution of the numerical grid in the shallow-water zone in comparison with the deep ocean. It turns out that both requirements can ideally be taken care of by introducing the following transformation of the coordinates:

$$x_2 = x, \qquad z_2 = \frac{\int_0^z N(s)\,ds}{\int_0^{-H(x)} N(s)\,ds}. \tag{4.2}$$

It takes into account the fluid stratification as well as the variability of the bottom

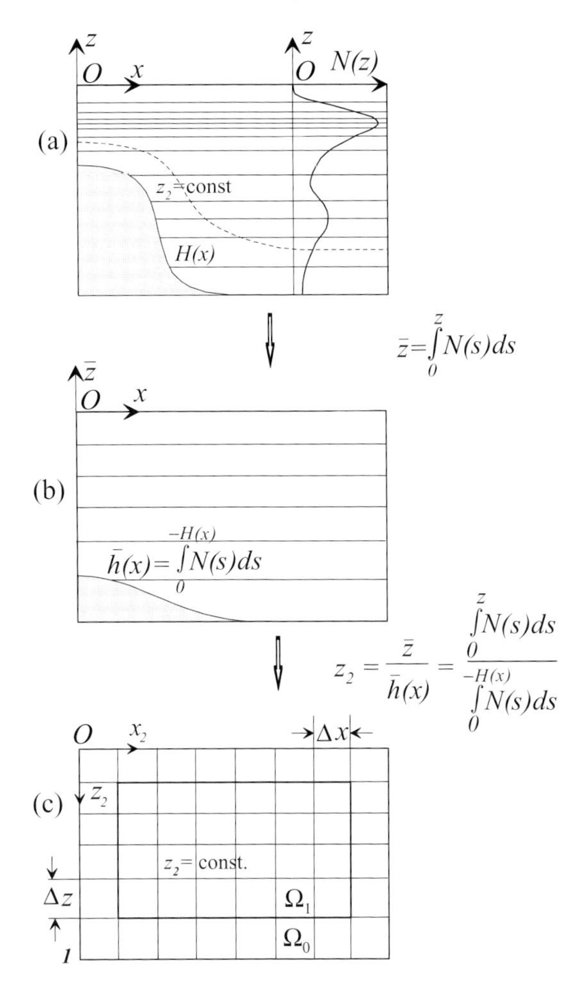

Figure 4.6. Transformation of the physical domain to a rectangle upon the application of the coordinate transformation (4.2). The dashed line in (a) represents in (c) a horizontal line.

topography. Such a replacement is similar to the σ-transformation,[2] which is frequently used in numerical models of oceanic circulation [115]. However, the basic difference is that the transformation (4.2) compresses a grid not only in the shallow-water areas, but also in the pycnocline layer. As a result, in the vertical plane the physical domain is transformed from an irregular area in the (x, z) coordinate system into a rectangle in the coordinates (x_2, z_2). Figure 4.6 illustrates this procedure, in which the transformation (4.2) is split into two steps to make it clearer.

[2] The σ-transformation $\sigma(x, z) = -z/H(x)$ maps the region from the undisturbed sea surface to the water depth onto the unit interval.

The above-introduced z_2-transformation makes the construction of a numerical scheme relatively simple, because rectangular grids can be used, but it complicates the governing equations. System (1.79), (1.81), after the application of (4.2), takes the form

$$\left.\begin{aligned} \omega_t + C(x_2, z_2)[J(\omega, \psi) - f v_{z_2}] &= g\mathcal{M}_1(\tilde{\rho})/\bar{\rho}_0 + A^{\mathrm{H}}\mathcal{M}_2(\omega) + A^{\mathrm{V}}\mathcal{M}_3(\omega), \\ v_t + C(x_2, z_2)[J(v, \psi) + f\psi_{z_2}] &= A^{\mathrm{H}}\mathcal{M}_2(v) + A^{\mathrm{V}}\mathcal{M}_3(v), \\ \tilde{\rho}_t + C(x_2, z_2)J(\tilde{\rho}, \psi) + \bar{\rho}_0/gN^2(x_2, z_2)\mathcal{M}_1(\psi) & \\ = K^{\mathrm{H}}\mathcal{M}_2(\tilde{\rho}) + K^{\mathrm{V}}\mathcal{M}_3(\tilde{\rho} + \rho_0), & \\ \omega = \mathcal{M}_2(\psi) + \mathcal{M}_3(\psi), & \end{aligned}\right\} \tag{4.3}$$

where

$$C(x_2, z_2) = N(x_2, z_2)r(x_2), \quad r(x_2) = 1/\bar{h}(x_2), \quad \bar{h}(x_2) = \int_0^{-H(x)} N(s)\,ds,$$

$$\mathcal{M}_1 = \frac{\partial}{\partial x_2} - z_2 p(x_2)\frac{\partial}{\partial z_2},$$

$$\mathcal{M}_2 = \frac{\partial^2}{\partial x_2^2} - 2z_2 p(x_2)\frac{\partial^2}{\partial x_2 \partial z_2} + z_2^2 p^2(x_2)\frac{\partial^2}{\partial z_2^2} + z_2(2p^2(x_2) - q(x_2))\frac{\partial}{\partial z_2},$$

$$\mathcal{M}_3 = r(x_2)\frac{dN}{dz}\frac{\partial}{\partial z_2} + r^2(x_2)N^2(x_2, z_2)\frac{\partial^2}{\partial z_2^2},$$

$$p(x_2) = r(x_2)\,d\bar{h}(x_2)/dx_2, \quad q(x_2) = r(x_2)\,d^2\bar{h}(x_2)/dx_2^2.$$

Equations (4.3) are written for constant vertical and horizontal turbulent exchange coefficients. We now construct the numerical scheme for system (4.3) and assume that it can be easily extended to the more general case, which indeed it can. The initial and boundary conditions (1.82) to either (1.85a) or (1.85b) must also be transformed when the transformation (4.2) is used.

For the finite difference approximation of the governing equations, we introduce in space-time (x_2, z_2, t) a rectangular grid $(\mathcal{I} \times \mathcal{J} \times \mathcal{N})$ with steps Δx $(i = 1, 2, \ldots, \mathcal{I})$, Δz $(j = 1, 2, \ldots, \mathcal{J})$ and Δt $(n = 1, 2, \ldots, \mathcal{N})$, respectively. To this end, it is assumed that all functions are known at the nth temporal step. To find the unknown functions at the $(n+1)$st step we use the alternative direction implicit method (ADIM); for a detailed outline of its application, see, e.g., ref. [115]. ADIM provides a means for solving (4.3) in two spatial dimensions with second-order accuracy.

Let us explain the idea of the ADIM by its application to the first equation of (4.3). For simplicity we denote the value of any function $y(x_2, z_2, t)$ at the grid node (i, j, n) by $y(i\Delta x, j\Delta z, n\Delta t) = y_{ij}^n$. Using Taylor series expansions for the function $y(x_2, z_2, t)$ at adjacent points $(i-1, j, n)$, $(i, j-1, n)$, $(i+1, j, n)$, and $(i, j+1, n)$, its first and second derivatives in the x_2- and z_2-directions can be computed to second-order accuracy, the result being

$$\frac{\partial y}{\partial x_2} = \frac{y^n_{i+1,j} - y^n_{i-1,j}}{2\Delta x} + o(\Delta x^2),$$

$$\frac{\partial y}{\partial z_2} = \frac{y^n_{i,j+1} - y^n_{i,j-1}}{2\Delta z} + o(\Delta z^2),$$

$$\frac{\partial^2 y}{\partial x_2^2} = \frac{y^n_{i+1,j} - 2y^n_{i,j} + y^n_{i-1,j}}{(\Delta x)^2} + o(\Delta x^2),$$

$$\frac{\partial^2 y}{\partial z_2^2} = \frac{y^n_{i,j+1} - 2y^n_{i,j} + y^n_{i,j-1}}{(\Delta z)^2} + o(\Delta z_2^2),$$

$$\frac{\partial^2 y}{\partial z_2 \partial x_2} = \frac{y^n_{i+1,j+1} + y^n_{i-1,j-1} - y^n_{i+1,j-1} - y^n_{i-1,j+1}}{4\Delta x \Delta z} + o(\Delta x^2, \Delta z^2, \Delta x \Delta z).$$

Then, the time integration of the first equation in (4.3) is split into two temporal semisteps, i.e. for the levels $n + \frac{1}{2}$ and $n + 1$. This yields the first semistep.

First semistep

$$\frac{\overbrace{\omega^{n+1/2}_{i,j}}^{\odot} - \omega^n_{i,j}}{0.5\Delta t} - C_{i,j} f \frac{v^n_{i,j+1} - v^n_{i,j-1}}{2\Delta z} + C_{i,j}$$

$$\times \left(\frac{(\omega^n_{i+1,j} - \omega^n_{i-1,j})}{2\Delta x} \frac{(\psi^n_{i,j+1} - \psi^n_{i,j-1})}{2\Delta z} - \frac{(\overbrace{\omega^{n+1/2}_{i,j+1}}^{\oplus} - \overbrace{\omega^{n+1/2}_{i,j-1}}^{\ominus})}{2\Delta z} \frac{(\psi^n_{i+1,j} - \psi^n_{i-1,j})}{2\Delta x} \right)$$

$$= \frac{g}{\bar{\rho}_0} \left(\frac{\rho^n_{i+1,j} - \rho^n_{i-1,j}}{2\Delta x} - z_{2j} p_i \frac{\rho^n_{i,j+1} - \rho^n_{i,j-1}}{2\Delta z} \right)$$

$$+ A^{\mathrm{H}} \left(\frac{\omega^n_{i+1,j} - 2\omega^n_{i,j} + \omega^n_{i-1,j}}{\Delta x^2} \right)$$

$$-2z_{2j} p_i A^{\mathrm{H}} \left(\frac{\omega^n_{i+1,j+1} + \omega^n_{i-1,j-1} - \omega^n_{i+1,j-1} - \omega^n_{i-1,j+1}}{4\Delta x \Delta z} \right)$$

$$+ A^{\mathrm{H}} \left[(z_{2j} p_i)^2 \frac{\overbrace{\omega^{n+1/2}_{i,j+1}}^{\oplus} - 2\overbrace{\omega^{n+1/2}_{i,j}}^{\odot} + \overbrace{\omega^{n+1/2}_{i,j-1}}^{\ominus}}{\Delta z^2} + z_{2j}(2p_i^2 - q_i) \frac{\overbrace{\omega^{n+1/2}_{i,j+1}}^{\oplus} - \overbrace{\omega^{n+1/2}_{i,j-1}}^{\ominus}}{2\Delta z} \right]$$

$$+ A^{\mathrm{V}} \left[r_i \left(\frac{dN}{dz} \right)_{i,j} \frac{\overbrace{\omega^{n+1/2}_{i,j+1}}^{\oplus} - \overbrace{\omega^{n+1/2}_{i,j-1}}^{\ominus}}{2\Delta z} + r_i^2 N^2_{i,j} \frac{\overbrace{\omega^{n+1/2}_{i,j+1}}^{\oplus} - 2\overbrace{\omega^{n+1/2}_{i,j}}^{\odot} + \overbrace{\omega^{n+1/2}_{i,j-1}}^{\ominus}}{\Delta z^2} \right],$$

in which the approximations of ω_{x_2}, $\omega_{x_2x_2}$, and $\omega_{x_2z_2}$ are written explicitly – that is at the time level n, where values of the vorticity are known. Consequently, only the three vorticity terms which are marked by overbraces with symbols $\oplus$, $\odot$, and $\ominus$ are unknown. Grouping similar terms, the above equation can be rewritten in the following tridiagonal form:

$$\mathscr{A}_i\omega_{i,j-1}^{n+1/2} + \mathscr{B}_i\omega_{i,j}^{n+1/2} + \mathscr{C}_i\omega_{i,j+1}^{n+1/2} = \mathscr{F}_i, \tag{4.4}$$

where $\mathscr{A}_i$, $\mathscr{B}_i$, and $\mathscr{C}_i$ are functions of Δx, Δz, etc. and known values of $\psi_{i,j}^n$; $\mathscr{F}_i$ is a function of known values at the nth time step of $\omega_{i,j}^n$, $\psi_{i,j}^n$, $\rho_{i,j}^n$, $v_{i,j}^n$. The system (4.4) with $j = 2, 3, \ldots, \mathscr{J}-1$ is not complete. It contains $\mathscr{J}-2$ equations, and thus two additional equations are required. With the boundary conditions[3] at the sea surface, $\omega_{i,1}^{n+1/2} = \mathscr{B}_1\omega_{i,2}^{n+1/2} + \mathscr{F}_1$, and the bottom, $\omega_{i,\mathscr{J}}^{n+1/2} = \mathscr{B}_{\mathscr{J}}\omega_{i,\mathscr{J}-1}^{n+1/2} + \mathscr{F}_{\mathscr{J}}$, the new system of linear equations becomes complete and can be solved by any standard procedure.

Second semistep

$$\begin{aligned}
&\frac{\overbrace{\omega_{i,j}^{n+1}}^{\odot} - \omega_{i,j}^{n+1/2}}{0.5\Delta t} - C_{i,j} f \frac{v_{i,j+1}^n - v_{i,j-1}^n}{2\Delta z} + C_{i,j} \\
&\times\left(\frac{(\overbrace{\omega_{i+1,j}^{n+1}}^{\oplus} - \overbrace{\omega_{i-1,j}^{n+1}}^{\ominus})}{2\Delta x}\frac{(\psi_{i,j+1}^n - \psi_{i,j-1}^n)}{2\Delta z} - \frac{(\omega_{i,j+1}^{n+1/2} - \omega_{i,j-1}^{n+1/2})}{2\Delta z}\frac{(\psi_{i+1,j}^n - \psi_{i-1,j}^n)}{2\Delta x}\right) \\
&= \frac{g}{\bar{\rho}_0}\left(\frac{\rho_{i+1,j}^n - \rho_{i-1,j}^n}{2\Delta x} - z_{2j}p_i\frac{\rho_{i,j+1}^n - \rho_{i,j-1}^n}{2\Delta z}\right) \\
&\quad + A^{\mathrm{H}}\left[\frac{\overbrace{\omega_{i+1,j}^{n+1}}^{\oplus} - 2\overbrace{\omega_{i,j}^{n+1}}^{\odot} + \overbrace{\omega_{i-1,j}^{n+1}}^{\ominus}}{\Delta x^2}\right] \\
&\quad - 2z_{2j}p_iA^{\mathrm{H}}\left[\frac{\omega_{i+1,j+1}^n + \omega_{i-1,j-1}^n - \omega_{i+1,j-1}^n - \omega_{i-1,j+1}^n}{4\Delta x\Delta z}\right] \\
&\quad + A^{\mathrm{H}}\left[(z_{2j}p_i)^2\frac{\omega_{i,j+1}^{n+1/2} - 2\omega_{i,j}^{n+1/2} + \omega_{i,j-1}^{n+1/2}}{\Delta z^2} + z_{2j}(2p_i^2 - q_i)\frac{\omega_{i,j+1}^{n+1/2} - \omega_{i,j-1}^{n+1/2}}{2\Delta z}\right] \\
&\quad + A^{\mathrm{V}}\left[r_i\left(\frac{dN}{dz}\right)_{i,j}\frac{\omega_{i,j+1}^{n+1/2} - \omega_{i,j-1}^{n+1/2}}{2\Delta z} + r_i^2N_{i,j}^2\frac{\omega_{i,j+1}^{n+1/2} - 2\omega_{i,j}^{n+1/2} + \omega_{i,j-1}^{n+1/2}}{\Delta z^2}\right].
\end{aligned}$$

[3] After transformation of variables from (x, z) to (x_2, z_2), boundary conditions can be written as Dirichlet (prescribe the value of the dependent variable on the boundary) or Neumann (prescribe the gradient of the variable normal to the boundary) conditions.

In contrast with equation (4.4), the approximations of $\omega_{x_2}, \omega_{x_2 x_2}$ are now implicit. Thus, the bias introduced by the first time step is now partly corrected. In complete analogy with (4.4), these equations can be regrouped as

$$\mathcal{A}_j \omega_{i-1,j}^{n+1} + \mathcal{B}_j \omega_{i,j}^{n+1} + \mathcal{C}_j \omega_{i+1,j}^{n+1} = \mathcal{F}_j, \tag{4.5}$$

where $\mathcal{A}_j, \mathcal{B}_j, \mathcal{C}_j$ are functions of Δx, Δz, etc. and known values of $\psi_{i,j}^n$, and $\mathcal{F}_j$ is a function of known values of $\omega_{i,j}^{n+1/2}$, $\omega_{i,j}^n$, $\psi_{i,j}^n$, $\rho_{i,j}^n$, and $v_{i,j}^n$ at the time steps n and $n+\frac{1}{2}$. System (4.5) with lateral boundary conditions at the left, $\omega_{1,j}^{n+1} = \mathcal{B}_1 \omega_{2,j}^{n+1} + \mathcal{F}_1$, and right, $\omega_{\mathcal{I}-1,j}^{n+1} = \mathcal{B}_{\mathcal{I}} \omega_{\mathcal{I},j}^{n+1} + \mathcal{F}_{\mathcal{I}}$, boundaries equally results in a tridiagonal matrix equation.

The value of the time step Δt, ensuring stability of the numerical scheme, must be taken in accordance with the Courant–Friedrichs–Levi (CFL) criterion, $\Delta t < \Delta x / c_{\max}$, where $c_{\max}$ is the maximum phase speed of the propagation of the baroclinic perturbations. For baroclinic tides this value is defined by the phase speed of the first baroclinic mode. Its magnitude is estimated from the linear problem (1.67) and, as a rule, does not exceed 1–2 m s^{-1}. This value may be slightly larger for nonlinear waves, a fact which is usually taken into account by choosing the time step Δt to be somewhat smaller than dictated by the CFL condition.

Next, the value of the stream function at the $(n+1)$st temporal step must be determined. It can be found from the final equation in (4.3) to be

$$\omega = \mathcal{M}_2(\psi) + \mathcal{M}_3(\psi). \tag{4.6}$$

This equation corresponds to $\omega = \Delta\psi$ and defines the vorticity in terms of the stream function in the new variables (x_2, z_2). Contrary to the "time-dependent" first three equations of system (4.3), equation (4.6) for the stream function is elliptic. It has the following finite difference form:

$$\begin{aligned}
\omega_{i,j}^n &= \frac{\psi_{i+1,j}^n - 2\psi_{i,j}^n + \psi_{i-1,j}^n}{\Delta x^2} \\
&\quad - 2 z_{2j} p_i \frac{\psi_{i+1,j+1}^n + \psi_{i-1,j-1}^n - \psi_{i+1,j-1}^n - \psi_{i-1,j+1}^n}{4\Delta x \Delta z} \\
&\quad + \left(z_{2j}^2 p_i^2 + r_i^2 N_{i,j}^2 \right) \frac{\psi_{i,j+1}^n - 2\psi_{i,j}^n + \psi_{i,j-1}^n}{\Delta z^2} \\
&\quad + \left[z_{2j} \left(2p_i^2 - q_i \right) + r_i \left(\frac{dN}{dz} \right)_{i,j} \right] \frac{\psi_{i,j+1}^n - \psi_{i,j-1}^n}{2\Delta z} =: \mathcal{G}(\psi_{i,j}^n)
\end{aligned}$$

or

$$\omega_{i,j}^n = \mathcal{G}(\psi_{i,j}^n),$$

where $\mathcal{G}$ is a short-hand representation of the right-hand side. To find the stream function at the $(n+1)$st time level, one ought to solve a large algebraic matrix equation and is computationally inconvenient.

Another way of finding $\psi_{i,j}^{n+1}$ is to employ the so-called pseudo-transient method (PTM). It converts an elliptic boundary value problem to an evolutional parabolic initial value problem by adding a term involving the time derivative of the dependent variable, i.e.

$$\omega_{i,j}^{n+1} = \mathcal{G}(\phi_{i,j}) + \frac{\partial \phi_{i,j}}{\partial \tau}. \tag{4.7}$$

This equation is solved for $\phi_{i,j}$ with initial value $\phi_{i,j} = \psi_{i,j}^{n}$ and integration in τ is continued until steady state is reached. The finite difference approximation of (4.7) is

$$\omega_{i,j}^{n+1} = \mathcal{G}[(\phi_{i,j})^{m}] + \frac{\phi_{i,j}^{m} - \phi_{i,j}^{m-1}}{\Delta\tau_m}, \tag{4.8}$$

in which $\Delta\tau_m$ ($m = 1, 2, \ldots, \mathcal{M}$) is the step size at integration level m for τ. Using the above-described ADIM, we start with $\Delta\tau_1 < 1$ and perform calculations to find the first value ($\phi_{i,j}^{1}$).

Setting the relative error, $0 < \epsilon < 1$, the following condition must be checked:

$$\max_{i,j} \left| \frac{\phi_{i,j}^{m} - \phi_{i,j}^{m-1}}{\phi_{i,j}^{m}} \right| < \epsilon.$$

If this inequality is not fulfilled, we consider the iteration sequence with

$$\Delta\tau_2 = \Delta\tau_1 \Bigg/ \left(\frac{1+\sqrt{\epsilon}}{1-\sqrt{\epsilon}} \right), \ldots, \Delta\tau_{\mathcal{M}} = \Delta\tau_{\mathcal{M}-1} \Bigg/ \left(\frac{1+\sqrt{\epsilon}}{1-\sqrt{\epsilon}} \right)^{\mathcal{M}-1},$$

where $\mathcal{M}$ is the number of iterations for which the inequality becomes valid and the iteration process is terminated. This procedure yields the final equation

$$\omega_{i,j}^{n+1} = \mathcal{G}(\phi_{i,j}^{\mathcal{M}}) + \underbrace{\frac{\phi_{i,j}^{\mathcal{M}} - \phi_{i,j}^{\mathcal{M}-1}}{\Delta\tau_{\mathcal{M}}}}_{\approx 0} \quad \Longrightarrow \quad \phi_{i,j}^{\mathcal{M}} = \psi_{i,j}^{n+1}, \tag{4.9}$$

for the stream function $\psi_{i,j}^{n+1}$ at the $(n+1)$st temporal step. Usually, the solution of problem (4.8) requires 10 to 20 iterations to reach an accuracy with ϵ in the range of 10^{-4}–10^{-5}.

Next, using the already found values of $\omega_{i,j}^{n+1}$ and $\psi_{i,j}^{n+1}$, the horizontal velocity components $v_{i,j}^{n+1}$ and the densities $\tilde{\rho}_{i,j}^{n+1}$ at the $(n+1)$st temporal layer are calculated again by using ADIM. This completes the computations at the $(n+1)$st level, and the whole procedure is repeated for $n+2, n+3, \ldots$

The next remark concerns the boundary conditions for the vorticity at the bottom, which was not previously defined. It is not clear what should be put for ω_0 at $z_2 = 1$ when equation (4.4) is solved for a viscous fluid when the no-slip condition (1.85a) is considered. To find a solution which accounts for the boundary layer at the rigid boundaries, the following procedure is usually applied. Apart from the computational domain Ω_0, an additional domain Ω_1 is introduced, the boundaries of which are shifted one spatial step inside Ω_0 (see Figure 4.6(c)). The boundary conditions for the vorticity $\omega = \omega_0$ are defined at the border of Ω_1 with the use of equation (4.9),

$$\omega_0 = \omega_{\Omega_1}^{n+1} = \mathcal{G}(\psi_{\Omega_1}^n),$$

where $\psi_{\Omega_1}^n$ is the value of the stream function at the bottom boundary Ω_1.

4.3 Qualitative analysis of the excitation mechanism

Estimations of the parameter ε_1 defined by (4.1) show that for many regions of the World Ocean its value is not small. Thus, for an adequate description of the generation mechanism under such conditions we should use the nonlinear model developed in the previous section. Before embarking on that, let us make some preliminary qualitative remarks about the tidal forcing, which will help us in interpreting the theoretical results.

Conceptually, the solution of the nonlinear boundary value problem (1.79), (1.80), or (1.81) with boundary conditions (1.82) to either (1.85a) or (1.85b) can be represented as a sum of a known barotropic tidal flux (external forcing) and an unknown baroclinic response. When substituted into the governing equations, this solution leads to the separation of terms which are responsible for the barotropic tidal flux on the one hand, and the pure baroclinic wave motions on the other. Some mixed terms also arise; they comprise the interaction effects. If we are not linearizing the system with respect to the background current, as was done in Section 3.1, the terms, which describe only the barotropic tidal flux and are responsible for the external forcing, can be written explicitly on the right-hand side of the equations. On this basis, the total time variation of the tidal forcing, F, in Euler variables is expressible as

$$\frac{dF}{dt} = \frac{\partial F}{\partial t} + U\frac{\partial F}{\partial x}, \tag{4.10}$$

where U represents the velocity of the basic flow in the x-direction.

A qualitative analysis of the specific input of the nonlinearity to the generation process, based on equation (4.10), was first performed by Bell [18] and more recently by Nakamura *et al.* [172], [173]. On the basis of a comparative analysis of different terms in the governing equations, it was shown that, over variable bottom topography of a stratified basin, several possible types of internal waves can be generated by the barotropic tidal flow. The first term on the right-hand side of (4.10), the local time derivative of the forcing, is responsible for the excitement of the internal tides. When the horizontal excursion of the fluid particles is small compared with the horizontal scale of the bottom topography, the oscillating tidal flow can generate internal waves of the same frequency as the basic flow. The length scales of these waves outside the source of generation depend only on the fluid stratification, the total depth, and the rotation, whilst their amplitudes and modal structure are defined by the shape of the obstacle and the intensity of the barotropic tidal flux.

The second term on the right-hand side of equation (4.10) represents the effect of advection. The nonlinear terms in the governing equations lead to a modification of the generated waves. In particular, the propagation from the source of generation can transform the baroclinic tidal wave into a *baroclinic bore*, which may disintegrate further to a sequence of solitary internal waves [73], [75], [129], [243], [257], [259]. For such a study, the Korteweg–de Vries theory is a good approximation [102], [130], [143].

When the advective term in (4.10) is *dominant*, and the local time derivative of the forcing (the first term) is also taken into account, *unsteady lee waves* are generated. These waves have properties quite different from internal tides. They are generated by fluxes on the lee side of bottom obstructions due to the advection effects. Usually, internal lee waves are much shorter than baroclinic tides. They have spatial scales of the same order as the bottom topographic changes. During the first half of the flood phase, when the tidal flux increases, lee waves are trapped at the downstream side of a bottom obstacle; they extract energy from the barotropic tidal flow and increase in amplitude. As the tidal flow slackens during the second half of the flood phase, the excited unsteady lee waves begin to propagate upstream.

In the *intermediate case* when local temporal derivatives and advection are of the same order, *mixed lee waves* are generated. Such waves have properties inherent in baroclinic tides as well as to lee waves.

The conditions of generation of different types of waves can be defined by the parameter kU_0/σ, where $U_0 = \Psi_0/H$ is the amplitude of the barotropic tidal flux and k is the horizontal wavenumber of the generated waves. Baroclinic tides are generated when $kU_0/\sigma \ll 1$. In fact, for baroclinic tides this condition means that the Froude number is much less than unity. In the opposite case, when the Froude number is large, $kU_0/\sigma \gg 1$, unsteady lee waves are produced. Mixed lee waves correspond to the intermediate case when $kU_0/\sigma = O(1)$ (see Table 4.1). Below we consider the process of internal wave generation at conditions of small as well

Table 4.1. *Regimes of internal wave generation.*

Baroclinic tides	$kU_0/\sigma \ll 1$
Mixed tidal–lee waves	$kU_0/\sigma = O(1)$
Unsteady lee waves	$kU_0/\sigma \gg 1$

as large Froude numbers, and we investigate the generation of both baroclinic tides and unsteady lee waves.

4.4 Generation mechanism at low Froude numbers: baroclinic tides

Let us analyze the predictions of the nonlinear model for small Froude numbers ($kU_0/\sigma \ll 1$) when primarily baroclinic tides are excited. The mechanism of their generation, described in ref. [243], can be clarified with the help of Figure 4.7, adapted from [261], which shows the density field during the two first tidal cycles after the beginning of the motion. The related computations were performed for $\Psi_0 = 30\ \mathrm{m^2\ s^{-1}}$ ($kU_0/\sigma = 0.53$ in a shallow-water zone and $kU_0/\sigma = 0.013$ off the shelf), and the other chosen input parameters (bottom profile, fluid stratification) are characteristic for the region of the New York Bight (latitude angle $\varphi = 40°$) where baroclinic tides are usually observed. Some typical data from this region were considered in Section 4.1.

A dome-like elevation of the pycnocline is formed over the slope during the flood phase of the tide from $t = 0$ to $t = 0.25T$, see Figure 4.7(a). The initial dome of the isopycnals is subsequently destroyed and separates into two progressive waves propagating in opposite directions, Figure 4.7(b). The barotropic tidal flux amplifies the process of separation and makes it more pronounced. Downward fluxes of water over the slope form steep trailing faces of propagating waves, shown by a vertical dashed arrow in Figure 4.7(c). By the instant $t = T$, the baroclinic bore is propagating away from the source of generation, more or less as it was formed.

The generated baroclinic bore propagates further during the second tidal cycle on the background of an oscillating barotropic tidal flux; therefore, it is exposed to accelerations and decelerations. However, the excited waves effectively leave the area of generation because their phase speed is at least twice as large as the velocity of the background flow.

In the center of the generation area, above the slope, the wave process repeats itself during the second tidal cycle ($T < t < 2T$, see Figures 4.7(e)–(h), whereas the waves that were excited at $t < T$ continue their propagation beyond the generation area. At the first stage of their evolution, at $0.5T < t < T$, the shoreward propagating long baroclinic tidal wave becomes more gently sloping at its leading face and steeper at its trailing edge. As a result, the nonlinear steepening of the

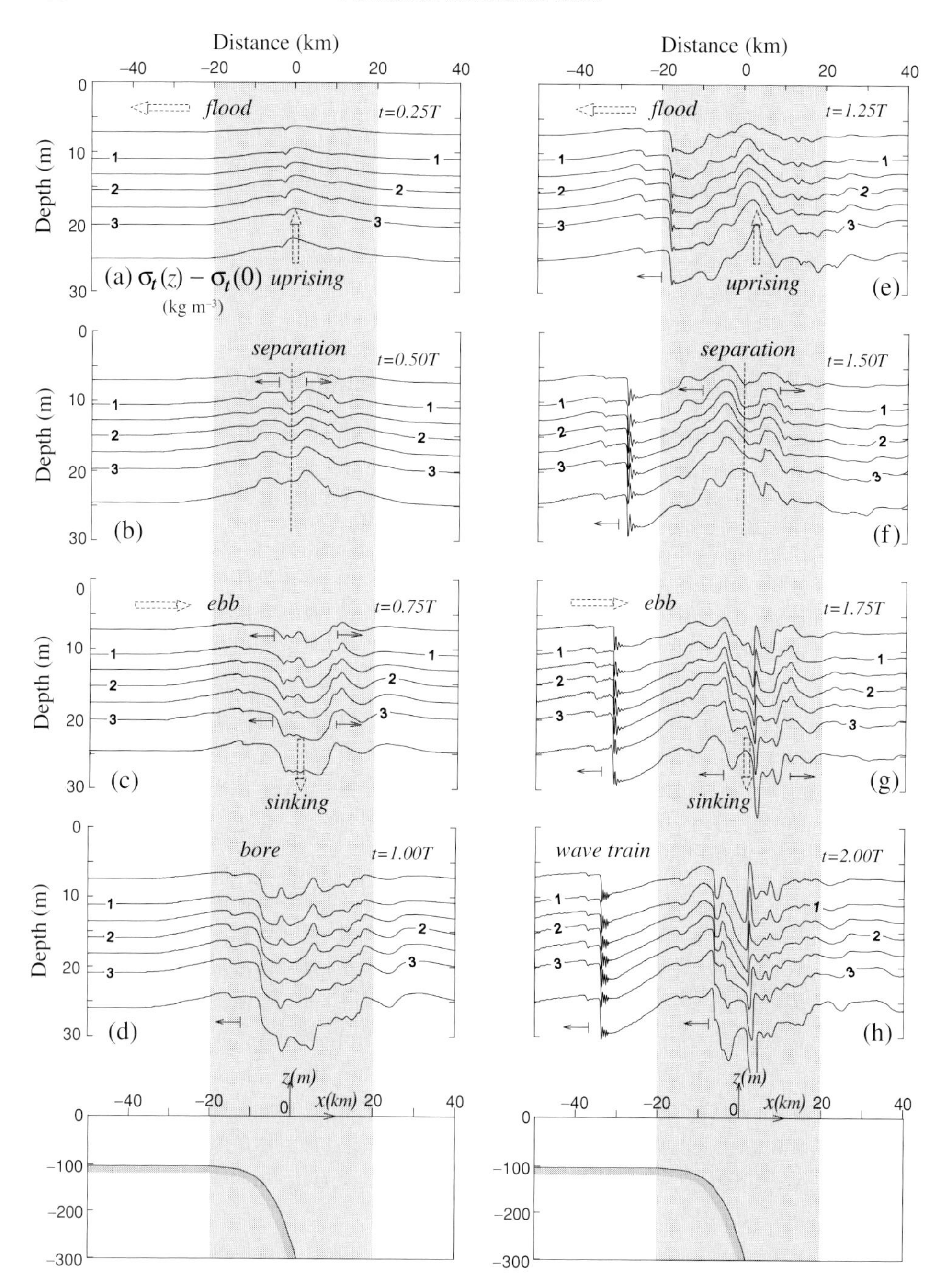

Figure 4.7. Predicted evolution of the density field $\sigma_t(z) - \sigma_t(0)$ (kg m^{-3}) in the slope-shelf area of the New York Bight ($\varphi = 40^\circ$) during two tidal cycles. The fluid stratification, the bottom topography, and the value of the vertically integrated tidal flux ($\Psi_0 = 30$ m^2 s^{-1}) were specified close to those observed in the ocean. $T = 12.4$ h is the tidal period.

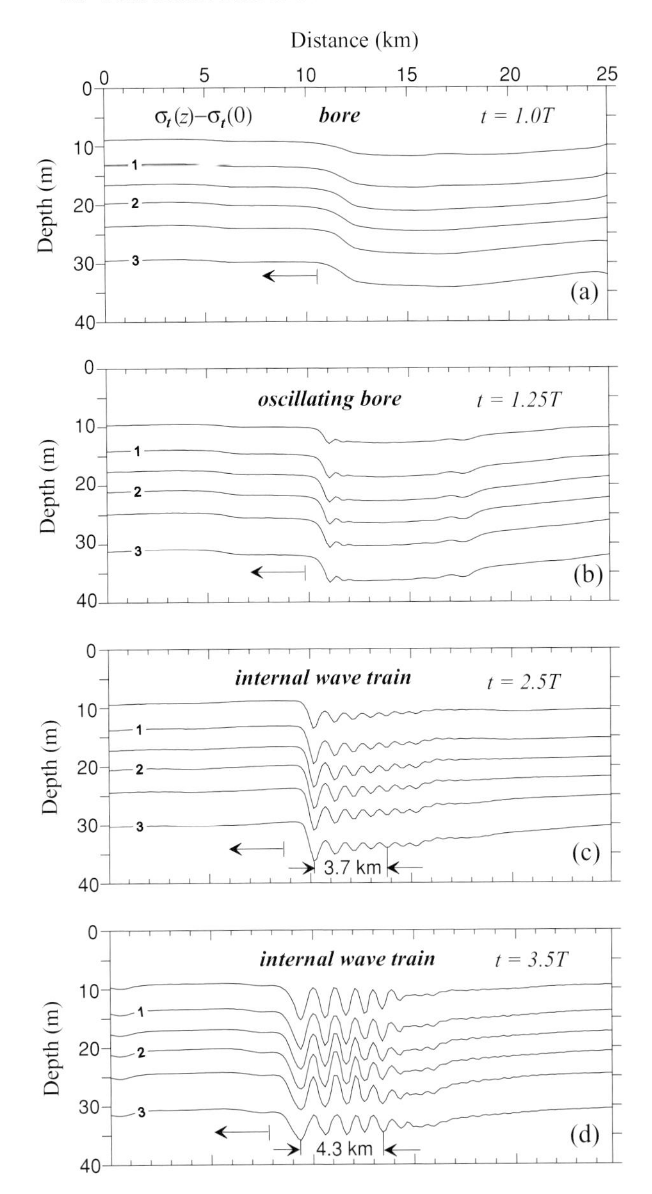

Figure 4.8. Nonlinear evolution of the shoreward propagating baroclinic tidal wave beyond the source of generation at a low Froude number. All input parameters are the same as for Figure 4.7.

baroclinic bore is well developed by the time $t = T$. Its further nonlinear evolution leads to the transformation into a packet of short internal waves. This process is shown in Figures 4.7 (e)–(h) and in greater detail in Figure 4.8. Evidently, the observed density field, measured in the New York Bight during JUSREX-92 and shown in Figure 4.4, represents an intermediate stage of this process.

The nonlinear steepening and disintegration occur because of the following instability mechanism described, e.g., in ref. [53]: due to the nonlinear dispersion, the shoreward propagating baroclinic bore transforms into a sequence of solitary waves arranged by amplitude, of which each one propagates with its own phase speed depending on its amplitude. As a result of this amplitude dispersion, the horizontal scale of the wave packet increases with time. This "stretching" of the wave packet is clearly seen when comparing Figures 4.8(c) and (d). The final stage of this nonlinear dispersion is shown in Figure 4.5, where the baroclinic bore is already transformed into a sequence of solitary internal waves.

Thus, from the above-described example one may infer that, for small Froude numbers, the dynamics of the baroclinic tides near bottom features can conventionally be divided into two stages:

- the generation over the inclined bottom topography, where nonlinear effects are not the dominating factor;
- the evolution beyond the source of generation, where the nonlinearity leads to a strong energy transfer from a long barotropic tidal wave to shorter scale internal wave packets and solitary waves.

An additional confirmation of the second statement, namely that the nonlinearity can play a decisive role at the second stage of evolution, is shown in Figure 4.9. Here we show the transformation of the profile of a long internal tidal wave propagating in a basin of *constant* depth under the action of only nonlinear advection.

The analytical solution (1.74)–(1.76), expressed as a linear first-mode periodic baroclinic tidal wave, was used to define the initial fields in the numerical model (1.79), (1.81), (1.84), (1.85b), in which $\Psi_0 = 0$ is assumed. It is evident that the analytical solution (1.74)–(1.76) of the BVP (1.65) is also the solution of the numerical model if the nonlinear and dissipative terms in (1.79), (1.81) are set equal to zero. In other words, in the linear case the initial wave profile of Figure 4.9(a) will indefinitely preserve its form during the wave evolution in a basin of constant depth.

The situation changes dramatically if the nonlinearities are "switched on" (Jacobian operators are taken into account in (1.79), (1.81)). The result of its influence is presented in Figures 4.9(b)–(d): first the propagating wave at the first stage becomes nonsymmetric, Figure 4.9(b), then the steep wave face transforms into a baroclinic bore which finally disintegrates into a sequence of solitary internal waves, Figures 4.9(c) and (d).

4.5 Influence of the intensity of the tidal forcing and dissipation

In this section we analyze the dependence of the characteristics of the tidally generated internal waves on the intensity of the external forcing [243] and study how far an initial disturbance can propagate from its source of generation [162].

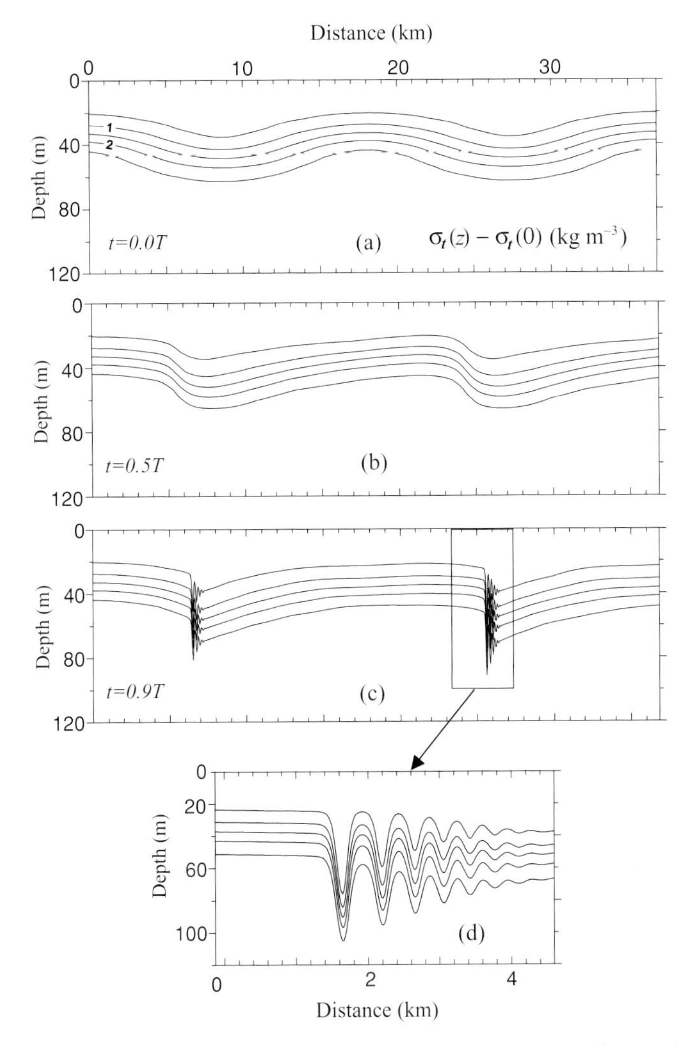

Figure 4.9. (a)–(c) Predicted nonlinear evolution of a first-mode internal periodic tidal wave (1.76) propagating in a basin of 1000 m depth. The isolines are conventional density anomaly lines with respect to the free surface: $\sigma_t(z) - \sigma_t(0)$ (kg m^{-3}). The initial wave amplitude is 10 m. Parameters of stratification were $N_{\mathrm{p}} = 0.02\,\mathrm{s}^{-1}$, $H_{\mathrm{p}} = 35$ m, $\Delta H_{\mathrm{p}} = 30$ m, $\varphi = 40^{\circ}$. (d) The internal wave train formed as a result of the nonlinear evolution.

As was shown above, the influence of the nonlinearity on the wave process basically consists of an energy transfer from large-scale to the small-scale motions: a long baroclinic tidal wave will become steeper during its propagation from the source of generation. At a certain definite stage of evolution, the wave will lose its form and disintegrate into a sequence of short nonlinear internal waves. Note that a linear periodic wave preserves its sinusoidal form during the propagation. The above-mentioned difference between linear and nonlinear waves is clearly seen in

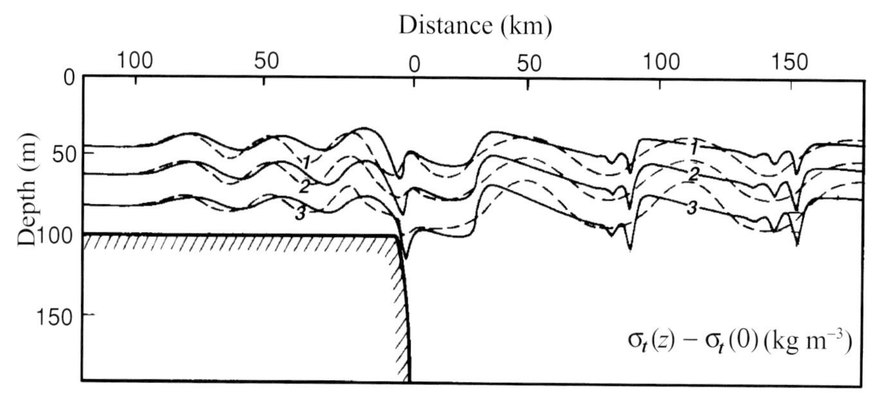

Figure 4.10. Model-predicted density field calculated for the Brazilian slope-shelf region in the linear (dashed lines) and nonlinear (solid line) cases. Isopycnals are plotted after three tidal cycles from the beginning of the motion. They are derived for $\Psi_0 = 70$ m^2 s^{-1}, $A^{\mathrm{H}} = K^{\mathrm{H}} = 10$ m^2 s^{-1}, and a realistic vertical stratification of the fluid.

Figure 4.10, where the results of the modeling of baroclinic tides near the Brazilian coast, Figure 3.2(a), are given for both the linear and nonlinear cases. This area is known as a "hot spot" of the World Ocean where the barotropic tidal energy efficiently transforms into baroclinic tides, and where strong nonlinear solitary waves are frequently observed. An example of such an observation of an internal tidal wave was shown in Figure 3.3; it represents strong internal waves propagating in the northeast direction from the shore. These measurements were conducted during Cruise 39 of the R/V *Akademik Vernadskiy* in June 1989. More recent measurements of strong internal tidal waves in this region are also reported, for instance in ref. [24].

If the conclusion formulated in Section 4.4 about the role of the nonlinearity is valid, the described mechanism of wave steepening and disintegration must become weaker with diminishing tidal forcing, or with a decrease of the wave amplitude. Figure 4.11 illustrates quite clearly how large the difference is between the linear (dashed lines) and nonlinear (solid lines) solutions at $\Psi_0 = 100$ m^2 s^{-1} and how this difference can substantially decrease with the weakening of the external forcing down to $\Psi_0 = 10$ m^2 s^{-1}. Estimations of the parameter ε_1 calculated according to formula (4.1) for the presented runs with $\Psi_0 = (100; 30; 10)$ m^2 s^{-1} give values of (0.28; 0.083; 0.028), respectively. It is evident from Figure 4.11 that, in the last case (solid and dashed lines numbered 3), the linear theory is a valid approach for the modeling of the baroclinic tides. It is also absolutely clear that the specific input of the nonlinearity is crucial at $\Psi_0 = 100$ m^2 s^{-1} (solid and dashed lines numbered 1) when the calculations must be performed by accounting for the nonlinear terms.

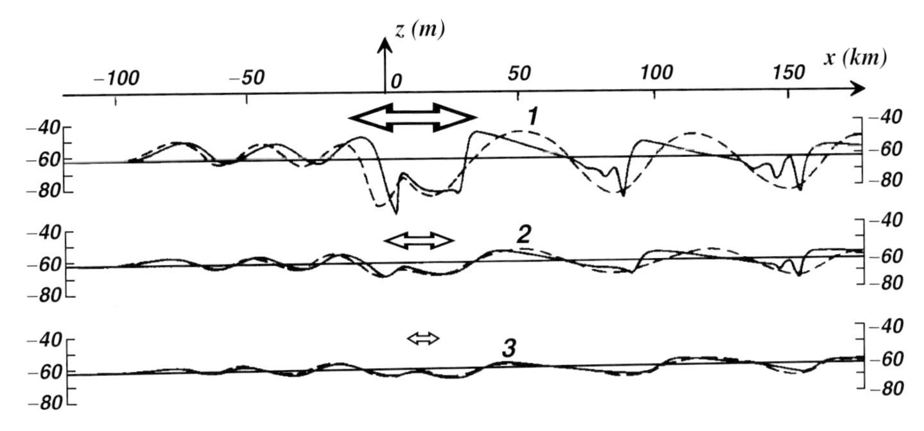

Figure 4.11. Vertical deviation of the pycnocline in the slope-shelf region after three tidal cycles from the beginning of the motion, derived for various values of the tidal discharge Ψ_0: (*1*) 100 $m^2\ s^{-1}$; (*2*) 30 $m^2\ s^{-1}$; (*3*) 10 $m^2\ s^{-1}$. The solid lines correspond to the nonlinear and the dashed lines correspond to the linear case. Calculations were performed for the slope (3.47) ($H_1 = 4$ km, $H_3 = 100$ m, $2l = 50$ km) at the following values of the parameters of the pycnocline (1.70): $N_p = 0.025\ s^{-1}$, $H_p = 60$ m, $\Delta H_p = 80$ m.

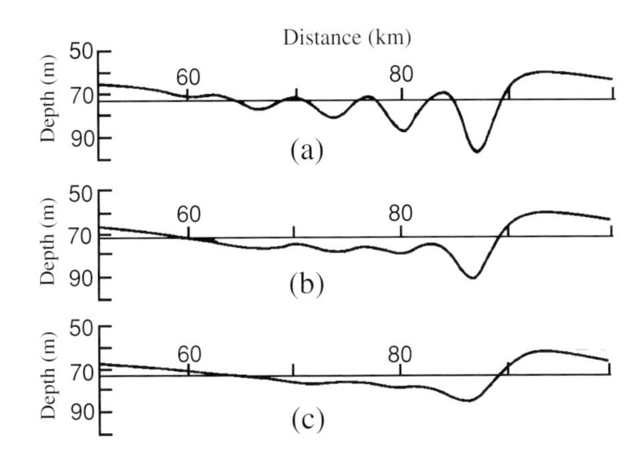

Figure 4.12. Snapshots of the topography of the pycnocline (isopycnal $\sigma_t(z) - \sigma_t(0) = 2\ \mathrm{kg\ m^{-3}}$) off the Brazilian continental slope after two tidal cycles from the beginning of the motion, obtained at $\Psi_0 = 30\ m^2\ s^{-1}$ for various coefficients of the horizontal turbulent exchange coefficients $A^H = K^H$: (a) 3 $m^2\ s^{-1}$, (b) 30 $m^2\ s^{-1}$, and (c) 300 $m^2\ s^{-1}$. All other parameters are the same as for Figure 4.11.

The other process which can play a key role in the dynamics of baroclinic tides is the background turbulent mixing. The values of the coefficients A^V, A^H, K^V, and K^H should be carefully chosen when performing computations because the quantitative (and sometimes also qualitative) characteristics are quite sensitive to the intensity of the subgrid mixing. The example shown in Figure 4.12 illustrates this statement quite clearly. With an increase of the coefficients $A^{\mathrm{H}} = K^{\mathrm{H}}$ from 3 to 300 $\mathrm{m^2\,s^{-1}}$, the amplitudes of the propagating waves decrease substantially. Moreover, due to the large value of $A^{\mathrm{H}} = K^{\mathrm{H}} = 30\ \mathrm{m^2\,s^{-1}}$, strong energy dissipation damps the formation of short-wave packets. As may clearly be seen from a comparison of Figures 4.12(a) and (b), the process of formation of short waves is completely blocked at $A^{\mathrm{H}} = K^{\mathrm{H}} = 300\ \mathrm{m^2\,s^{-1}}$ (Figure 4.12(c)). In the last case, the steepening of the long wave and its disintegration and fission are compensated for by the energy leakage due to dissipation.

The problem of the correct choice of the coefficients of the background turbulence can be solved by including a turbulence closure scheme in the model, see, e.g., ref. [151]. However, such a procedure solves the problem only partly because other adjustable parameters arise, which can only be identified by *in situ* or laboratory measurements. Thus, the information from the field experiments about the internal wave attenuation is very important and has a high priority for the calibration of the theoretical models. One such example is considered below.

The Mascarene Ridge in the Indian Ocean has provided a wealth of empirical data on internal wave dynamics; see, e.g., refs. [161], [162], [206], and [208]. We present here a short summary from refs. [161], [162], and [254], since the waves reported in these papers are closely connected to those considered in this section.

For the analysis we use the data obtained in different years by current and temperature meters at the moorings M1, M2 , . . . , M6 presented in Figure 4.13. The

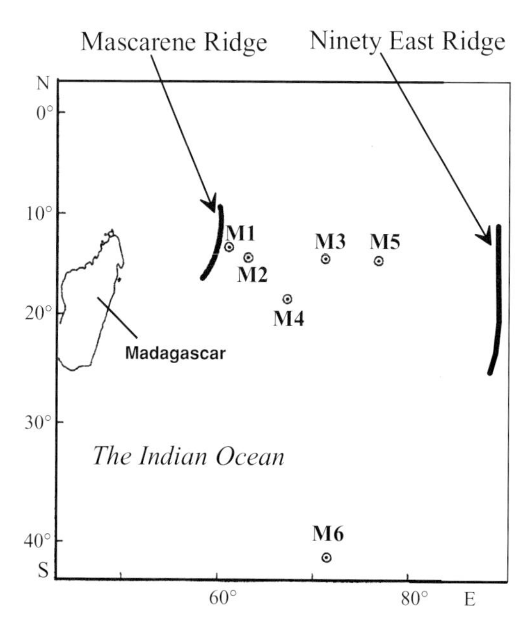

Figure 4.13. Schematic diagram of the location of moorings with current and temperature meters near the Mascarene Ridge.

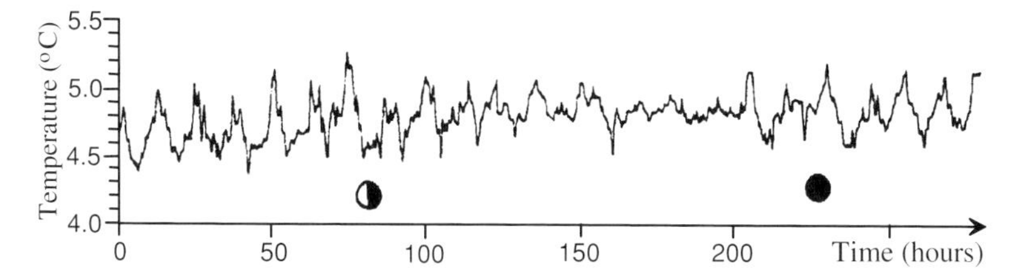

Figure 4.14. Temperature–time series measured at mooring M3 of Figure 4.13, 300 km off the ridge at the 1200 m depth level. Moon phases are shown by circles.

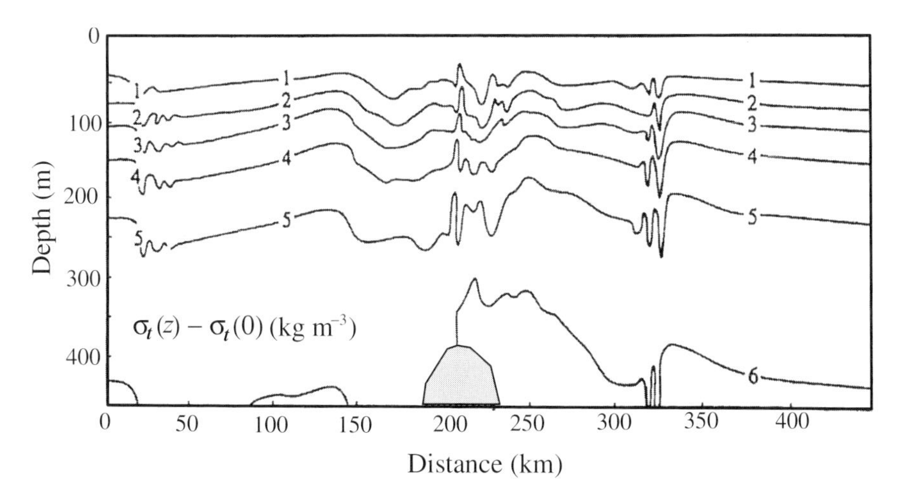

Figure 4.15. Isopycnal surfaces (seasonal thermocline) in the neighborhood of the Mascarene Ridge, computed for two tidal cycles using the nonlinear tidal model. Numbers at the curves correspond to the excess density relative to that at the surface (in kg m^{-3}) [254].

opportunity to use the data from different experiments follows from the quasistationary character of the source of generation. The barotropic tidal flux moving through the ridge topography is subjected to a variability associated with a fortnightly repetition, which has quasistationary character. Consequently, the field of the internal tidal waves is quasistationary, changing in accordance with the variability of the barotropic tide. This neap–spring tidal variability is clearly seen in Figure 4.14. We analyze below the recorded data together with the results of the theoretical modeling of baroclinic tides performed on the basis of the previously described nonlinear mathematical model. The result of the modeling is shown in Figure 4.15. It represents the density field in the neighborhood of the ridge and in the abyssal zone surrounding it. The values of the coefficients of horizontal turbulent exchange were set equal to 10 m^2 s^{-1}, and the water discharge in the barotropic tidal flow, Ψ_0, was chosen to be 174 m^2 s^{-1}. All other input parameters (bottom topography, water stratification) were close to those measured in the considered area.

As was expected from the examples previously considered, the nonlinear baroclinic tidal wave in close proximity to the ridge is characterized by a gently sloping leading edge and a steep trailing edge, which transforms into a packet of short internal waves during its propagation from the ridge. These fragments are clearly seen in Figure 4.15.

Observations from a drifting ship near the Mascarene Ridge confirm this mechanism. The temperature sensors lowered from the ship showed passages of packets of short internal waves. Their amplitudes of the vertical displacements at a depth of 90 m were about 20 m, and the phase speed of these waves was estimated to be 2.2 m s^{-1}. Both values are very close to those found for the near surface layer from the numerical runs. At the same time, the mooring measurements carried out in the principal thermocline (depth between 1000 and 1200 m), where the vertical displacements of the first baroclinic mode assumes a maximum, revealed isopycnal deflections up to 80 m. This is also in good agreement with the theoretical results.

Analysis of the temperature–time series and temperature spectra indicates that wave amplitudes decrease while the wave propagates in the ocean. In our analysis of the experimental data, we assume that the waves are generated over the Mascarene Ridge and propagate roughly towards the east and southeast. We neglect the additional generation of internal tides in the near bottom areas at large depths. We make this assumption because there are no significant bottom topographic features along the wave trajectory from the Mascarene Ridge to the Central Basin of the Indian Ocean. No doubt, internal waves can propagate from other sources to the points where moorings were located; the Ninety East Ridge is the closest (see Figure 4.13). However, waves generated over this ridge can be shown to have significantly smaller amplitudes than those propagating from the Mascarene Ridge. Nevertheless, their influence should be noted, especially in the region of mooring M5.

To analyze the wave attenuation while the waves propagate into the ocean, we consider the time series of vertical displacements and spectral functions, calculated on the basis of these time series. Temperature measurements at moored buoy stations were converted into vertical displacements by dividing the values of the time series by the mean vertical temperature gradient.[4]

Time series for such vertical displacements for several moored stations, located at increasing distance from the Mascarene Ridge at levels between 1200 and 1300 m, are shown in Figure 4.16.

[4] More appropriate would be to construct isotherm–depth–time series; however, these time series may be interrupted whenever they leave the depth range of the thermistor chains.

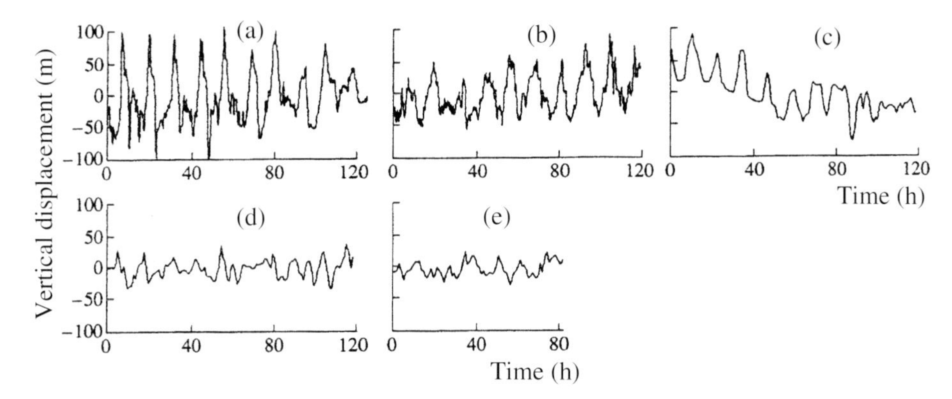

Figure 4.16. Time series of the vertical displacements caused by internal tides at the 1200 m level; (a)–(e) correspond to moorings M1–M5 in Figure 4.13, respectively.

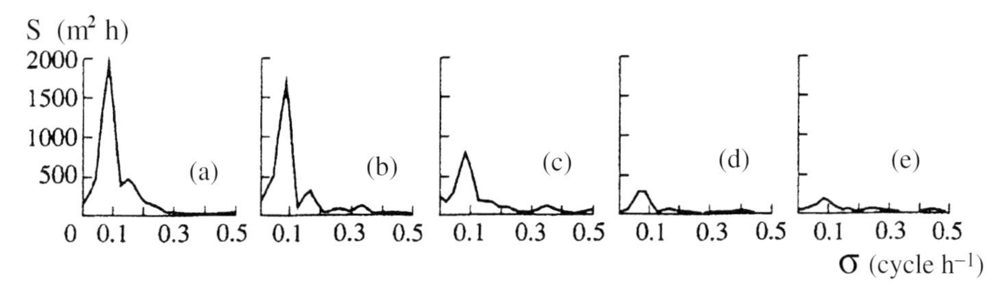

Figure 4.17. Spectra of the time series of the vertical displacements displayed in Figure 4.16.

The corresponding spectra of the vertical displacements for the same stations are shown in Figure 4.17, where one can clearly see how the spectral peak at the semidiurnal frequency is decreasing with progressing distance from the ridge while the wave propagates from a powerful source.

The dependence of the amplitude of the internal tidal waves on the distance is shown in Figure 4.18. These data indicate a conspicuous decrease in wave amplitude with distance. At short distances from the source, the effect is stronger than at greater distances. It is caused by a significant influence of the nonlinearities in the governing equations and various interactions with other processes. The maximum wave amplitude reaches a value of 80 m (mooring M1) in the region of wave generation at a distance of 100 km from the ridge, while the mean amplitude is 62 m. At a distance of 300 km, the mean amplitude is reduced by one-quarter and is equal to 45 m (mooring M2). Analysis of observations carried out within another time interval indicates that, at a distance of 900 km from the ridge, the mean amplitude is equal to 20–25 m (M4) and, at a distance of 1200 km from the

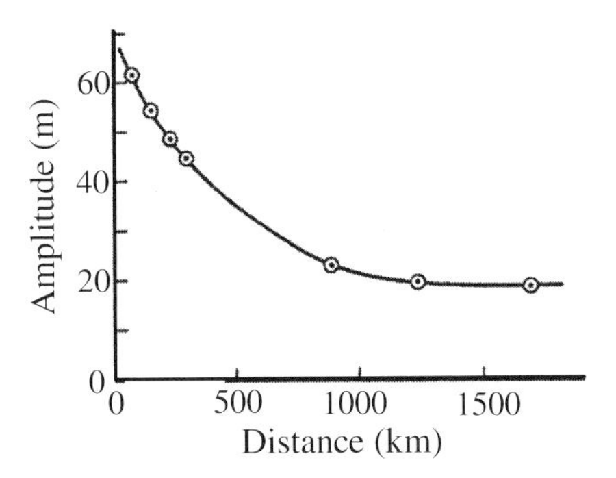

Figure 4.18. Decay of the mean amplitude of internal tidal waves with distance from the Mascarene Ridge.

ridge, it is close to 20 m (M3). At a distance of 1700 km, the amplitude never exceeded 20 m (M5). Mooring M5 is closer to the Ninety East Ridge than to the Mascarene Ridge. According to the results of the measurements, the amplitudes of the vertical displacements at this station are very small. Even if the waves from the Ninety East Ridge contribute to the amplitude, their influence is obviously very small. The joint contribution of the two wave systems leads to oscillations which slightly exceed the background level. On the basis of these measurements, one can estimate that the wave amplitude decreases by 10% over one wavelength or after one period of the wave.

In addition to the considered results, the measurements in the Crozet Basin (M6) and in the Madagascar Basin southwest of the Mascarene Ridge indicate that semidiurnal oscillations there have an even lower value (12–15 m), which is close to the background level of the Garrett–Munk model spectrum [69].[5] In some spectral functions, the spectral peak is weak or absent; this provides support for the assumption that only the background wave field is registered there and no waves from distant sources are observed. These measurements allow us to obtain an estimate of the utmost distant tide propagation equal to 2000–2500 km or 12–15 wavelengths.

These estimations are in agreement with the TOPEX/Poseidon altimeter data on the surface manifestation of internal tides near the Hawaiian Ridge. As was shown by Ray and Mitchum [204], [205], in many places of the World Ocean the spatially coherent signal of baroclinic tides at the free surface was detected more than 1000 km away from the source of their generation. Theoretical results obtained by St. Laurent and Garrett [225] confirm the idea that first-mode baroclinic tides radiate much of their energy over $O(1000$ km) distances.

[5] The mean spectra of horizontal wavenumbers are used extensively for the description of the spectra of internal waves. By comparing the Garrett–Munk model spectra with the mean spectra of frequencies and horizontal wavenumbers, we can describe the mean state of the field of internal waves in the investigated region.

4.6 Critical Froude numbers: excitement of unsteady lee waves

Quite different from the characteristics considered in Section 4.4 are those exhibited by waves when the nonlinearity of the wave process is relatively strong. As mentioned in Section 4.3, at critical Froude numbers not only baroclinic tides but also lee waves can be excited. Such a situation is presented in Figure 4.19, where time series showing the evolution of the density field during two tidal cycles are calculated for $\varepsilon_1 = 0.4$. In this run the value of the tidal flow Ψ_0 was equal to 60 $m^2\ s^{-1}$, and the appropriate Froude number, estimated for the shallow-water zone, was equal to 1.45. Fluid stratification was specified by formula (1.70) with $N_p = 0.015\ s^{-1}$, $H_p = 30$ m, $\Delta H_p = 35$ m. Other input parameters were $\varphi = 10°$, $H_1 = 300$ m, $H_3 = 100$ m, $2l = 10$ km; and $T = 12.4$ h is the tidal period. The values and directions of the barotropic tide are shown by the horizontal arrows.

During the first half of the tidal period, not shown in Figure 4.19, the barotropic current flows leftward and produces upward fluxes over the slope. These vertical flows elevate the isopycnals and generate two waves propagating in opposite directions, as can be seen in the two uppermost panels in Figure 4.19. During the second half of the tidal period ($0.5T < t < T$), the barotropic tidal flux changes its direction and now flows rightward. So, the disturbances moving leftward now propagate upstream, whereas deep-water waves (moving rightward) propagate downstream.

As can be seen from Figure 4.19, the position of the wave A_1 relative to the bottom is almost stationary during the ebb tidal phase (compare the wave position at the vertical section a–a at $0.625T < t < 0.825T$). Taking into account the existence of the background tidal flow, one can conclude that these disturbances, in fact, represent internal waves propagating upstream. Thus, these waves are trapped by the topography, and, as was mentioned above, can extract energy from the mean flow. As a result, their amplitudes increase with time. By the end of the first half of the tidal period, the amplitude of wave A_1 is more than 30 m.

During the next half of the tidal period ($T < t < 1.5T$) when the tidal flow changes again to the opposite direction, the generated wave A_1 is released and propagates leftward as a free wave from the source of generation. Estimation of the phase speed, obtained from Figure 4.19, without influence of the tidal flow, yields a magnitude of 0.33 m s^{-1}. This value is almost 1.5 times smaller than the maximum velocity of the tidal flux in the shallow-water zone. Thus, the generated internal waves are really trapped waves.

The influence of the background tidal flow on the propagation of the generated waves A_1 and B_1, manifested as accelerations and decelerations, is clearly seen during the second tidal cycle (compare the positions of the waves A_1 and B_1 during the time intervals $T < t < 2T$). Even a change of the direction of propagation of the wave A_1 at $1.625T < t < 1.875T$ can be observed when the downstream tidal flux

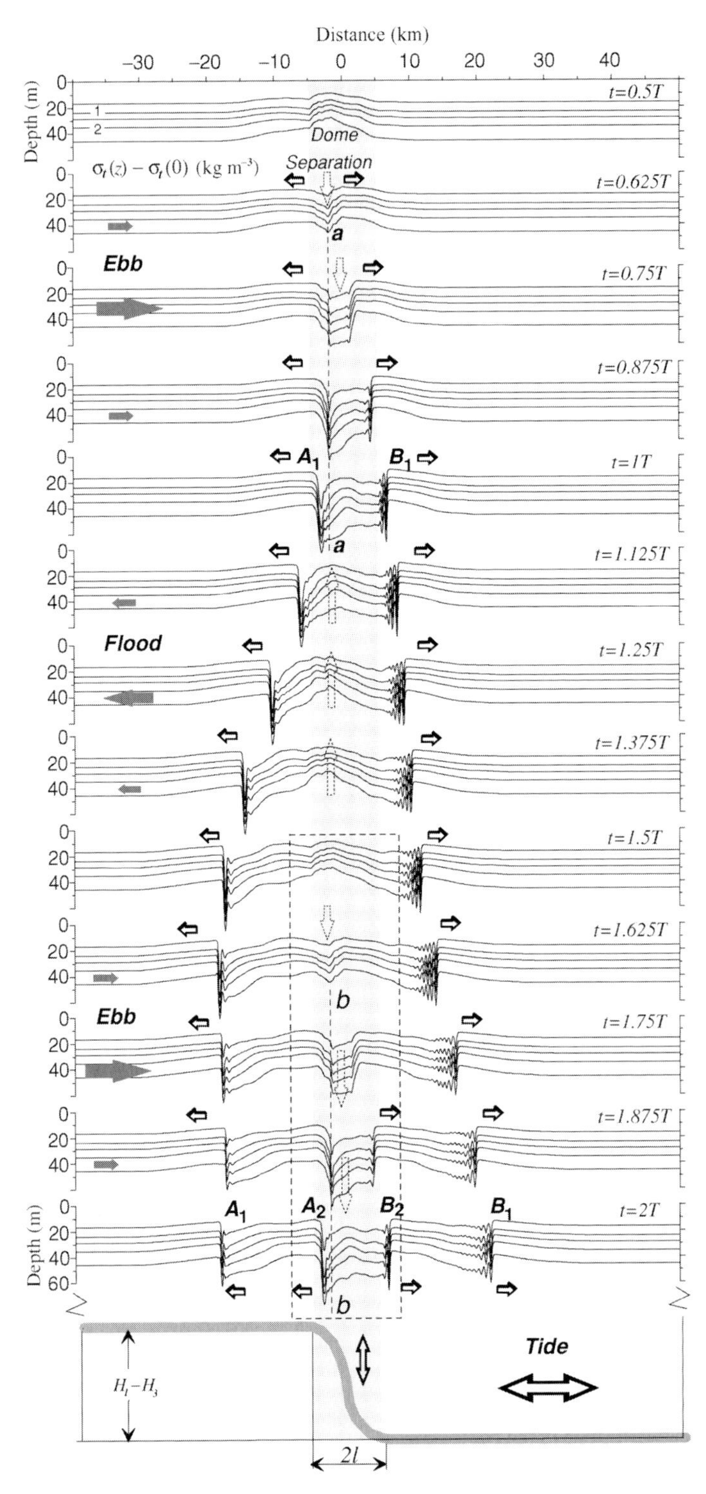

Figure 4.19. Evolution of the density field $\sigma_t(z) - \sigma_t(0)$ (kg m^{-3}) above the continental slope (3.47) caused by a tidal flux. Waves moving in opposite directions are marked by A_1, A_2 and B_1, B_2, respectively.

in the shallow-water zone is sufficiently strong to capture the upstream propagating wave. In this area, the local Froude number is larger than unity. The parameter kU_0/σ, estimated for the wave A_1, has the value 15 if one takes its length scales to be about 1 km; this is much larger than unity and thus these waves can be considered to be unsteady lee waves.

We conclude from Figure 4.19 that the intensive small-scale disturbances, generated above the slope and propagating shoreward, have the characteristics of unsteady lee waves rather than internal tides. As shown above by the qualitative analysis of equation (4.10), the spatial variability of the forcing is responsible for the generation of such waves and not its local time variation. In other words, the nonlinear advective terms in the governing equations (1.79) are much more effective for the generation of internal waves than the local time derivatives.

The described generation mechanism of unsteady lee waves is further clarified by Figure 4.20, in which the portion depicted by the dashed rectangle in Figure 4.19 is shown in greater detail. It is clearly seen that the wave A_2 generated during the ebb phase of the tide at the lee side of the slope does not propagate upstream when $1.625T < t < 1.875T$. During this time period its amplitude increases to almost 40 m (compare waves at section b–b in Figure 4.20). Only at the final stage of the ebb ($t > 1.875T$) when the tidal flux slackens does this wave start to propagate shoreward. At the same time, the waves propagating rightward are not trapped by the tidal flux. They leave the generation area and effectively evolve into a sequence of solitary waves according to the mechanism considered above. This is so because the velocity of the barotropic tidal flux in the deep part of the ocean is usually small, only several centimeters per second, and the subcritical regime of its generation exists here. One can recognize only weak accelerations and decelerations of the wave packet B_1 in Figure 4.19.

Thus, the considered examples demonstrate the manifestation of strong nonlinearities in the mechanism of the generation of internal tides. Figure 4.21 illustrates how large the difference of the generated fields can be between the linear and nonlinear cases. The interesting feature is that the distance between the two successive wave trains propagating to the deep part is equal to the wavelength of the first baroclinic tidal mode: the values of λ_3 and L_3 (see Figure 4.21) coincide reasonably well with one another, with an error of less than 3%. At the same time, the difference between the values λ_1 and L_1 in the shallow-water zone is more pronounced, reaching 20%. This discrepancy may be explained by the same fact as was found above when analyzing Figures 4.19 and 4.20: waves propagating seaward, to the right, have characteristics of nonlinear baroclinic tides because the Froude number in the deep part of the basin equals about 0.4, which is much less than unity. This is why these waves evolve according to the mechanism described above in the low-Froude-number regime; see Figure 4.9.

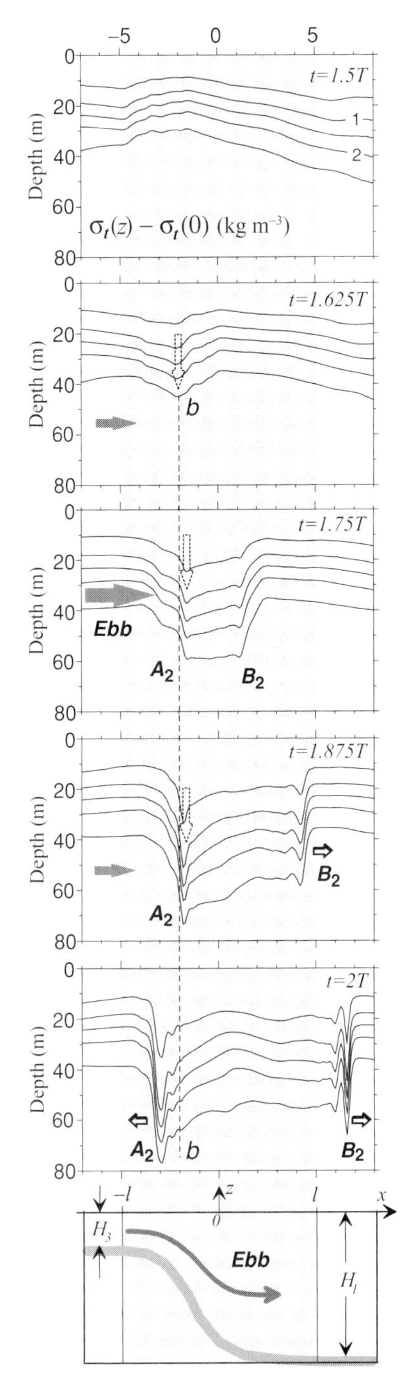

Figure 4.20. Portion marked in Figure 4.19 by the dashed rectangle. Formation of the baroclinic lee wave A_2, arrested by the tidal flux at the lee side of the bottom topography, is shown by the vertical dashed arrows. All model parameters are the same as in Figure 4.19.

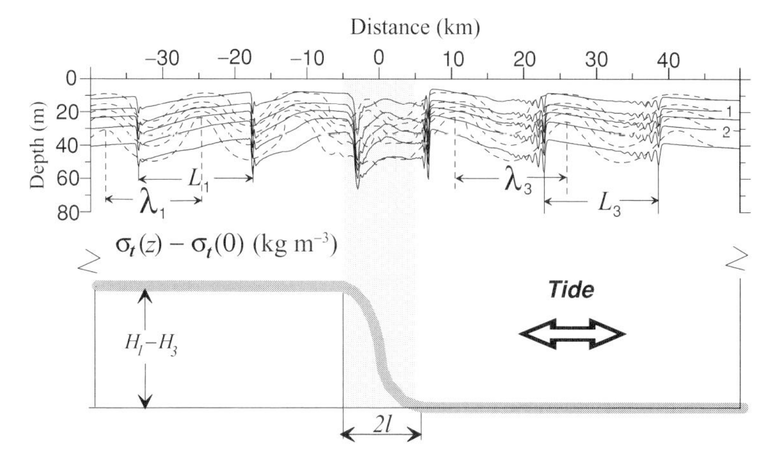

Figure 4.21. Density anomaly field $\sigma_t(z) - \sigma_t(0)$ (kg m^{-3}), calculated by using the linear (dashed lines) and nonlinear (solid lines) theories, after three tidal periods. All input parameters are the same as in Figure 4.19.

The waves propagating shoreward, to the left, exhibit the peculiarities inherent in lee waves, which are usually trapped by the flux at the lee side of the topography. For this reason, their phase speed must be related to the maximum velocity of the background flow but not to the phase speed of the baroclinic tide. Thus, as the wave regime at the shelf is critical (the Froude number is equal to unity), the distance L_1 between the two waves is much larger than the wavelength λ_1.

The next interesting finding from Figure 4.21 is that the waves generated in the shallow-water region are exposed to a dispersion not so remarkable as it is for deep-water waves. The wave trains here contain two to three waves, whereas the deep-water counterparts are well developed and consist of six to seven waves. This is not surprising if one takes into account the water depths in zones I and III. In fact, we encounter here a situation with two limiting cases: the shallow-water theory onto the shelf – in the limit without rotation it corresponds to nondispersive waves – and the finite-depth theory in zone III, where the dispersion effects are the usual phenomena.

5

Evolutionary stages of baroclinic tides

In this chapter we consider the evolutionary stage of baroclinic tides, i.e. the behavior of the tidally generated internal waves beyond the source of generation. During propagation, the long baroclinic tidal waves which are radiated from this area of generation – usually bottom features – are subjected to nonlinear effects, as shown in Chapter 4. If the nonlinearity is sufficiently strong, these waves are usually transformed into a sequence of solitary internal waves or wave trains. So, we shall now concentrate on the dynamic structure of solitary internal waves. First, we will give a brief summary of a number of analytical theories and consider the stationary solutions of weakly nonlinear equations. Then, we will move on to consider strong solitary internal waves; even though any analytical theory fails to describe their structure and dynamics, nevertheless they are a common feature of the real ocean. Using several numerical procedures, the governing equations can be handled and a physical understanding can be extracted. Using this approach, we will study the spatial–temporal structure of strong waves and indicate their differences from strict analytical solutions of equations describing weaker nonlinear interactions. Finally, we will present the wave transformation appropriate over variable bottom topography that includes strong effects such as wave overturning and breaking.

5.1 Analytical models for the evolution of baroclinic tides

Different weakly nonlinear theories of internal waves in a continuously stratified fluid are usually developed on the basis of the following equations:

$$\left.\begin{aligned}(\nabla^2\psi)_t + J(\nabla^2\psi, \psi) - f v_z &= b_x,\\ v_t + J(v, \psi) + f\psi_z &= 0,\\ b_t + J(b, \psi) + N^2(z)\psi_x &= 0,\end{aligned}\right\} \tag{5.1}$$

where $b = g\tilde{\rho}/\bar{\rho}_0$. This system can be derived from (1.79), (1.81) if the effects related to diffusion and dissipation are neglected. Sometimes this system can be justified on the sole ground that wave attenuation and water mixing are not of primary interest. The boundary conditions are $\psi = b = v_z = 0$ at $z = 0$ and $z = -H$.

We use the scales

$$\left.\begin{aligned} (x, z, t) &= \{\lambda\bar{x}, H\bar{z}, (\lambda/c_0)\bar{t}\}, \\ (\psi, b, v) &= \left\{(-a_\xi c_0)\bar{\psi}, (a_\xi c_0^2/H)\bar{b}, (a_\xi c_0/H)\bar{v}\right\}, \end{aligned}\right\} \tag{5.2}$$

in which the fluid depth, H, is used as a vertical scale and a typical horizontal wavelength, λ, represents the horizontal scale. Moreover, the horizontal convective time scale is taken to be λ/c_0, where $c_0 = HN_\mathrm{p}$ is the velocity scale for the baroclinic wave speed and N_p is the maximum value of the buoyancy frequency. Furthermore, a_ξ is a typical wave amplitude. The dimensionless variants of equations (5.1), after dropping bars, are

$$\left.\begin{aligned} \psi_{zzt} + \mu\psi_{xxt} + \varepsilon J(\psi_{zz}, \psi) + \varepsilon\mu J(\psi_{xx}, \psi) - (\lambda f/c_0)v_z &= b_x, \\ v_t + \varepsilon J(v, \psi) + (\lambda f/c_0)\psi_z &= 0, \\ b_t + \varepsilon J(b, \psi) + \bar{N}^2(z)\psi_x &= 0. \end{aligned}\right\} \tag{5.3}$$

Here, $\varepsilon = a_\xi/H$ is a dimensionless parameter measuring the wave amplitude, $\mu = (H/\lambda)^2$ is a measure of dispersion, and $(\lambda f/c_0)$ is the ratio of the wavelength to the Rossby radius of deformation, which defines the specific input of the rotational dispersion.

If the wave amplitude is small, $\varepsilon \ll 1$, and the lengths of the waves are relatively large, $\mu \ll 1$, an asymptotic expansion can be pursued and system (5.3) may be reduced to the Korteweg–de Vries (K–dV) equation, which is a first-order equation; this transformation was originally performed by Benney [20] for the nonrotational case (later, Lee and Beardsley [134] extended this perturbation approach to second order). Following these references, the separable asymptotic solution for $f = 0$, $\varepsilon \ll 1$, and $\mu \ll 1$, can be shown to be expressible as

$$\begin{aligned} \psi(x, z, t) = {} & A(x, t)\Phi(z) + \varepsilon A^2(x, t)\Phi^{(1,0)}(z) + \mu A_{xx}(x, t)\Phi^{(0,1)}(z) \\ & + \varepsilon\mu\left[\left(A(x, t)A_{xx}(x, t) - \tfrac{1}{2}A_x^2(x, t)\right)\Phi_a^{(1,1)}(z) + \tfrac{1}{2}A_x^2(x, t)\Phi_b^{(1,1)}(z)\right] \\ & + \varepsilon^2 A^3(x, t)\Phi^{(2,0)}(z) + \mu^2 A_{xxxx}(x, t)\Phi^{(0,2)}(z) + \cdots, \end{aligned} \tag{5.4}$$

$$\begin{aligned} b(x, z, t) = {} & A(x, t)\frac{N^2(z)}{c}\Phi(z) + \varepsilon A^2(x, t)B^{(1,0)}(z) + \mu A_{xx}(x, t)B^{(0,1)}(z) \\ & + \varepsilon\mu\left[\left(A(x, t)A_{xx}(x, t) - \tfrac{1}{2}A_x^2(x, t)\right)B_a^{(1,1)}(z) + \tfrac{1}{2}A_x^2(x, t)B_b^{(1,1)}(z)\right] \\ & + \varepsilon^2 A^3(x, t)B^{(2,0)}(z) + \mu^2 A_{xxxx}(x, t)B^{(0,2)}(z) + \cdots, \end{aligned} \tag{5.5}$$

where $A(x, t)$ is the wave profile. Superscripts (n, m) indicate that this term represents the $O(\varepsilon^n, \mu^m)$ term of the expansion. To second order, the term related to $O(\varepsilon\mu)$ consist of two parts, denoted by subscripts a and b. The linear eigenvalue problem for $\Phi(z)$,

$$\Phi_{zz} + \frac{N(z)^2}{c^2}\Phi = 0, \tag{5.6}$$

$$\Phi(-H) = \Phi(0) = 0, \tag{5.7}$$

defines a set of solutions $(\Phi_j(z), c_j)$, where j is the number of the baroclinic mode.

The high-order corrections of the vertical structure of internal waves are defined by functions $\Phi^{(n,m)}(z)$, $\Phi_a^{(n,m)}(z)$, $\Phi_b^{(n,m)}(z)$, $B^{(n,m)}(z)$, $B_a^{(n,m)}(z)$, $B_b^{(n,m)}(z)$. They are the solutions of an inhomogeneous BVP, essentially equations (5.6) and (5.7), where the right-hand side of (5.6) is a known function from the solution of the lower-order problems (for more details, see ref. [130]).

If only the first-order terms in the expansions (5.4) and (5.5) are retained, then for $f = 0$, $v = 0$, and with an accuracy of $O(\varepsilon, \mu)$, the K–dV type equation

$$A_t + cA_x + \alpha_1 AA_x + \beta A_{xxx} = 0 \tag{5.8}$$

can be derived from (5.3), which describes all possible unidirectional waves. The parameters α_1, β, and c are coefficients of nonlinearity, dispersion, and linear phase speed of long internal waves, respectively. In the Boussinesq approximation, α_1 and β are determined as

$$\left.\begin{aligned} \alpha_1 &= \frac{3}{2}c\frac{\int_{-H}^{0}(d\Phi/dz)^3\,dz}{\int_{-H}^{0}(d\Phi/dz)^2\,dz}, \\ \beta &= \frac{c}{2}\frac{\int_{-H}^{0}\Phi^2\,dz}{\int_{-H}^{0}(d\Phi/dz)^2\,dz}. \end{aligned}\right\} \tag{5.9}$$

In the more common case when the amplitudes of the internal waves are relatively large, the second-order terms in expansions (5.4) and (5.5) can introduce an essential correction to the first-order solution.

Generally speaking, the contributions of the various higher-order terms depend on the model parameters, and sometimes several of them may be equally important. For instance, in a two-layer model of fluid stratification, the coefficient of quadratic nonlinearity, α_1, is determined by the pycnocline location: it is negative when the pycnocline lies near the free surface, but positive when it is closer to the bottom, see, e.g., ref. [53]. Therefore, when the pycnocline location is near or at the mid-depth, the value of the coefficient α_1 is very small or zero, and the next cubic nonlinearity terms become the major source of nonlinearity. It is evident that in such zones, called "turning points," the quadratic and cubic nonlinearities may be comparable and have the same influence as the linear terms. To take into account this effect, the second-order shallow-water theories leading to the so-called extended Korteweg–de Vries equation (eK–dV) were built. For the two-layer stratification see, for instance, refs. [53], [92], [114], [155], and the more recent publications [39] and [182]; for a continuously stratified fluid, the appropriate theories are described, e.g., in refs. [72], [83], [84], [103] and [130].

Omitting all intermediate computations, we can write here the final form of the eK–dV equation as

$$A_t + cA_x + \alpha_1 AA_x + \alpha_2 A^2 A_x + \beta A_{xxx} = 0, \tag{5.10}$$

in which the coefficient of cubic nonlinearity is given by [103]

$$\alpha_2 = -\frac{\alpha_1^2}{c} + 3c\frac{\int\limits_{-H}^{0} \left\{\left[\Phi_z^4 - (N\Phi/c)^4\right] + \left[c\Phi_z^2 + N^2\Phi^2/c - 2\alpha_1\Phi_z/3\right]\mathcal{Q}_z\right\} dz}{\int\limits_{-H}^{0} \Phi_z^2\, dz}. \tag{5.11}$$

Here, the function $\mathcal{Q}(z)$ is the solution of the following boundary value problem:

$$\left.\begin{aligned} \mathcal{Q}_{zz} + (N/c)^2\mathcal{Q} &= (\alpha_1 N^2/c^4)\Phi + (\mathcal{Q}_z^2/c^3)\Phi^2, \\ \mathcal{Q}(-H) &= \mathcal{Q}(0) = 0\,, \end{aligned}\right\} \tag{5.12}$$

in which Φ is given by (5.6), (5.7).

The eK–dV equation (5.10) is integrable and can describe various nonlinear waves, such as solitons, breathers, or dissipationless shock waves. It is valid to second order in the wave amplitude and to first order in the wavelength, and can be applied for an ocean of constant depth when rotational effects are negligible.

For a continuously stratified fluid, rotation was included for the first time in the first-order nonlinear theory of internal waves by Ostrovsky [181], but only as a "small" effect (the frequency of the considered waves was much larger than the Coriolis parameter). This assumption allowed Ostrovsky to derive a wave-type evolution equation, now called the "Ostrovsky equation." This equation was subsequently obtained in a different way by Leonov [135], who showed that strictly form-preserving solitons do not exist for this equation, and Grimshaw [82] extended it to include transverse variations.

However, soliton solutions can be obtained by intermediate asymptotics in the process of the wave evolution. As shown by Gerkema and Zimmermann for two-layer stratification [73], [75], rotation influences the nonlinear internal wave evolution by reducing the number of generated solitons. Solitary waves of preserved form exist because of the balance between advective nonlinearities and dispersion. Rotation increases the wave dispersion, and thus the balance with the nonlinearity is modified. This leads to a reduction of the number of solitons. This effect will be considered in Chapter 6.

Holloway *et al.* [102] incorporated the Coriolis parameter into the second-order theory; their so-called rotated-modified-extended Korteweg–de Vries equation has the following form:

$$\left(A_t + cA_x + \alpha_1 AA_x + \alpha_2 A^2 A_x + \beta A_{xxx}\right)_x = Af^2/2c. \tag{5.13}$$

The next important factor influencing the oceanic internal wave evolution is the variability of the bottom topography. The effect of slowly varying depth can be accounted for by inserting an additional term, $\mathscr{S}$, into (5.13) [102] [272], leading to

$$\left(A_t + cA_x + \alpha_1 AA_x + \alpha_2 A^2 A_x + \beta A_{xxx} - (c/\mathscr{S})\mathscr{S}_x A\right)_x = Af^2/2c, \tag{5.14}$$

where

$$\mathscr{S} = \left(\frac{c_{\text{ref}}^3 \int_{-H}^0 (d\Phi_{\text{ref}}/dz)^2\, dz}{c^3 \int_{-H}^0 (d\Phi/dz)^2\, dz}\right)^{1/2}, \tag{5.15}$$

in which Φ is again a solution of (5.6), (5.7). The subscript "ref" indicates that the values of c_{ref} and Φ_{ref} are known for some initial reference point x_{ref}. In general, the coefficients of nonlinearity and dispersion α_1, α_2, and β in equation (5.14) are not constant but depend on the total ocean depth. Moreover, depending on the water depth and stratification, the values of the coefficients α_1, α_2 can be positive or negative. Qualitative estimation of soliton dynamics on the basis of equation (5.14) is not evident, because usually the coefficients α_1, α_2, and β can be calculated only numerically.

However, there are several important particular laws of fluid stratification for which the BVP (5.6), (5.7) has analytical solutions. The simplest one is a two-layer fluid stratification with water densities ρ_+, ρ_- and thicknesses h_+, h_- for the upper and lower layers, respectively. In this case the phase speed c and parameters β, α_1, and α_2 are expressible by the following formulas:

$$\left.\begin{aligned} c &= \left(\frac{g(\rho_- - \rho_+)}{\rho_-}\frac{h_- h_+}{h_- + h_+}\right)^{1/2}, \quad \beta = c\frac{h_- h_+}{6}, \\ \alpha_1 &= \frac{3c}{2}\frac{(h_- - h_+)}{h_- h_+}, \quad \alpha_2 = -\frac{3c}{8h_-^2 h_+^2}(h_-^2 + h_+^2 + 6h_- h_+). \end{aligned}\right\} \tag{5.16}$$

It is seen that the value of α_2 is always negative while the value of α_1 may be of either sign, depending on the location of the interface of the layers.

The BVP (5.6), (5.7) can also be solved analytically for some other interesting particular cases of fluid stratification including three-layer models with constant density in the upper and lower layers and piecewise linear or smoothed sigmoidal density change between the two [83], [105].

In the more common case of a continuously stratified fluid the analytical solution can be found for the following smoothed form of the pycnocline [245], [256]:

$$N(z) = N_p \left(\left[\frac{2(z + H_p)}{\Delta H_p} \right]^2 + 1 \right)^{-1} . \tag{5.17}$$

Here, H_p represents the depth where the Brunt–Väisälä frequency has its maximum value N_p, and ΔH_p is a vertical length scale characterizing the variation of the buoyancy frequency; see Figure 5.1. This law of fluid stratification is very similar to those considered in Section 1.5 for baroclinic tides; see formula (1.70) and Figure 1.6.

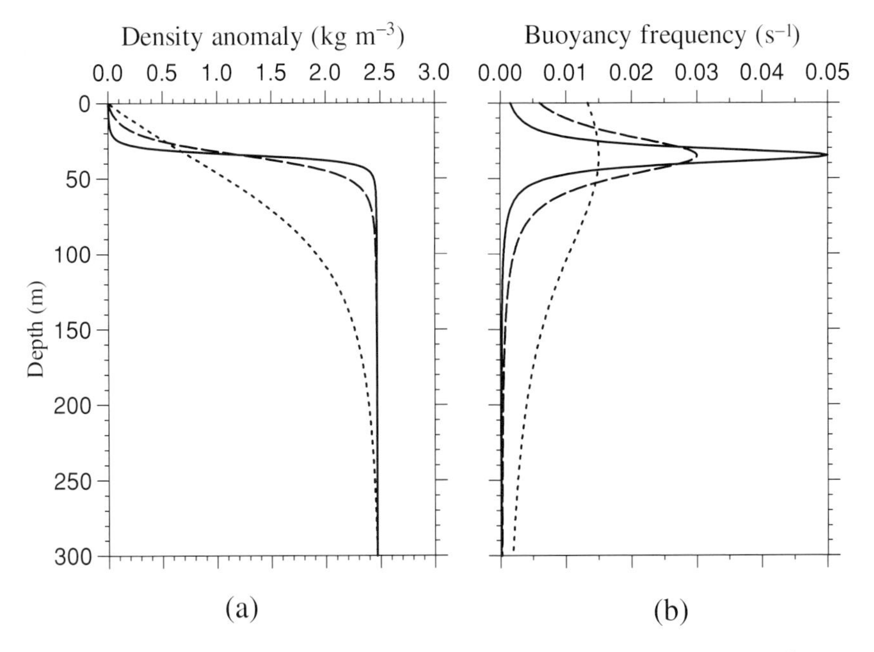

Figure 5.1. Vertical profiles of (a) density anomalies $\rho(z) - \rho(0)$ (kg m^{-3}) and (b) buoyancy frequency $N(z)$ (s^{-1}) characterizing the three density stratifications (type 1, solid line; type 2, short-dashed line; type 3, dashed line). The buoyancy frequency profiles were computed using equation (5.17) with the three sets of parameters listed in Table 5.1, and the water depth was chosen to be 300 m.

Table 5.1. *Values of the parameters characterizing three types of water stratifications given by formula (5.17).*

Type of stratification[a]	N_p (s^{-1})	H_p (m)	ΔH_p (m)
1 ———————	0.05	35	12.35
2 – – – – – – – – – –	0.03	35	35
3 ----------------------	0.015	35	196.5

[a] Solid, dashed, and short-dashed lines refer to Figure 5.1.

These stratifications represent typical water-mass distributions on a continental shelf, characterized by the presence of a seasonal pycnocline of different strength. Note that the three different density profiles plotted in Figure 5.1(a) have the same values of the density at the surface and at the bottom.

The advantage of this law of fluid stratification over piecewise linear or sigmoidal density profiles, often used in theory and in practice for the description of pycnocline spreading [123], is that it is more general, yet allows a relatively simple analytical solution of the BVP (5.6), (5.7). For the stratification (5.17), the eigenfunctions $\Phi_j(z)$ are given by

$$\Phi_j(z) = \left[c_a(z/H + c_b)^2 + 1\right]^{1/2} \sin(j\pi z_3), \tag{5.18}$$

and the eigenvalues c_j of the jth baroclinic mode listed in (5.19) possess the simple square root trigonometric form [245], [256]:

$$c_j = N_p H \left[(j\pi/h)^2 - c_a\right]^{-1/2}. \tag{5.19}$$

Here

$$\left.\begin{aligned} c_a &= (2H/\Delta H_p)^2, \quad c_b = H_p/H, \\ z_3 &= [\arctan (c_a)^{1/2} (z/H + c_b) - \arctan (c_a)^{1/2}]/ (c_a)^{1/2} h, \\ h &= [\arctan (c_a)^{1/2} (c_b - 1) - \arctan (c_a)^{1/2}]/c_a. \end{aligned}\right\} \tag{5.20}$$

5.2 Solitary internal waves as manifestations of the coherent structure of baroclinic tides

The weakly nonlinear equations presented above describe the evolutionary stage of baroclinic tides. They are useful when studying the transformation of a baroclinic tidal bore into a series of relatively short nonlinear internal waves. The specific feature of these tidally produced short internal waves is their ability to propagate very long distances, i.e. several hundreds of kilometers [5], [24], [78], [143]. Thus, such shape-preserving waves very likely possess the structure of steady state solutions of weakly nonlinear equations.

To find a stationary solution of the K–dV equation (5.8), we employ coordinates that move with the phase speed c_p ($\theta = x - c_p t$, $z = z$). The wave form A in this system depends only on the variable $\theta = x - c_p t$, and (5.8) can be rewritten as

$$(c - c_p)A' + \frac{\alpha_1}{2}(A^2)' + \beta A''' = 0, \tag{5.21}$$

where the primes denote derivatives with respect to θ, and $c_p - c$ is the nonlinear correction of the phase speed. This equation can be integrated and rewritten in the form

$$(c - c_p)A + \frac{\alpha_1}{2}A^2 + \beta A'' = 0. \tag{5.22}$$

The constant of integration was assumed to be zero. This follows because, at infinity, A, A', and A'' all vanish, which holds provided wave motions in "quiet" water are considered without any background noise.

If (5.22) is multiplied by A', the first integral

$$\frac{3\beta}{\alpha_1}\left(\frac{dA}{d\theta}\right)^2 = -\left[A^3 + \frac{3(c - c_p)}{\alpha_1}A^2 - E_0\right] \tag{5.23}$$

can be constructed, where E_0 is the constant of integration, the value of which is related to the energy (amplitude) of the internal wave.

Solutions of equation (5.23) represent solitary internal waves or periodic cnoidal waves, depending on the roots of the equation $A^3 + [3(c - c_p)/\alpha_1]A^2 - E_0 = 0$. The exact analytical solution of (5.22) describing an internal soliton has the form

$$A(\theta) = \frac{3(c_p - c)}{\alpha_1}\left\{\cosh\left[\left(\frac{c_p - c}{4\beta}\right)^{1/2}\theta\right]\right\}^{-2}. \tag{5.24}$$

The amplitude a_ξ of the wave profile in (5.24) is equal to $3(c_p - c)/\alpha_1$. This allows immediate determination of the relation between the nonlinear phase speed and the wave amplitude:

$$c_p = c + \frac{\alpha_1 a_\xi}{3}. \tag{5.25}$$

Taking into account (5.25), the multiplier $\left[(c_p - c)/4\beta\right]^{1/2}$ in the argument of the cosh function of (5.24) can be rewritten as $\left[(\alpha_1 a_\xi)/12\beta\right]^{1/2}$. The latter also defines the soliton width λ as follows:

$$\lambda = \left(\frac{12\beta}{\alpha_1 a_\xi} \right)^{1/2}. \tag{5.26}$$

With such identifications the soliton profile takes the final form

$$A(x, t) = a_\xi \cosh^{-2} \left(\frac{x - c_\mathrm{p} t}{\lambda} \right). \tag{5.27}$$

As can be seen from (5.25) and (5.26), the phase speed of a soliton increases with growing amplitude, but at the same time it becomes shorter. In the simpler case of a two-layered fluid, the coefficients of nonlinearity and dispersion, α_1 and β, are explicitly expressed by (5.16), so that c_p and λ take the forms

$$\left. \begin{aligned} c_\mathrm{p} &= c \left[1 + \frac{a_\xi}{2} \frac{(h_- - h_+)}{h_- h_+} \right], \\ \lambda &= \left[\frac{4}{3a_\xi} \frac{(h_- h_+)^2}{(h_- - h_+)} \right]^{1/2}. \end{aligned} \right\} \tag{5.28}$$

It is clear that this two-layer, first-order theory for internal solitons is invalid if the density jump is located exactly in the middle between the bottom and the free surface, i.e. when $h_- = h_+$ ("turning point"). In this case, the width of the soliton would be infinite.

The next extreme case is linear density stratification with constant value of the buoyancy frequency N_0. In this case the eigenfunction Φ of the BVP (5.6), (5.7) has sinusoidal form, and thus $\alpha_1 = 0$ in (5.9). Then from (5.25) and (5.26) we have

$$c_\mathrm{p} = c = N_0 H/\pi, \quad \lambda \to \infty. \tag{5.29}$$

For this case, a correct account of the nonlinear effects can be accomplished only by the inclusion of cubic nonlinear terms.

The importance of these standard solutions is evident by virtue of the fact that, according to the current understanding of stratified fluid dynamics, many initial disturbances ultimately tend to a train of solitary waves. However, the procedure described above is not flawless. One of the major shortcomings is the restriction of values of the parameters of nonlinearity ε and dispersion μ. The requirement $\varepsilon, \mu \ll 1$ is invalid for many solitary waves that were measured in the ocean. An attempt to broaden the range of applicability for such waves under consideration was described above, and it demands that approximations of (5.4) and (5.5) are taken into account which are higher than the first-order expansions. This, in turn, leads to considerable mathematical difficulties, the major one being the sophisticated law of vertical fluid stratification, which requires a numerical solution of the BVP (5.6), (5.7), as well as a numerical estimation of the nonlinear and dispersion coefficients.

Such numerical procedures for the solution of the evolution equation (5.10), (5.13) or (5.14) were developed and applied, for instance, in refs. [103] and [130].

The situation becomes somewhat simpler if we consider stationary solutions of solitary waves, which preserve their form due to the balance between the nonlinear and dispersive terms. In this case, we can assume that $\varepsilon = \mu$. Moreover, for stationary waves propagating in a nonrotating ocean with phase speed c_{p}, the governing system (5.1) can be reduced to a single equation. This was demonstrated by Long [157].

5.2.1 Long's equation

The system of equations describing internal waves in a nonrotating ocean of constant depth has the form

$$(\nabla^2\psi)_t + J(\nabla^2\psi, \psi) = g\rho_x/\bar{\rho}_0, \quad \rho_t + J(\rho, \psi) = 0. \tag{5.30}$$

Here the full density ρ is used as a variable. Assuming that the stream function, ψ, and the density, ρ, in stationary waves moving with the phase speed c_{p} depend only on two variables, namely z and $\theta = x - c_{\mathrm{p}}t$,

$$\psi = \psi(\theta, z), \quad \rho = \rho(\theta, z), \tag{5.31}$$

we can rewrite the second equation (5.30) as

$$-c_{\mathrm{p}}\rho_\theta + \psi_z\rho_\theta - \psi_\theta\rho_z = 0. \tag{5.32}$$

Equation (5.32) transforms to the equation

$$J(\rho, \Theta) \stackrel{\mathrm{def}}{=} \begin{pmatrix} \rho_\theta & \rho_z \\ \Theta_\theta & \Theta_z \end{pmatrix} = 0, \tag{5.33}$$

if the new, modified stream function $\Theta = \psi - c_{\mathrm{p}}z$ is intoduced. It requires that ρ is an arbitrary function of Θ,

$$\rho = \rho(\Theta) = \rho(\psi - c_{\mathrm{p}}z). \tag{5.34}$$

Substituting (5.34) into the first equation of (5.30) and taking into account that $g\Theta_\theta = -J(gz, \Theta)$, the following equation can be deduced:

$$J(\Delta\Theta, \Theta) + \vartheta'(\Theta)J(gz, \Theta) = 0,$$

where $\vartheta'(\Theta) = \rho_\Theta/\bar{\rho}_0$, and the prime denotes differentiation with respect to Θ and $\Delta = \partial^2/\partial\theta^2 + \partial^2/\partial z^2$. The preceding equation can also be rewritten as

$$J\left\{\Delta\Theta + \vartheta'\Theta gz, \Theta\right\} = 0$$

or, alternatively, as

$$\Delta\Theta + \vartheta'\Theta g z = \Phi_0'(\Theta), \tag{5.35}$$

where $\Phi_0'(\Theta)$ is an arbitrary function. Returning to the initial variables, $\psi = \Theta + c_p z$, (5.35) takes the form

$$\Delta\psi + \vartheta'(\psi - c_p z) g z = \Phi_0'(\psi - c_p z). \tag{5.36}$$

This equation contains two functions, $\vartheta(\Theta)$ and $\Phi_0(\Theta)$, which make its analysis not so obvious. It can, however, be simplified if waves propagating in an undisturbed fluid only are considered. To this end, we rewrite equation (5.36) in terms of the new independent variable $s = z - \psi/c_p$ (instead of $\Theta = \psi - c_p z$) and we take into account the following relations:

$$\left\{\vartheta', \Phi_0'\right\} = \frac{d\left\{\vartheta, \Phi_0\right\}}{d\Theta} = \frac{d\left\{\vartheta, \Phi_0\right\}}{d(\psi - c_p z)} = \frac{1}{c_p}\frac{d\left\{\vartheta, \Phi_0\right\}}{d(z - \psi/c_p)}.$$

For undisturbed conditions at infinity ($\Delta\psi|_\infty = 0$), we then find

$$\frac{gz}{c_p}\frac{d\vartheta}{d(z - \psi/c_p)} = -\frac{1}{c_p}\frac{d\Phi_0}{d(z - \psi/c_p)}. \tag{5.37}$$

Since $\vartheta'(\Theta) = \rho_\Theta/\bar{\rho}_0 = -N^2(\Theta)/g$, we immediately obtain $d\Phi_0/dz = -zN^2(z)$. Substituting these formulas into (5.36), we obtain the following equation:

$$\Delta\psi + N^2\left(z - \frac{\psi}{c_p}\right)\frac{\psi}{c_p^2} = 0. \tag{5.38}$$

This is Long's equation.

5.2.2 First-order weakly nonlinear theory

To construct the theory of solitary internal waves (SIWs) we use Long's equation (5.38) along with the "rigid lid" assumption and the streamline boundary conditions at the free surface and at the bottom, respectively,

$$\psi(z = 0) = \psi(z = -H) = 0. \tag{5.39}$$

Introducing the dimensionless variables

$$\zeta = (x - c_p t)/\lambda, \quad \eta = z/H, \quad F = \psi/(Hc_p), \tag{5.40}$$

equation (5.38) takes the form

$$\mu F_{\zeta\zeta} + F_{\eta\eta} + \varkappa\bar{N}^2(\eta - F)F = 0, \tag{5.41}$$

where $\mu = (H/\lambda)^2$, $\varkappa = (HN_p/c_p)^2$, and $\bar{N}$ is the dimensionless buoyancy frequency. Suppose that the parameter μ is small; then the solution of equation (5.38) can be written in the form of the asymptotic expansions [136]

$$F = \mu F_1 + \mu^2 F_2 + \cdots, \qquad \varkappa = \varkappa_0 + \mu\varkappa_1 + \mu^2\varkappa_2 + \cdots. \tag{5.42}$$

As was already noted for stationary internal waves, the nonlinear wave steepening is compensated for by dispersion, $a_\xi/H = (H/\lambda)^2$, and expansions (5.42) are valid for weak nonlinearity and weak dispersion. If $a_\xi/H \gg (H/\lambda)^2$, then overturning of internal waves will take place. Alternatively, when $a_\xi/H \ll (H/\lambda)^2$, the linear theory is valid.

Expanding the function $\bar{N}^2(\eta - F) = \bar{N}^2[\eta - (\mu F_1 + \mu^2 F_2 + \cdots)]$ into a Taylor series near the point $\mu = 0$, i.e.

$$\bar{N}^2(\eta - F) = \bar{N}^2(\eta) - \frac{d\left[\bar{N}^2(\eta)\right]}{d\eta}F + \frac{1}{2!}\frac{d^2\left[\bar{N}^2(\eta)\right]}{d\eta^2}F^2 + \cdots,$$

and substituting this, as well as (5.42), into (5.41), we find the first-order and second-order problems for the determination of F_1 and F_2.

- **To first order**

$$\left.\begin{aligned} F_{1\eta\eta} + \varkappa_0 F_1 \bar{N}^2(\eta) = 0, \\ F_1(\eta = 0) = F_1(\eta = -1) = 0. \end{aligned}\right\} \tag{5.43}$$

- **To second order**

$$\left.\begin{aligned} F_{2\eta\eta} + \varkappa_0 F_2 \bar{N}^2(\eta) = -\left(F_{1\zeta\zeta} + \varkappa_1 F_1 \bar{N}^2 - \varkappa_0 F_1^2 \frac{d\bar{N}^2}{d\eta}\right), \\ F_2(\eta = 0) = F_2(\eta = -1) = 0. \end{aligned}\right\} \tag{5.44}$$

Problem (5.43) admits a solution of the form

$$F_1(\zeta, \eta) = W(\eta)U_1(\zeta), \tag{5.45}$$

in which $W(\eta)$ must solve the eigenvalue problem

$$\left.\begin{aligned} d^2W/d\eta^2 + \varkappa_0 W \bar{N}^2(\eta) = 0, \\ W(\eta = 0) = W(\eta = -1) = 0. \end{aligned}\right\} \tag{5.46}$$

The homogeneous form of (5.44) is identical to (5.43). Therefore, to obtain a bounded solution of problem (5.44), the function $W_j(\eta)$ must be orthogonal to the right-hand side of (5.44) (see, e.g., ref. [80]),

$$\int_{-1}^{0} W_j(\eta)\left(F_{1\zeta\zeta} + \varkappa_1 F_1 \bar{N}^2 - \varkappa_0 F_1^2 \frac{d\bar{N}^2}{d\eta}\right) d\eta = 0.$$

This condition leads to the stationary K–dV equation for $U_1(\zeta)$,

$$d^2U_1/d\zeta^2 + \nu_1 U_1^2 + \nu_2 \varkappa_1 U_1 = 0, \tag{5.47}$$

where

$$\left.\begin{aligned} \nu_1 &= -\varkappa_0 \int_{-1}^{0} W_j^3 \frac{d\bar{N}^2}{d\eta} d\eta, \\ \nu_2 &= \int_{-1}^{0} W_j^2 \bar{N}^2 d\eta, \end{aligned}\right\} \tag{5.48}$$

and $W_j(\eta)$ is normalized in such a way that

$$\int_{-1}^{0} W_j^2 \, d\eta = 1. \tag{5.49}$$

If the buoyancy frequency $N(z)$ has an arbitrary form, the vertical structure of the soliton $W(\eta)$ and the coefficients ν_1 and ν_2 of equation (5.47) must be determined numerically. However, we can obtain the solution of problem (5.46) analytically if $N(z)$, given by formula (5.17), is used. In this case, the solution has the form [245]

$$\left.\begin{aligned} W_j(\eta) &= A_0 \left[c_a(\eta + c_b)^2 + 1\right]^{1/2} \cdot \sin(j\pi z_3), \\ \varkappa_0 &= (j\pi/h)^2 - c_a. \end{aligned}\right\} \tag{5.50}$$

Here j is the identifier of the mode, A_0 is a constant which can be determined from condition (5.49), and the parameters c_a, c_b, h, and the function $z_3 = z_3(z)$ are the same as in (5.20).

The existence of an analytical solution of problem (5.46) allows us to find from (5.48) the explicit expressions for the coefficients ν_1 and ν_2. After simple mathematical transformations, we obtain

$$\nu_1 = \frac{96H^2(j\pi A_0)^3 \left[\sin\mathscr{R} - \sin(\mathscr{P} + \mathscr{R} + j\pi)\right]}{\Delta H_{\mathrm{p}}^2 \mathscr{P}\left[\mathscr{P}^2 - (3j\pi)^2\right]},$$

$$\nu_2 = A_0^2 \mathscr{P}/(2c_a^{1/2}),$$

where

$$\mathscr{R} = \arctan(c_a^{1/2}), \qquad \mathscr{P} = \arctan\left[c_a^{1/2}(c_b - 1)\right] - \mathscr{R}. \tag{5.51}$$

The solution of equation (5.47), $U(\zeta)$ can be found in the same way as for the K–dV equation (5.22). This will not be repeated here. In summary, and with (5.45), we can write the first-order solution of the weakly nonlinear problem (5.43) in the form

$$\left.\begin{aligned}\psi_1(x,z) &= -a_\xi c_p U_1(x,z) W_j(z),\\ U_1(x,t) &= \cosh^{-2}\left[(x-c_p t)/\lambda\right],\\ W_j(z) &= A_0\left[c_a(z/H+c_b)^2+1\right]^{1/2}\sin(j\pi z_3),\end{aligned}\right\}\tag{5.52}$$

in which the phase speed, c_p, and the wavelength, λ, are defined by

$$\left.\begin{aligned}c_p &= HN_p/(\varkappa_0+2a_\xi\nu_1/3\nu_2 H)^{1/2},\\ \lambda &= (-6H^3/a_\xi\nu_1)^{1/2}.\end{aligned}\right\}\tag{5.53}$$

It is seen that the speed of the soliton and its length depend on the amplitude in a qualitatively expected way: the wavelength decreases with growing amplitude as $\sim a_\xi^{-1/2}$, while the phase speed grows due to the negative value of ν_1.

Now let us use the first-order weakly nonlinear theory of solitary internal waves for the prediction of the parameters of solitons and comparison with real oceanic solitary waves. The observational data, collected in the New York Bight during the JUSREX-92 campaign (Section 4.1), will be used for this purpose. The amplitudes of the registered solitary internal waves were in this case about 8 m. Taking into account that the depth H on a shelf is 80 m, the nonlinearity parameter, $\varepsilon = a_\xi/H$, was close to 0.1.

The measurements of the density fields, showing the wave trains, are summarized in Figure 5.2. The towings a, b, and c as indicated in Figure 4.1 have the following designations: the first letter J designates the month July, the next two numbers denote the day of the month, and the last letter identifies the ship track. To show the episodes which look like solitons, additional numbers are used in Figure 5.2.

Cross-sections of the density fields of the SIWs J15c, J21c-1,2,3,4, J24b-1,2, and J24c-1,2 are shown in Figure 5.3, which shows the J15c, J21c-1,2,3,4, J24b-1,2, and J24c-1,2 waves from Figure 5.2 obtained from JUSREX-92 with superimposed solitons, constructed with the help of the first-order classical K–dV theory (5.52).

Inspection of Figure 5.3 leads to the conclusion that the wave profiles of the SIWs measured in the New York Bight are qualitatively in good agreement with the theoretical curves; no larger quantitative divergence can be discerned between them. Existing differences can be explained by the apparently chaotic background of the internal waves arising at the shelf with amplitudes of 1–3 m (see Figure 5.2). The agreement between the measured and calculated solitons is even more explicitly seen in Figure 5.4. All measured vertical profiles of the horizontal velocity of the solitons were normalized with the value of their smoothed amplitude.

Let us analyze now the data of the vertical wave velocities registered at the moorings during the passage of the waves. The time series $w(t)$ of the six SIWs measured at the 29 m depth at mooring M3 (see Figure 4.1) are shown in

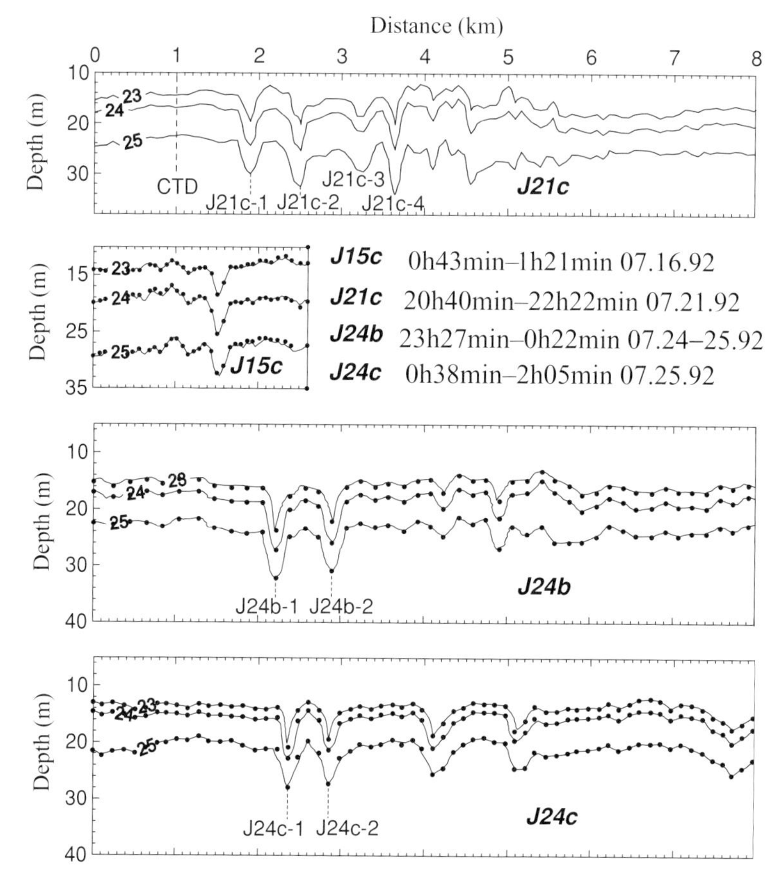

Figure 5.2. Isopycnal displacements for solitary internal waves measured on four transects in the New York Bight during the JUSREX-92 campaign; see also Figure 4.1.

Figure 5.5. They are all normalized by the maximum value of the computed vertical orbital velocity. The time series of the vertical velocity of the theoretical soliton is shown by the thick line. The observational curves $w(t)$ have the same characteristic shape as the K–dV counterpart; they lie overall close to the theoretical curve.

Thus, the inference which can be drawn from a judicious comparison of Figures 5.3, 5.4, and 5.5 is that the measured solitary wave profiles of both horizontal and vertical velocities, as well as their time series, are in excellent agreement with the first-order K–dV theory. Such excellent agreement occurs when the wave amplitude, i.e. the parameter of the nonlinearity, ε, is relatively small (we considered solitary waves, for which $\varepsilon < 0.1$). The situation will become much more complicated for strong waves and greater values of ε.

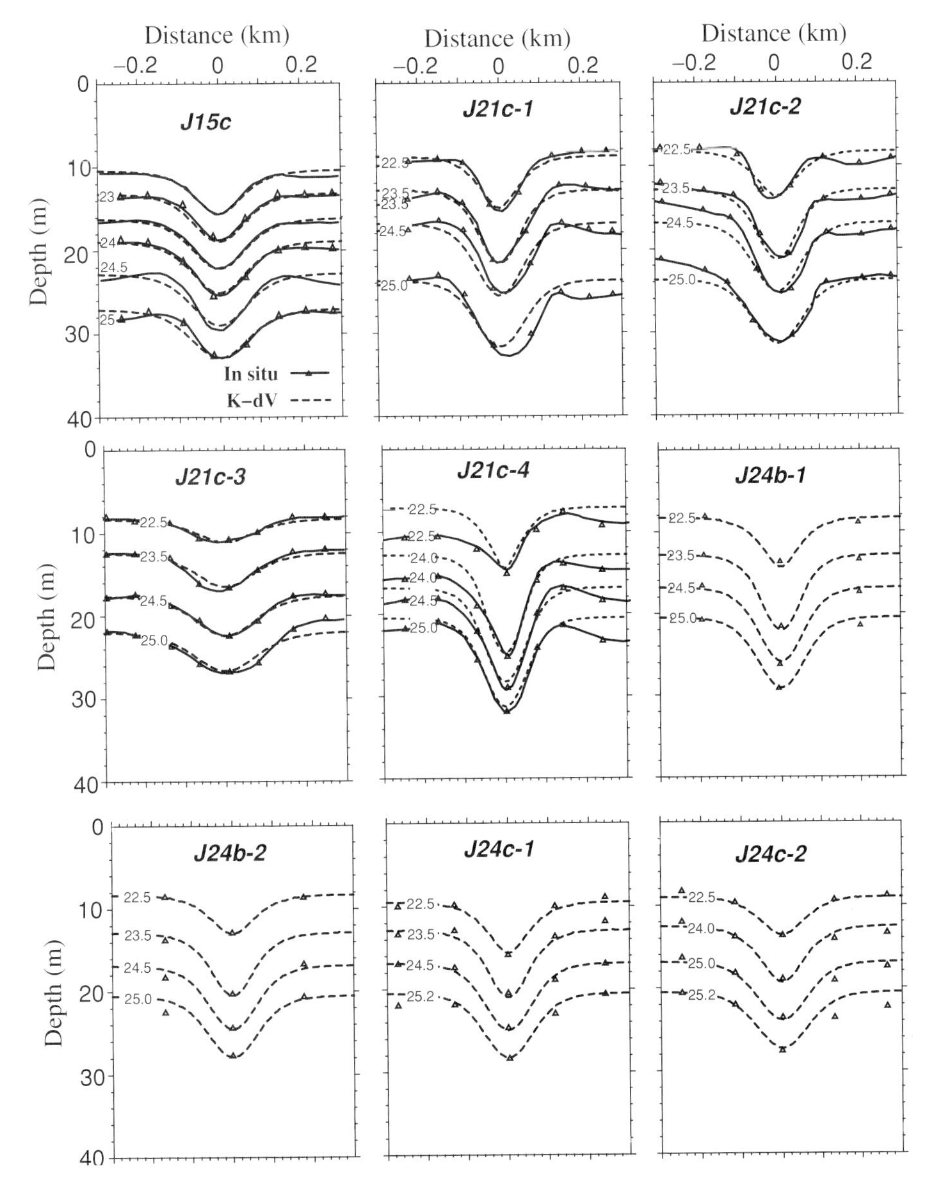

Figure 5.3. Enlarged wave profiles from Figure 5.2 shown by triangles and solid lines, predicted by the first-order K–dV solitons given by the dashed lines.

5.2.3 Second-order weakly nonlinear theory

Following ref. [245], consider now a second-order correction to a soliton with mode number $j = n$ as obtained from the first-order theory and expressed by the function $F_1(\zeta, \eta)$. The second-order term $F_2(\zeta, \eta)$ of the expansion (5.42) can be found once $F_1(\zeta, \eta)$ is substituted into equations (5.44). The second-order function

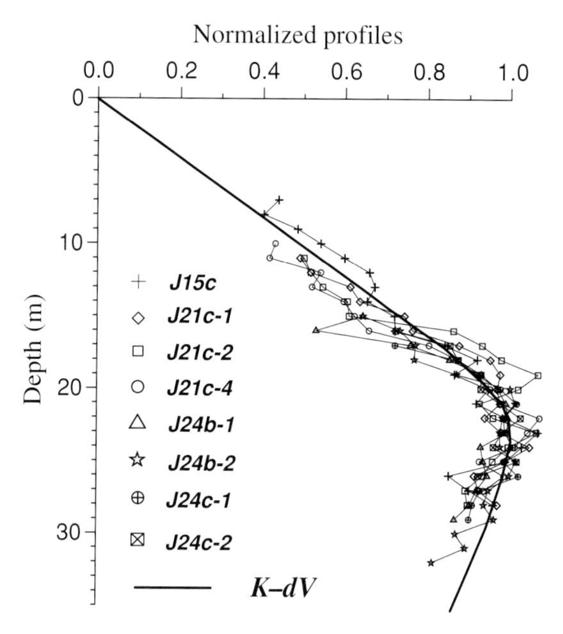

Figure 5.4. Normalized vertical wave profiles of the SIWs measured in the New York Bight (triangles and thin solid lines) and predicted first-mode profile for the K–dV soliton (thick line).

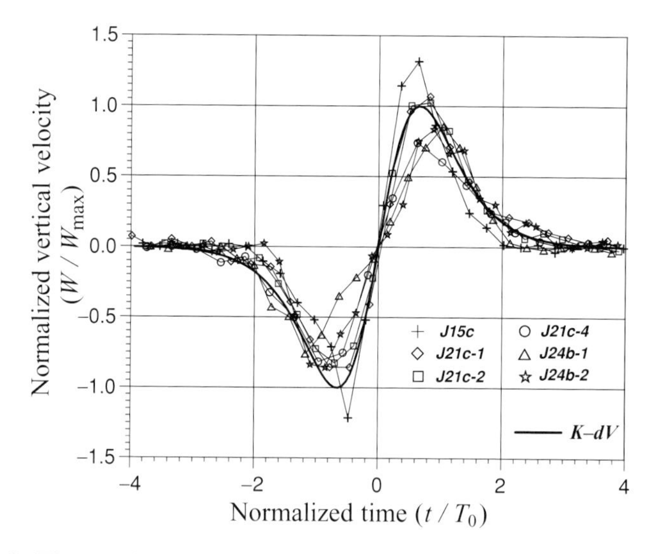

Figure 5.5. Time series of the vertical velocity of SIWs measured in the New York Bight at mooring M2 during JUSREX-92 (see Figure 4.1) normalized with the peak vertical velocity derived from the tow-yo data. The profile for the K–dV soliton is shown by thick solid line. The time scale T_0 is found from the expression $T_0 = \lambda / c_{\mathrm{p}}$.

F_2 is represented as a series of products, i.e.

$$F_2(\zeta, \eta) = \sum_{j=1}^{\infty} U_{2j}(\zeta) W_j(\eta), \tag{5.54}$$

in which $W_j(\eta)$ are the eigensolutions of the BVP (5.46), and the functions $U_{2j}(\zeta)$ are still to be determined. In this case, it is reasonable to find a solution when the spectral parameter $\varkappa_0$ in (5.44) is the eigenvalue of the BVP (5.46) [175].

Substituting (5.54) into (5.44), we obtain

$$\begin{aligned}\frac{(\pi/h)^2}{[c_a(\eta + c_b) + 1]^2} &\sum_{j=1}^{\infty} (n^2 - j^2) U_{2j}(\eta) W_j(\eta) \\ &= -\left(F_{1\zeta\zeta} + \varkappa_1 F_1 \bar{N}^2 - \varkappa_0 F_1^2 \frac{d\bar{N}^2}{d\eta} \right).\end{aligned} \tag{5.55}$$

Multiplying (5.55) by $W_k(\eta)$ with $k = 1, 2, 3, \ldots$ and integrating from $\eta = -1$ to $\eta = 0$, we find

$$(\pi/h)^2 \sum_{j=1}^{\infty} (n^2 - j^2) U_{2j}(\zeta) I_1^{kj} = -(U_{1\zeta\zeta} I_2^{kn} + \varkappa_1 U_1 I_3^{kn} - A_0 \varkappa_0 U_1^2 I_4^{kn}), \tag{5.56}$$

in which I_1^{kj}, I_2^{kn}, I_3^{kn}, and I_4^{kn} are coefficients which arise after the integration procedure. For the model pycnocline (5.17), these integrals have the following simple forms:

$$\left.\begin{aligned} I_1^{kj} &= \begin{cases} 0 & \Rightarrow k \neq j, \\ h/2 & \Rightarrow k = j, \end{cases} \\ I_3^{kn} &= \begin{cases} 0 & \Rightarrow k \neq n, \\ h/2 & \Rightarrow k = n, \end{cases} \\ I_2^{kn} &= h \int_0^1 \sin(n\pi z_3) \sin(k\pi z_3) \cos^{-4}(z_3 \mathscr{P} + \mathscr{R})\, dz_3, \\ I_4^{kn} &= \mathscr{P}[\sin(\mathscr{P} + \mathscr{R} + k\pi) - \sin \mathscr{R}] \\ &\times \left\{ \frac{(k+2n)\pi}{\mathscr{P}^2 - [(k+2n)\pi]^2} + \frac{(k-2n)\pi}{\mathscr{P}^2 - [(k-2n)\pi]^2} \frac{2k\pi}{\mathscr{P}^2 - (k\pi)^2} \right\}, \end{aligned}\right\} \tag{5.57}$$

where $\mathscr{P}$ and $\mathscr{R}$ are defined in (5.51). To calculate the integral I_2^{kn}, quadrature formulas must be used.

Relation (5.56) forms an algebraic system of equations with an infinite number of unknown functions $U_{2j}(\zeta)$. The integer n is the number of soliton modes in the first approximation, k is the counting index of the equations, and j is the counting index of the term on the left-hand side of (5.56).

Let us analyze (5.56) more closely. If the integer k of the normalized function $W_k(\eta)$ coincides with n, then the left part of (5.56) vanishes and the resulting equation turns into the K–dV equation (5.47). If $k \neq n$, then all terms in (5.56) vanish except the kth ones, which enter into the system on the left-hand side of the kth equation. This circumstance permits us to find the function $U_{2k}(\xi)$:

$$U_{2k}(\zeta) = \frac{2h}{\pi^2(n^2 - k^2)} U_1(\zeta) \times \left\{\left[A_0 \varkappa_0 I_4^{kn} + \nu_1 I_2^{kn}\right] U_1(\zeta) - a_\xi \varkappa_1 I_2^{kn}\right\}, \quad (5.58)$$

for $k = 1, 2, 3, \ldots, k \neq n$.

In summary, according to (5.40) and (5.42), the final second-order solution for the stream function ψ_2 can be written as

$$\psi_2 = \mu^2 c_p H F_2(x, \eta) = \frac{2a_\xi^2 c_p \nu_1 \mathscr{P} \Delta H_p}{3(\pi H)^2} \sum_{\substack{j=1 \\ j \neq n}}^{\infty} \frac{U_1(\zeta)}{n^2 - j^2}$$
$$\times \left\{\left[A_0 \varkappa_0 I_4^{nj} + \nu_1 I_2^{nj}\right] \frac{3U_1(\zeta)}{2\nu_1} + I_2^{nj}\right\} W_j(\eta). \quad (5.59)$$

The first-order approximation of the internal wave field (5.52) that goes together with this solution consists of the single-mode solitary wave with mode number n and amplitude a_ξ. Expression (5.59) represents the corrections to the basic solution and depends quadratically on the wave amplitude, $\sim a_\xi^2$. Moreover, the second-order correction (5.59) consists of a sum of modes. Thus, this solution shows a conspicuous soliton of finite amplitude and multimodal structure. The modes with mode numbers close to n, i.e. $n + 1, n - 1$, have the largest amplitudes. The contributions to the solution of the waves with increasing distance from n decrease to both sides ($n \pm 2, n \pm 3, \ldots$). This decrease is largely governed by the multiplier $1/(n^2 - j^2)$.

It is true that the expression in curly brackets in (5.59) also contributes to the dependence of the amplitude on the mode number, because the coefficients I_2^{nj}, I_4^{nj} also depend on j. However, this is less obvious because they must be determined numerically. Nevertheless, a qualitative analysis shows that I_2^{nj}, $I_4^{nj} \sim 1/j$ when $j \to \infty$.

Consider an example for which the second-order terms exhibit a quantitative influence on the characteristics of the solitary internal wave. Figure 5.6 displays the spatial structure of the SIWs of the first mode with amplitude a_ξ equal to 20 m. The soliton has a wavelength $\lambda \cong 600$ m, and thus the parameters of nonlinearity ε and dispersion μ for the depth $H = 185$ m both have the value ~ 0.1.

Figure 5.6 shows that the structure of the soliton with the second-order corrections is qualitatively very similar to the soliton obtained from (5.52), as the plots in these two cases are very similar. In both cases, the flow is characterized by a depression of the pycnocline, by maximum and minimum values of the horizontal

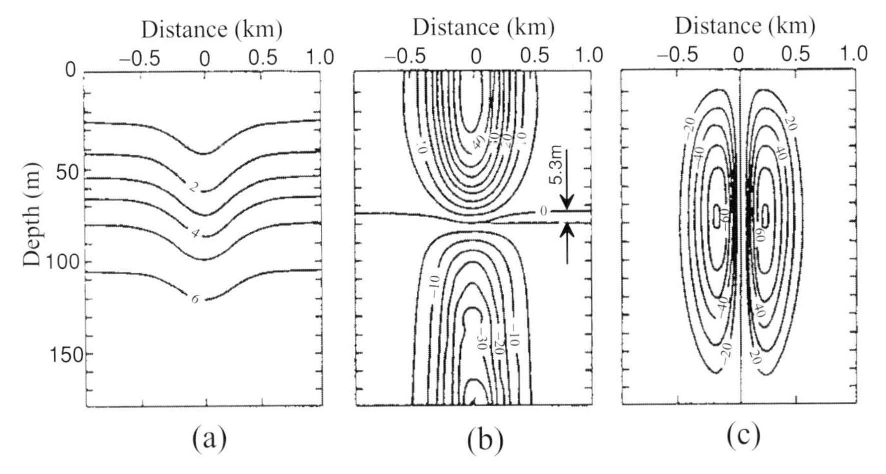

Figure 5.6. The combined fields of the first- and second-order solutions for a solitary internal wave: (a) reference density $\sigma_t(z) - \sigma_t(0)$ (kg m^{-3}), (b) horizontal (cm s^{-1}) and (c) vertical (mm s^{-1}) velocity isotachs. The model parameters are $H = 185$ m, $N_p = 0.03$ s^{-1}, $H_p = 60$ m, and $\Delta H_p = 100$ m for the model pycnocline (5.17).

velocity at the free surface and the bottom, respectively, and by two extrema of the vertical velocity in the layer of the density jump. The single evident qualitative change is a depression of the zero isopleth to a depth of 5.3 m just in the soliton center.

The velocity fields of second order are shown in Figure 5.7. Whereas the first-order soliton has a maximum value of the horizontal velocity of +53.2 cm s^{-1} at the free surface, and its minimum value is equal to −34.2 cm s^{-1} at the bottom (Figure 5.6), the appropriate corrections for the velocity are −4.2 cm s^{-1} and −2.1 cm s^{-1}, respectively, and they are in the range of 5–10%. The maximum value of the horizontal velocity at $z = -70$ m is equal to +6.7 cm s^{-1} and is located exactly where the first-order horizontal velocity changes its sign. This is why the second-order corrections are so important here, and why the zero isopleth in Figure 5.6 is depressed.

The appropriate corrections of the amplitude of the vertical speed of the first-order solution can also be estimated to be about 5–10%. The maximum value of the first-order solution (5.52) is equal to 68.5 mm s^{-1}, located at 75 m depth, whereas its second-order corrections are at a depth of 110 m and amount to ± 6.1 mm s^{-1}.

5.3 Structure of large-amplitude solitary internal waves

The weakly nonlinear theories presented above – whether they are first- or second-order – have several evident limitations in their applicability to real oceanic SIWs:

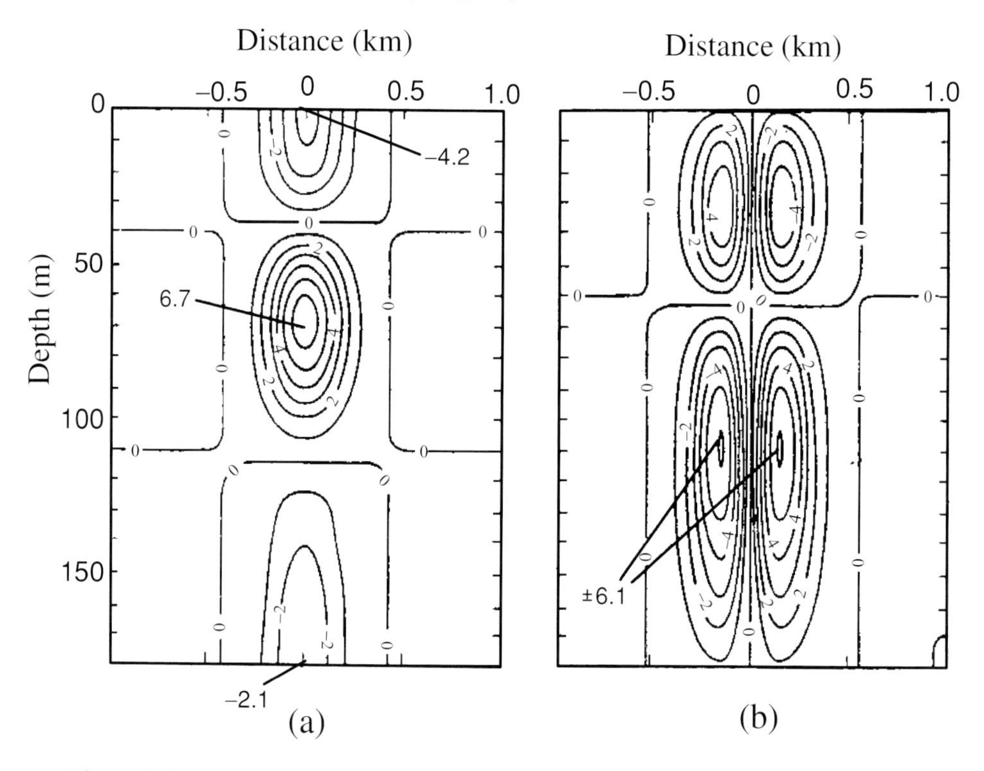

Figure 5.7. Fields of the second-order corrections for (a) horizontal (cm s^{-1}) and (b) vertical (mm s^{-1}) velocity components obtained by formula (5.59). The input parameters are the same as in Figure 5.6.

the capability of the K–dV models to reproduce observational results decreases as the wave amplitude increases. For example, in the Mediterranean Sea, in the area of the Strait of Messina, where $H = 300$ m and $H_p = 35$ m, SIWs with amplitudes $a_\xi = 50$–60 m and wavelengths λ of order 250 m were observed [23]. The first-order theory [21], [53], which uses the K–dV equation, demands the formal condition $H/\lambda \ll 1$ to hold, but that is not met in this case. The asymptotic theories [19], [179] developed for deep water are equally invalid because the wavelength, λ, practically coincides with the total depth, H, of the basin, which is in contradiction with the condition of a small value of wave dispersion: $(H/\lambda)^2 \ll 1$. The application of the asymptotic theory developed for finite depth basins [112], [124] is more appropriate in this case. However, the vertical scale, a_ξ, of the considered waves is almost twice as large as the scale of the stratification, H_p, and the basic condition for the application of this theory [215] is not satisfied either.

There are several other sites of the World Ocean where such exceptional SIWs with amplitudes of 100 m and more occur: such waves were registered in the equatorial Atlantic [24], [128]; in the Sulu Sea [5]; in the Andaman Sea [180]; and close to the Mascarene Ridge in the Indian Ocean [120].

Recognition of these limitations in the applicability of the K–dV models led to the development of evolution equations containing higher-order nonlinearities, as was presented in Sections 5.1 and 5.2. However, all these theories are not free of the formal restriction, namely that $a_\xi/H,\ H/\lambda \ll 1$.

Note also that for a two-layer stratification it is possible to build some asymptotic expansions which permit consideration of large-amplitude waves; see, for instance, refs. [39], [158], and [182]. In such models, which have been successfully used to explain laboratory observations of large-amplitude solitary internal waves [153], the full nonlinearity of the shallow-water theory is included, i.e. no limitation is imposed on the wave amplitude. Unfortunately, idealized stratification conditions do not permit application of these results to real oceanic conditions.

In order to overcome these limitations in the description of large-amplitude SIWs, a numerical integration of the complete set of the nonlinear, nonhydrostatic equations was also performed. Two different approaches were used:

- integration of the stationary Long equation (5.38), [26], [235];
- integration of the nonstationary equations [130], [229].

With the first approach, a large number of solitary wave solutions can be found, including waves characterized by a zone of closed circulation cells where unstable stratification may exist. The second approach consists of producing solitary internal waves as a result of the evolution of initial disturbances; it tends to filter out unstable waves that, nevertheless, can be solutions of the stationary problem. Thus, although both approaches can be used to study characteristics of large-amplitude SIWs, the first one emphasizes the mathematical complexity of nonlinear dispersive waves, whereas information inherent in the realistic manifestation in the ocean can more easily be inferred from the second. In this section the structure of large-amplitude SIWs is studied by analyzing results of a nonstationary numerical model and by studying high-resolution *in situ* data acquired north and south of the Strait of Messina [256].

5.3.1 Numerical model for stationary wave solutions

The scheme described in Section 4.2 was used for the numerical implementation of solitons. The calculations were carried out in a nonrotating ocean with the stratification given by formula (5.17) and by assuming constant depth. The finest resolution, with spatial grid steps $\Delta x = \Delta z = 1$ m, was used.

In the model, the analytical solution (5.52) was initialized with the first-mode K–dV SIW. Solution (5.52) is valid when the wave amplitude a_ξ is relatively small in comparison with the depth H ($a_\xi/H \ll 1$). If $a_\xi \sim O(H)$, the solitary internal wave inserted into the nonstationary system (1.79), (1.81) as an initial profile will evolve toward a new stationary solitary wave solution. Such a transitional process is illustrated in Figure 5.8. The modification of the K–dV soliton is continued until the new stationary solution is reached. It is tempting to use this method to produce solitary internal waves with large amplitudes. In order to guarantee that the new SIWs are stationary waves, the following strategy is applied: the changes in the shape of the SIW, its amplitude, velocities, and energy ought to be less than 0.1% over a time period of $50T_0$ ($T_0 = \lambda/c_p$ is the time scale for the internal solitons). If this is so, then the wave will be considered as a stationary solitary wave. The greatest pycnocline depression in the lowest panel in Figure 5.8 ($t = 25T_0$) clearly represents the formation of such a wave.

5.3.2 *Characteristics of large waves*

Figure 5.9(a) shows the density field of a large-amplitude SIW obtained by the numerical model (solid lines) and calculated using the analytical K–dV model (dashed lines) for type 2 density stratification (see Figure 5.1 and Table 5.1). It is seen that the two waves have approximately the same values for their amplitudes. In the near-surface layer, down to ~40 m, the isopycnal displacements of the numerical SIW exceed those of the K–dV solitary internal wave; in the lower layer, the opposite situation occurs. The most conspicuous difference between the numerical and K–dV solitary internal waves is in the greater wavelength of the numerical than the K–dV solitary internal wave. The dependence of the wavelength of the numerical SIW on the density stratification and wave amplitude is, however, nontrivial and will be discussed later.

Figure 5.9(b) shows the horizontal velocity field. The form of the locus of zero horizontal velocity (heavy solid line) differs remarkably from the horizontal straight line that would result from the K–dV model. Note that the minimum of the horizontal velocity is not located at the bottom, as in the K–dV solution. This indicates that numerically obtained SIWs are essentially multimodal waves. Figure 5.9(c) displays the typical down- and upwelling also exhibited by the K–dV solution.

Figure 5.10 shows the vertical structure of the numerical and K–dV SIWs of different amplitudes for the density stratification discussed above. It is seen that the numerical SIWs possess several characteristics which differ from the K–dV internal solitons. Indeed, the depth of the maximum isopycnal displacement (Figure 5.10(a)), the depth of the zero horizontal velocity (Figure 5.10(b)), and the depth of the maximum vertical velocity (Figure 5.10(c)) are functions of the wave amplitude: whereas the first parameter decreases, the second and third increase with increasing wave amplitude. In general, the deviation of the numerical

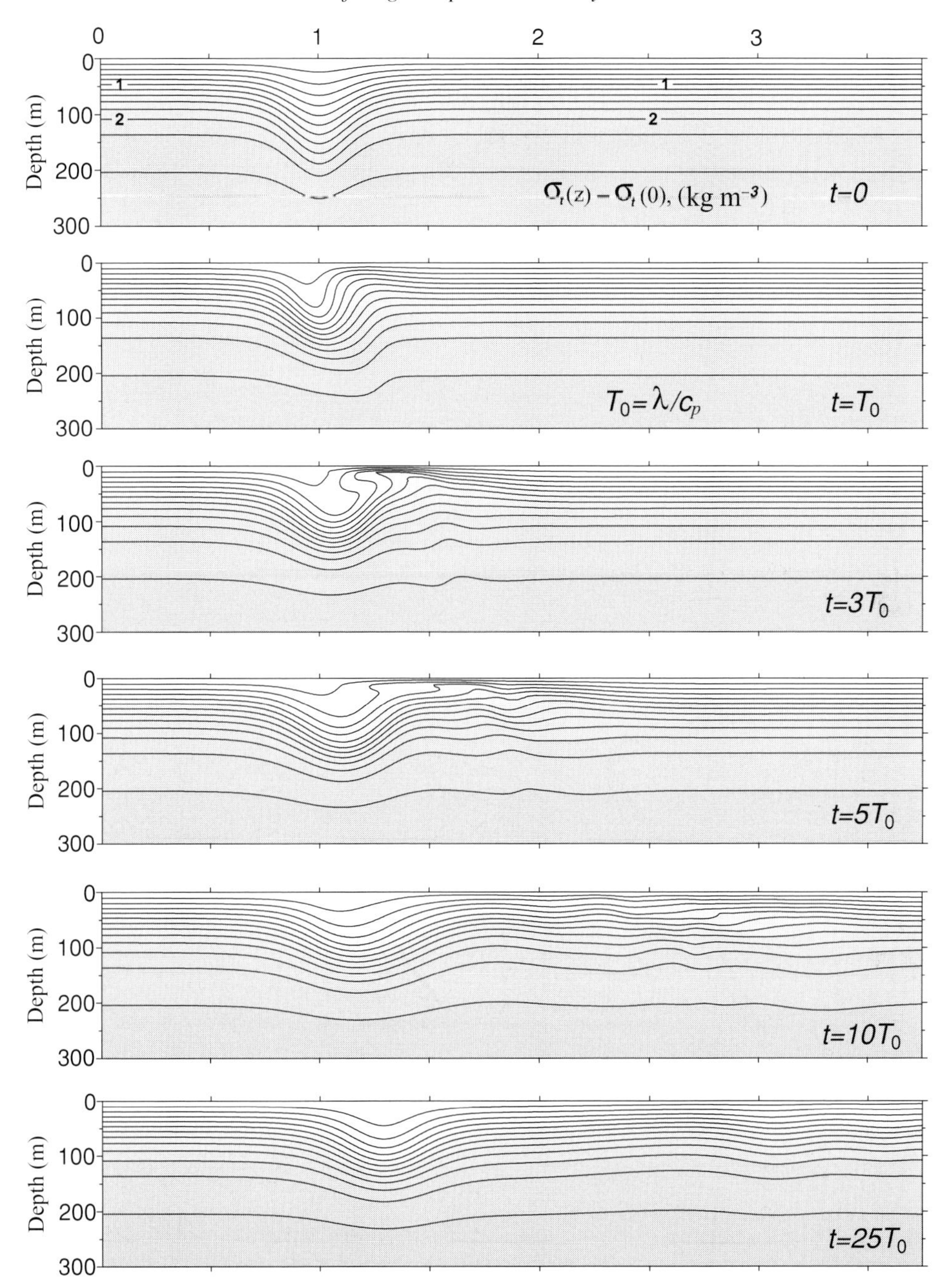

Figure 5.8. Process of the adjustment of a SIW (shown in a coordinate system moving with speed c_p) defined by the solution (5.52) of the K–dV equation (5.47) to the ambient conditions. The SIW amplitude is $a_\xi = 80$ m, the stratification of the fluid is defined by formula (5.47), with the parameters of the stratification corresponding to type 2 in Figure 5.1 and Table 5.1. The water depth is $H = 300$ m.

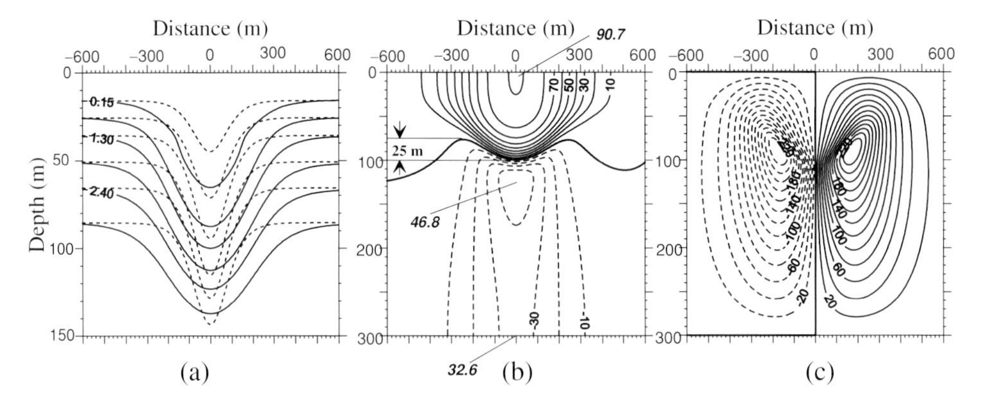

Figure 5.9. Fields of (a) density anomaly (kg m^{-3}), (b) horizontal (cm s^{-1}), and (c) vertical (mm s^{-1}) velocities of a large-amplitude solitary internal wave for stratification 2. (a) Solid lines refer to the results of the numerical model and the dashed lines refer to the results of the K–dV model (solution (5.52)); (b) and (c) refer solely to the results of the numerical model. Solid lines denote positive and dashed lines denote negative velocities.

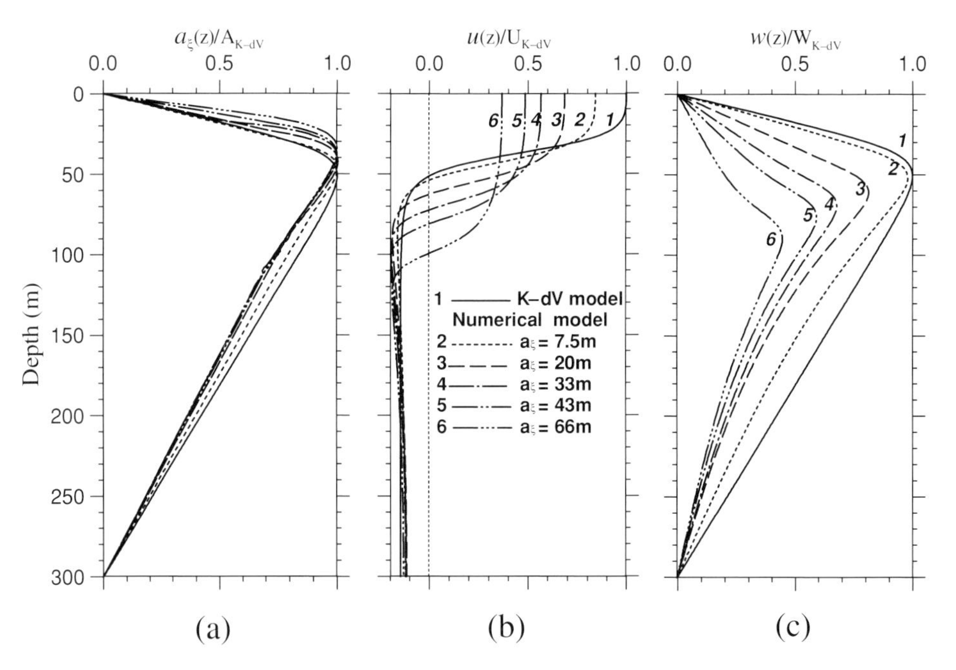

Figure 5.10. Vertical profiles of (a) the normalized isopycnal displacements, (b) the normalized horizontal velocities both calculated at the wave center, and (c) the normalized vertical velocities calculated at the position of maximum vertical velocity as simulated by the numerical model and by the K–dV model for a stratification of type 2 (Table 5.1) and different wave amplitudes. The normalizations are done with the amplitude $A_{\text{K–dV}}$, maximum horizontal velocity $U_{\text{K–dV}}$, and maximum vertical velocity $W_{\text{K–dV}}$ of the K–dV soliton.

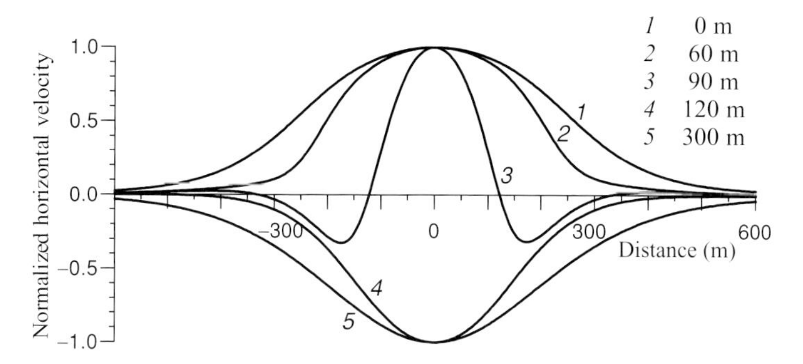

Figure 5.11. Normalized horizontal velocities calculated at different depths across the SIW shown in Figure 5.9.

from the K–dV SIWs increases with increasing amplitude. Moreover, the maximum horizontal velocity, and also the vertical velocities, of the numerical SIWs are smaller than those of the K–dV waves, independently of the wave amplitude. This is a consequence of the fact that, for this stratification, the nonlinear effects in the K–dV model are overestimated.

Such distinctive, perhaps unexpected, features of the horizontal velocity field are inherent in large SIWs; they can lead to confusing conclusions in interpretations of data of *in situ* measurements. Figure 5.11 clearly illustrates this fact. Here, the horizontal profiles of the velocity normalized by the maximum value across the SIW are displayed for various depths. Obviously, the curve labeled "3" exhibits quite a complex structure with one maximum and two additional minima. If the measurements of the velocity are conducted at this depth, they may be interpreted as the passage of three waves with different polarities, whereas actually this structure belongs to a single large SIW. Knowledge of the structure of the K–dV wave only will invariably lead to such a misinterpretation. The mathematics of weakly nonlinear waves is therefore intolerably insufficient. Figure 5.11 further shows that the width of the intensive soliton is different at different depths. Comparison of two pairs of normalized profiles, for example 1 and 2 (depths 0 m and 60 m, respectively), or profiles 4 and 5 (depths 120 m and 300 m, respectively), confirms this statement.

The wavelength of a soliton was introduced by Koop and Butler [122] as

$$\lambda(z) = \frac{1}{2a_\xi(z)} \int_{-\infty}^{\infty} \xi(x, z)\, dx, \tag{5.60}$$

in which, depending on the value of z, $\xi(x, z)$ represents the displacements of the isopycnals whose undisturbed depth is z and $a_\xi(z)$ the amplitude of the soliton at

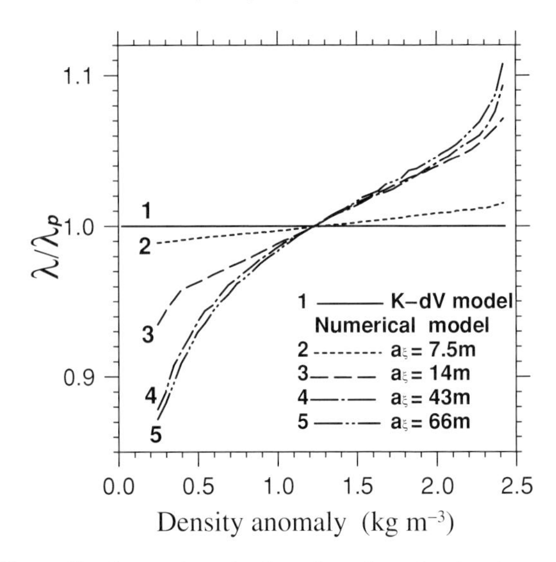

Figure 5.12. Normalized wavelength plotted against the density anomaly, as obtained by the numerical model (lines 2–5) and the K–dV model (solid line 1) for different wave amplitudes. The normalization used is explained in the text.

the wave center. This definition is meaningful as long as ξ is only of one sign, but becomes questionable otherwise.

The wavelength λ, normalized by the wavelength λ_p for different amplitudes, is shown in Figure 5.12 for the numerical and K–dV SIWs as a function of the density anomaly $(\rho(z) - \rho(0))$ for the stratification of type 2 (Table 5.1). The normalization was carried out using the value of the wavelength, λ_p, defined at the depth of the pycnocline $z = -H_p$. Because of this normalization, the resulting wavelength of the K–dV SIW does not depend on depth and is also independent of the wave amplitude. For SIWs with large amplitudes, the wavelength increases with increasing depth, and this dependence grows with growing amplitude. The dependence of different structures of the simulated large-amplitude SIWs on the density stratification is illustrated in Figure 5.13 for the three types of density stratification identified in Table 5.1. Figures 5.13(a)–(c) hold for a sharp pycnocline; by contrast, Figures 5.13(g)–(i) show the structure of a large SIW propagating through a fluid with a wide pycnocline, and Figures 5.13(d)–(f) correspond to an intermediate case.

If we compare the density fields in these three cases, no essential dependence of the horizontal structure of SIWs on the stratification is discernible. Qualitatively, the density wave profiles in Figures 5.13(a), (d), and (g) look the same, and the pycnocline can be roughly represented by the function $\cosh^{-2}(x)$.

By contrast, the u-field changes are much more pronounced among the three different stratifications. The "double hump" peculiarity in the complex velocity structure of the zero u-isopleth alluded to above becomes less pronounced with

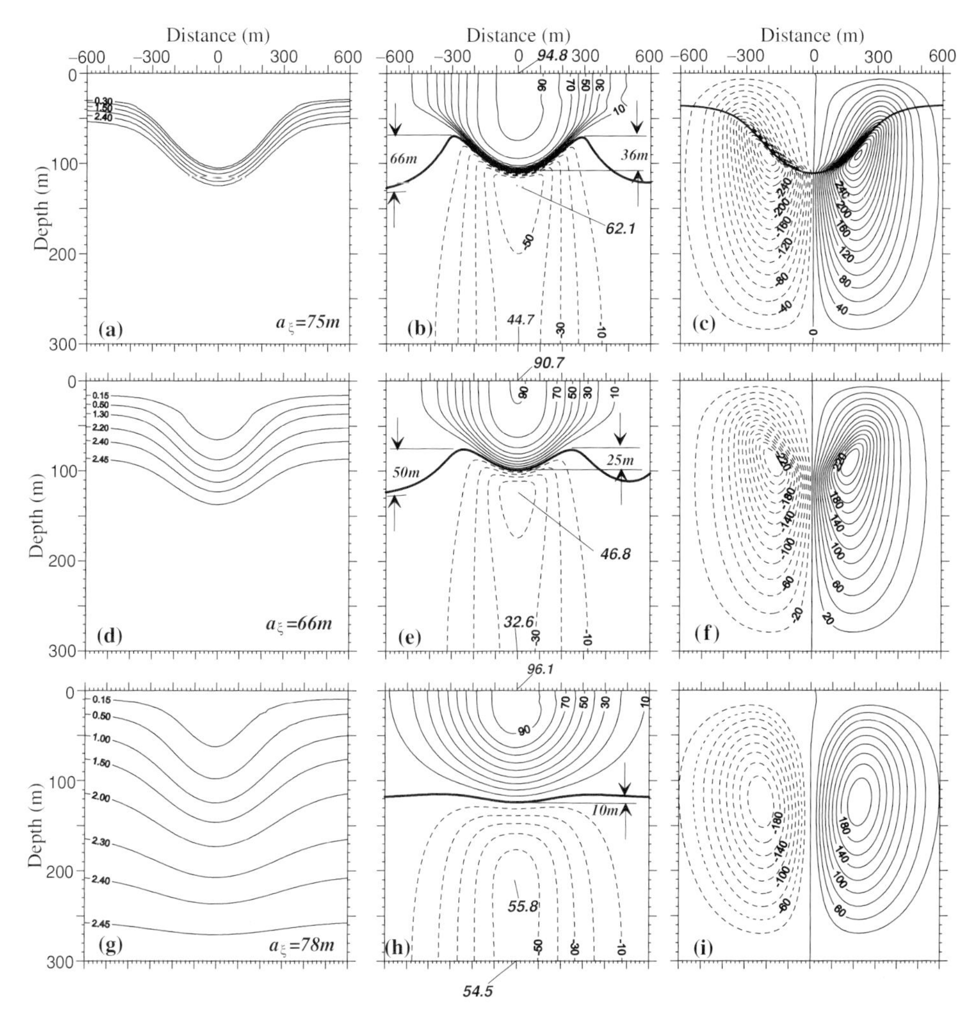

Figure 5.13. Fields of density anomaly (kg m^{-3}) ((a), (d), (g)), horizontal (cm s^{-1}) ((b), (e), (h)) and vertical (mm s^{-1}) ((c), (f), (i)) velocities of large-amplitude solitary internal waves for stratification 1, 2, and 3 in Figure 5.1 (top, middle, and bottom rows, respectively).

the widening of the pycnocline. For the narrow pycnocline, the zero isopleth at the wave center is located 36 m deeper than at its periphery, Figure 5.13(b). For the intermediate pycnocline this value equals 25 m, and for the wide pycnocline it is only 10 m, and the zero isopleth is practically horizontal, Figure 5.13(h).

The qualitative change of the w-velocity field with the pycnocline widening is shown in Figures 5.13(c), (f), and (i). The regular ellipsoidal type w-speed isolines for the wide pycnocline are considerably deformed for the narrow pycnocline. This deformation occurs especially in the layers where the pycnocline is located.

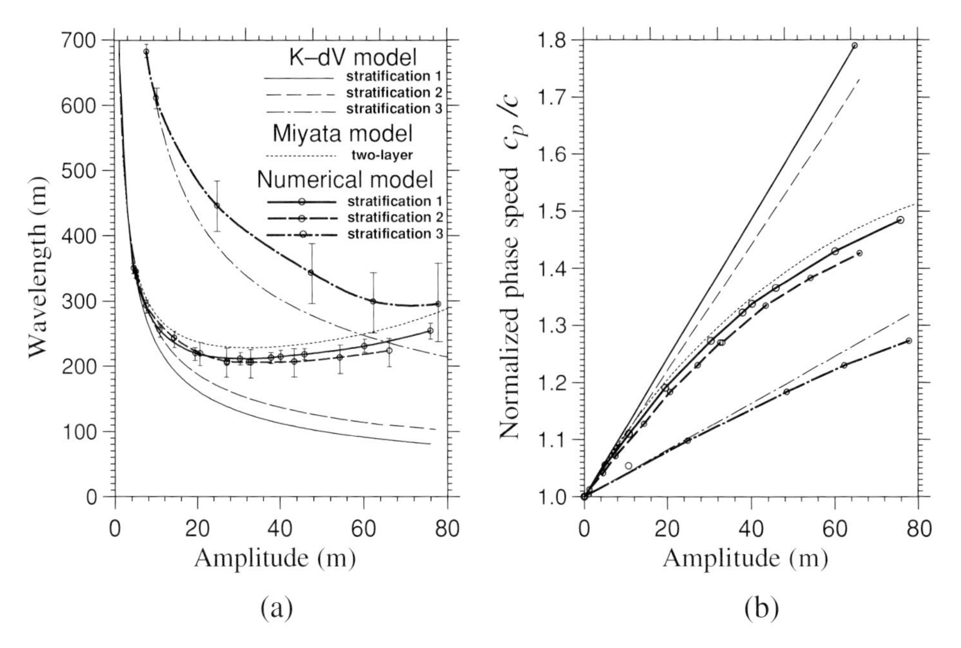

Figure 5.14. Relationship between (a) wavelength and wave amplitude and (b) nondimensional phase speed and wave amplitude of solitary internal waves, as simulated by the numerical model and calculated by the K–dV model for stratifications 1–3 in Figure 5.1, and as calculated by the Miyata model for the two-layer stratification. The nonlinear phase speed c_p is normalized with the linear phase speed c.

The wavelength–amplitude relationship and the phase-speed–amplitude relationship are plotted in Figures 5.14(a) and (b) for different theoretical models and density stratifications. Results are presented for the numerical, the K–dV, and the Miyata two-layer model [158]. The calculations were performed for the three density stratifications listed in Table 5.1; those referring to the Miyata model were determined for a two-layer stratification with the method outlined in Section 3.3. The wavelengths presented in Figure 5.14 refer to the isopycnal for which the undisturbed depth coincides with the pycnocline. The dependences of the wavelength of the numerical solitary internal waves on the depth are given by the bars in Figure 5.14(a). For small wave amplitudes, the difference in wavelength between the numerical and the K–dV wave is small and increases with an increase of the amplitude. In contrast with the results of the K–dV model, the wavelength of the numerical solitary internal waves for large-amplitude solitons increases with increasing amplitude. This corresponds to the known characteristics of the Miyata model (see also refs. [39] and [182]). While in the presence of weak vertical density gradients the difference in phase speed between the numerical and the K–dV wave is small even for large wave amplitudes, for a sharp pycnocline it is small only for small wave amplitudes. For these stratifications a much better agreement can be observed between the numerical and Miyata solitary internal waves.

5.3.3 Observational evidence of large waves

Several characteristics of large-amplitude SIWs discussed in Section 5.3.2 can be observed in the ocean. In this section, the recorded data of SIWs measured north and south of the Strait of Messina (Mediterranean Sea, Figure 5.15(a)) during the Atlantic Ionian Stream 1995 cruise and during the Rapid Response 1997 cruise from aboard the R/V *Alliance* are compared with results obtained by applying the numerical and the K–dV model [256]. The hydrographic data were obtained using a CTD probe and a towed CTD chain, and the current data were obtained using a 75 kHz acoustic Doppler current profiler (ADCP) mounted on the vessel. Two large-amplitude solitary internal waves were measured, at about 25 km north and 45 km south of the strait sill with water depths of about 300 m and 1000 m, respectively; see Figure 5.15(a). The observed waves resulted from the disintegration of internal bores generated by the interaction of the barotropic semidiurnal tidal flow with the strait sill according to the mechanism described in Chapter 4. We feel confident that our data refer to almost stationary solitary internal waves because of the large distance traveled by the waves as compared with the typical length scale of the front of the parent internal bore [23].

Figure 5.15(b) shows the undisturbed density profiles, measured by the CTD chain near the wave fronts north and south of the strait. The corresponding buoyancy frequency profiles, together with the smoothed profiles used in the numerical simulations, are shown in Figure 5.15(c). Using these profiles, the numerical model was run until a stationary solitary wave solution was reached. After several model runs, initialized by using the K–dV solitary internal waves of different amplitudes, the numerical SIWs were produced having almost the same amplitudes as the measured waves. The measured and numerical SIWs are shown in Figure 5.16 as density fields ((a) and (d)), horizontal ((b) and (e)), and vertical ((c) and (f)) velocities. Figures 5.16(a)–(c) refer to SIWs to the north and Figures (d)–(f) refer to SIWs south of the Strait of Messina.

Several characteristics of the observed and computed SIWs are very similar. Among these are the shape of the isopycnal displacements, the locus of the zero horizontal velocity, the positions of the extrema in the vertical velocity, and the values of the maximum horizontal and vertical velocities. They confirm that the numerical model is indeed capable of describing the main physical processes governing the dynamics of large-amplitude SIWs in the Strait of Messina. This fact is clearly seen from Table 5.2, where typical values referring to the observed and simulated solitary internal waves are listed.

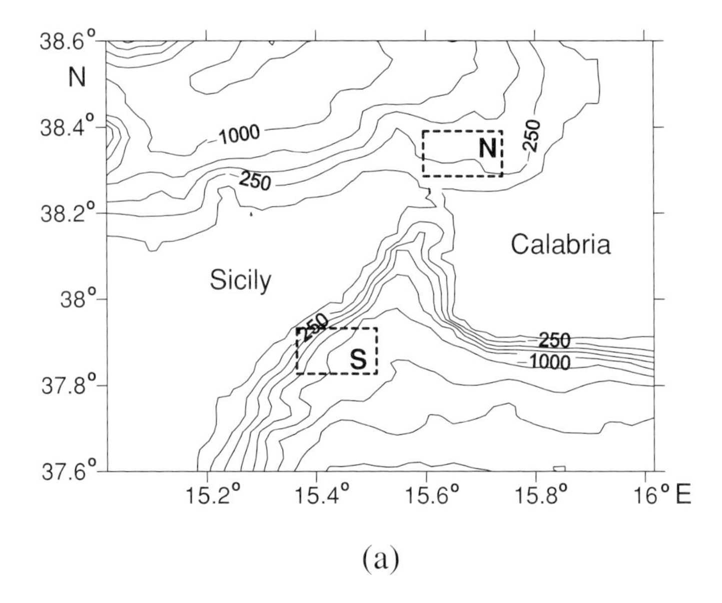

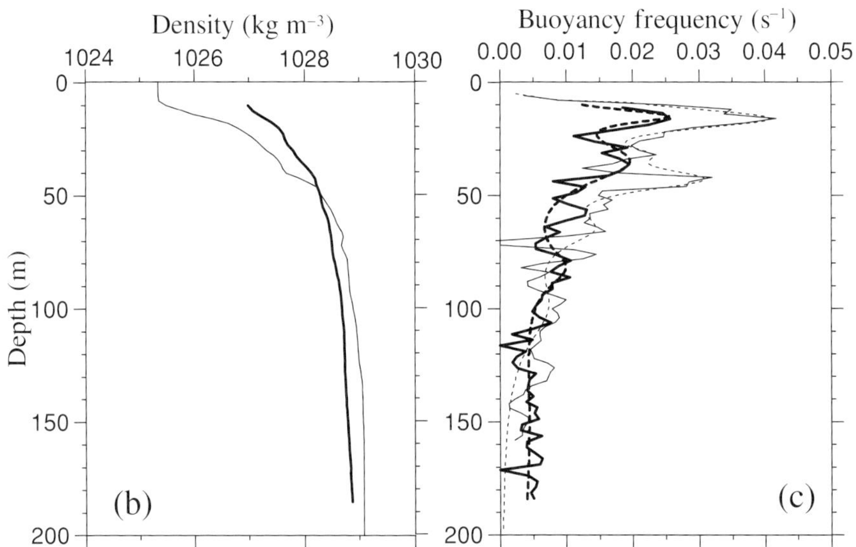

Figure 5.15. (a) Map of the Strait of Messina. Measurements of SIWs to the north and south of the strait are shown by boxes N and S, respectively. Undisturbed vertical profiles of (b) densities and (c) buoyancy frequencies measured by the CTD chain near the Strait of Messina. Thick and thin lines correspond to the measurements to the north and south of the strait, respectively. The dashed lines in (c) refer to the smoothed buoyancy frequency profiles used in the numerical simulations.

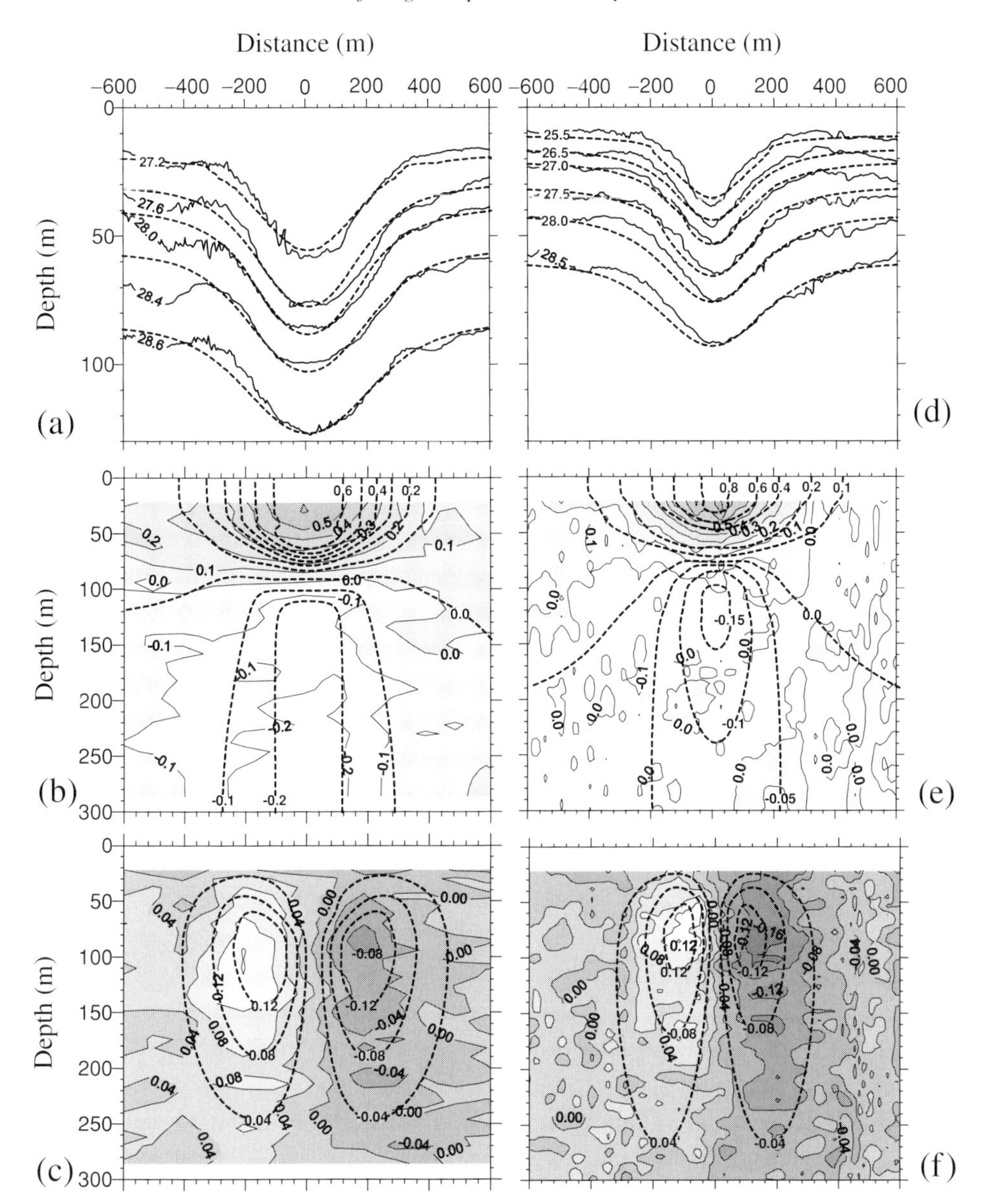

Figure 5.16. Fields of (a), (d) conventional density anomaly (kg m^{-3}), (b), (e) horizontal (cm s^{-1}), and (c), (f) vertical (mm s^{-1}) velocities of two large-amplitude solitary internal waves as measured by the CTD chain and the ADCP north ((a)–(c)) and south ((d)–(f)) of the Strait of Messina and as simulated by the numerical model (dashed lines).

Table 5.2. *Values of several characteristic quantities of large-amplitude solitary internal waves north and south of the Strait of Messina as measured by the CTD chain and the ADCP, and as simulated by the numerical model.*

Characteristic	Measured (north)	Simulated (north)	Measured (south)	Simulated (south)
Amplitude (m)	55	50	34	33
Phase speed (m s^{-1})	1.0	1.0	1.2	1.1
Maximum horizontal velocity (m s^{-1})	0.61	0.67	0.81	0.83
Maximum vertical velocity (m s^{-1})	0.16	0.16	0.16	0.15
Minimum vertical velocity (m s^{-1})	−0.11	−0.16	−0.19	−0.15
Depth of zero horizontal velocity at the wave center (m)	94	91	85	77

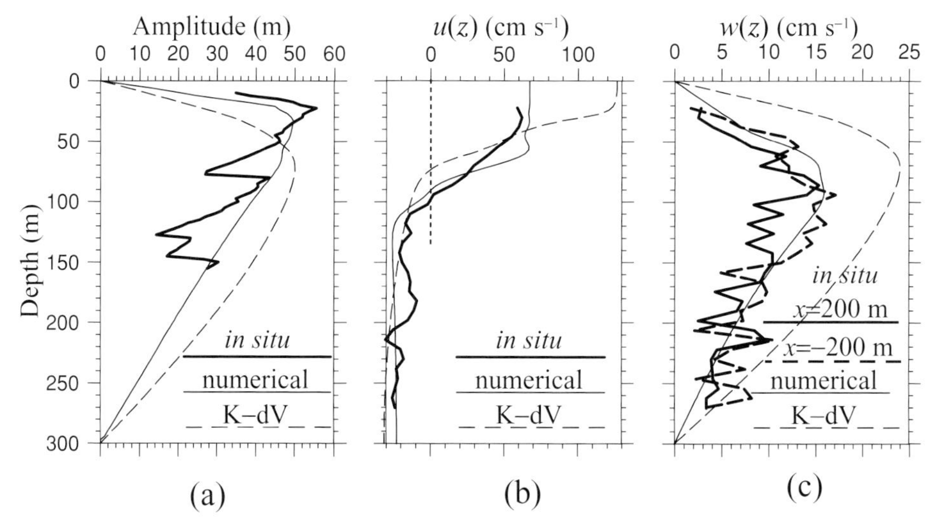

Figure 5.17. Vertical profiles of (a) isopycnal displacements at the wave center, (b) horizontal velocities at the wave center, and (c) vertical velocities at the position of their extremal values at $x \pm 200$ m in Figure 5.16 *north* of the Strait of Messina. *In situ* data are shown by thick lines; theoretical curves are represented by thin lines.

A comparison between characteristic quantities of the two SIWs, as measured by the CTD chain and the ADCP probe, and as simulated by using the numerical and the K–dV model, is given in Figures 5.17, north, and 5.18, south of the Strait of Messina. Figure 5.17(a) shows vertical profiles of the isopycnal displacements at the wave center referring to the solitary internal wave north of the Strait of

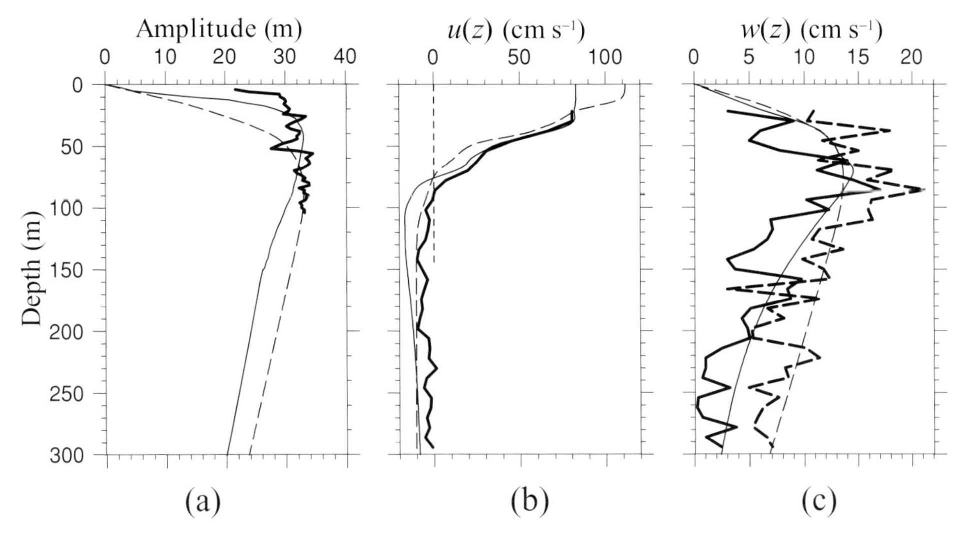

Figure 5.18. As for Figure 5.17, but for measurements *south* of the Strait of Messina.

Messina. Note the good agreement between the depth of the maximum isopycnal displacement measured by using the CTD chain and that simulated by the numerical model. In the measured data, several local extrema are, however, present which are not captured by the model. They may be the result of the superposition of different subscale internal waves existing in the area. In Figure 5.17(b) the vertical profiles of the horizontal velocity at the wave center for the solitary internal wave north of the Strait of Messina are illustrated. Comparison between the values measured by the ADCP probe and simulated by the models shows very good agreement in the maximum horizontal velocity and the depth of vanishing horizontal velocity. Figure 5.17(c) shows vertical profiles of the vertical velocity at the positions of extreme vertical velocity (i.e. at the position where the vertical velocity has a maximum, in the rear of the wave, and in the position where it is minimum, at the front of the wave), also for the solitary internal wave north of the Strait of Messina. Although the measured data are, in general, noisy, satisfactory agreement between the measured and simulated profiles is nevertheless observed.

Comparison between the characteristics of the solitary internal waves south of the Strait of Messina as measured with the CTD chain and the ADCP probe and as simulated by the numerical model reveals equally good agreement (see Figure 5.18).

It can clearly be seen from Figures 5.17 and 5.18 that the K–dV model is not able to capture several characteristics of the measured solitary internal wave, but they can be adequately simulated by the numerical model. This is particularly evident for the SIW measured north of the Strait of Messina, which has a larger amplitude than that measured south of it.

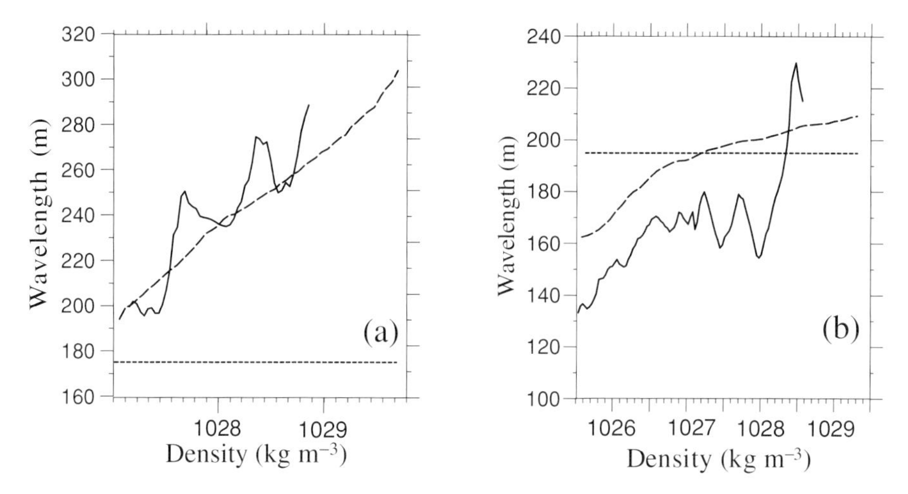

Figure 5.19. Wavelength as a function of the density of two large-amplitude solitary internal waves as measured by the CTD chain (a) north and (b) south of the Strait of Messina (solid lines), and as simulated by the numerical model (long-dashed lines) and the K–dV model (short-dashed lines).

The inadequacy of the K–dV model for the simulation of large-amplitude SIWs becomes dramatically evident if the wavelengths, as measured at different depths by the CTD chain north and south of the Strait of Messina, are compared with those simulated by using the numerical model and the K–dV model, Figure 5.19. It is seen that the K–dV model predicts wavelengths which are much smaller than those measured by the CTD chain and simulated by the numerical model. Moreover, the wavelengths of K–dV solitary internal waves are independent of depth.

5.4 Interaction of large-amplitude SIWs with bottom topography

Solitary internal waves propagating for a long distance from the source of generation must inevitably encounter bottom features like oceanic ridges, banks, or continental slopes. In this section we concentrate on identifying the mechanism of energy transfer from large-scale internal waves to short scales during the process of wave–topography interaction.

Evolution of nonlinear internal waves over variable bottom topography has been studied in many papers.[1] The weakly nonlinear K–dV theory is usually used as a tool to interpret the *in situ*, laboratory, and observational data. Here, the process of wave–topography interaction will be considered without a formal restriction on the amplitude of the propagating wave.

[1] See refs. [5], [83], [84], [91], [92], [93], [102], [103], [140], [142], [143].

As an example for the model application, topographic features from Lake Constance are used.[2] Several different scenarios of the nonlinear wave evolution over bottom topography are investigated. They indicate possibly important sinks of wave energy from the large-scale motions to the small scales and to turbulence. Even though they are obtained for lake conditions [253] and as such stand as examples, they apply equally well to oceanic solitary waves.

5.4.1 Scenarios of wave–topography interaction

It is a commonly held belief that the energy influx extending into depths of lakes is supplied by wind acting on the free surface: it drives the surface water and generates internal waves in the form of basin-scale standing waves [106] or propagating nonlinear waves [58]. After storms, internal waves in lakes may take the form of an internal surge or packets of internal solitons, generated by the nonlinear steepening of a basin-scale finite-amplitude wave [58], [230], [231], [267]. Since these solitons are much shorter in length than the wind-induced initial large-scale thermocline displacements, their generation results in a transfer of energy within the internal wave field from large to small scales.

Here, we consider the evolution of plane SIWs in a variable-depth channel, typical of a lake of variable depth. As was mentioned above, the vertical fluid stratification, wave amplitudes, and bottom parameters are taken to be close to those values observed in Lake Constance, a typical mountain lake; see Figure 5.20(a).

The numerical model described above is applied to the study of the interaction of SIWs with a variable bottom topography. It is assumed that a plane wave of depression with amplitude a_ξ, of which the vertical structure is defined by the density profiles, propagates from the deep part to an underwater ridge or a slope region near the shore. The objective of our modeling efforts is to investigate how the process of evolution of a SIW reacts to the variable bottom topography.

To simulate real conditions more closely, we selected a region of Mainau Island where a sill acts as a constriction, Figure 5.20(b). Since Lake Constance is elongated in one direction, we can assume that the lake axis is the preferable direction of the propagation of internal waves. So, the idea is to investigate the evolution of an internal wave propagating through the constriction near Mainau Island. Such waves can encounter three qualitatively different typical profiles of the bottom topography, and these are presented in Figure 5.21 for the cross-sections 1–3 depicted in Figure 5.20.

Profile 1, roughly following the lake thalweg, is characterized by a gentle slope and a smooth transition from the deep part over the Sill of Mainau ($\sim$80 m deep).

[2] Lake Constance is situated on the borders of Germany, Switzerland, and Austria (see Figure 5.20).

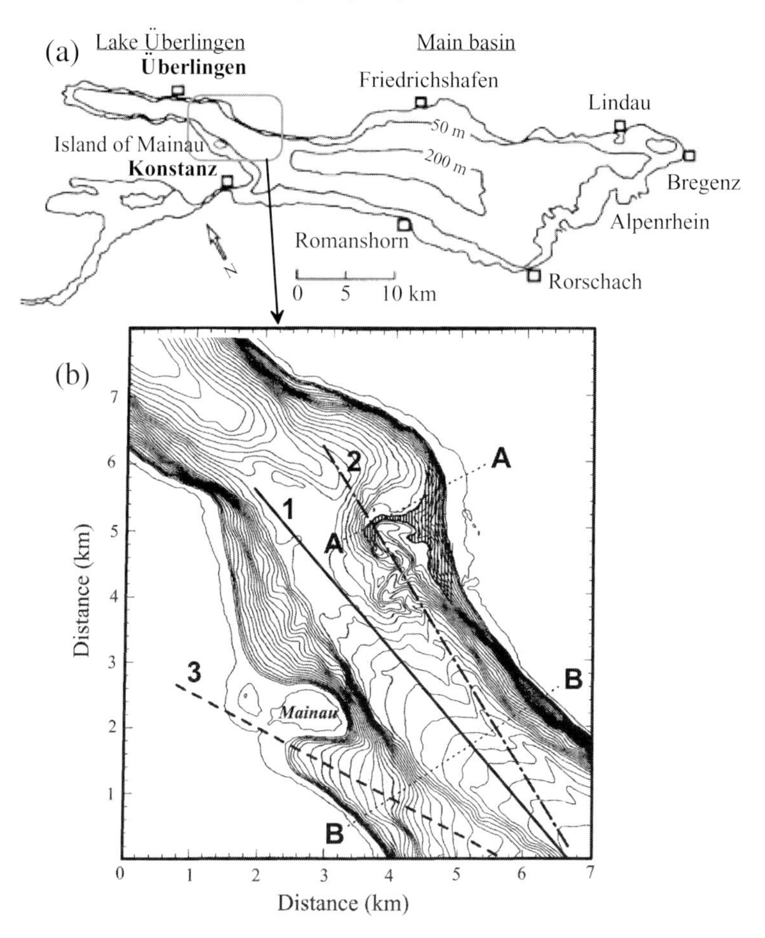

Figure 5.20. (a) Lake Constance. (b) Local bathymetric map in the region of Mainau Island. Contour intervals = 5 m. Cross-sections 1–3 show the selected directions of two-dimensional wave propagation.

Internal waves propagating more closely to the northern shore of the lake (profile 2) encounter in section A–A a higher sill, with its top at about 50 m below the surface, followed by a relatively abrupt drop and rapid increase in the water depth. A qualitatively different profile is encountered by waves propagating in direction 3; this possibility should also be taken into account because almost 40% of cross-section B–B can be described by such a bottom profile. This slope is shorter and steeper in comparison with profiles 1 and 2. Moreover, unlike the first two profiles, it ends on the shore. Most likely, the wave evolution here will differ considerably from the dynamical processes in the other two places.[3]

[3] The reader is cautioned not to consider these profiles, and the two-dimensional strongly nonlinear processes derived by using them, as realistic scenarios for the region close to Mainau Island. Rather, the selection should be regarded simply as a typical scenario from which qualitative results can be deduced.

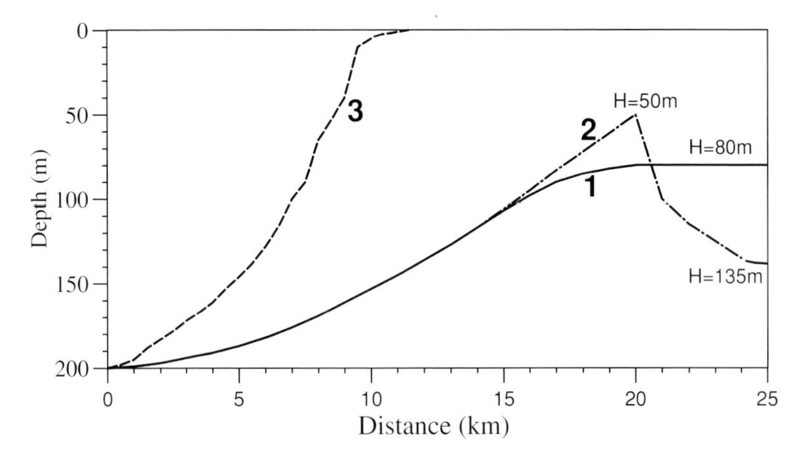

Figure 5.21. Bottom profiles used in the model: 1–3 correspond to cross-sections 1–3 in Figure 5.20.

The fluid stratification used in this study is shown in Figure 5.22. The two profiles of the buoyancy frequency presented here reflect typical vertical distributions for the summer and autumn seasons. In summer, the density jump is more pronounced and lies above the similar autumn maximum. As a consequence, the maximum of the wave displacements in summer is located closer to the free surface than in autumn. This leads, in turn, to a conspicuous difference between the wave behavior over bottom features in different seasons, as we shall see in the following.

Amplitudes of internal waves measured in Lake Constance usually lie in the range of 0–15 m. To identify the influence of the nonlinearity of a wave process on the mechanism of the internal wave transformation over variable bottom topography, i.e. to determine the dependence of the wave structure on the wave amplitudes, four different incoming solitary internal waves of depression were taken as initial conditions for each season:

- $a_\xi = 2$ m, weakly nonlinear case;
- $a_\xi = 5$ m, intermediate nonlinearity;
- $a_\xi = 10$ and 15 m, strong nonlinearity.

The approach used here to obtain the initial fields (at $t = 0$) for the incident wave was the same as that used in the previous section. At the initial stage we considered a basin with constant depth of 200 m (deep part of Lake Constance). The K–dV solution is used as an initial profile for the interaction of the intense SIW with the bottom topography.

Numerical runs were performed for every SIW listed above and every profile shown in Figures 5.20(b) and 5.21, for both summer and autumn seasons. In the following we describe the three different scenarios of the wave evolution in the

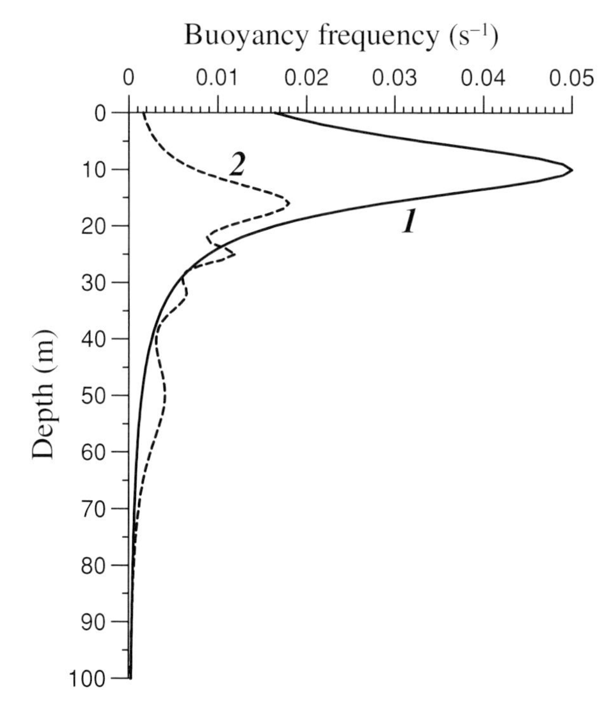

Figure 5.22. Buoyancy frequency profiles measured in Lake Constance during summer (1) and autumn (2) [178].

considered area which were found for different bottom profiles. As will be shown below, the wave evolution over variable bottom topography strongly depends on the parameter $a_\xi/(H - H_\xi)$, which in fact is the ratio of the wave amplitude, a_ξ, to the distance from the depth, $z = -H_\xi$, where this maximum of the isopycnal deflection occurs to the bottom.

Scenario 1: Wave adjustment when $a_\xi/(H - H_\xi) \ll 1$

This scenario was found for profile 1. As already mentioned above, this profile is characterized by a gentle slope, i.e. by a weak change of the bottom depth from 200 m in the deep part to ∼80 m depth at the Sill of Mainau. Taking into account that for summer and autumn stratifications the maximum amplitude of the SIW lies very far from the bottom, H_ξ equals 20 m and 35 m depth, respectively, internal waves under conditions of a weakly variable bottom are not destroyed and do not transform into a secondary wave packet during their evolution. These SIWs – weak or strong – simply adjust permanently and passively to the local ambient conditions, i.e. stratification and depth, by preserving their initial form. One example of such adiabatic behavior for autumn stratification is presented in Figure 5.23, which shows the density and the u and w velocity fields for an incident

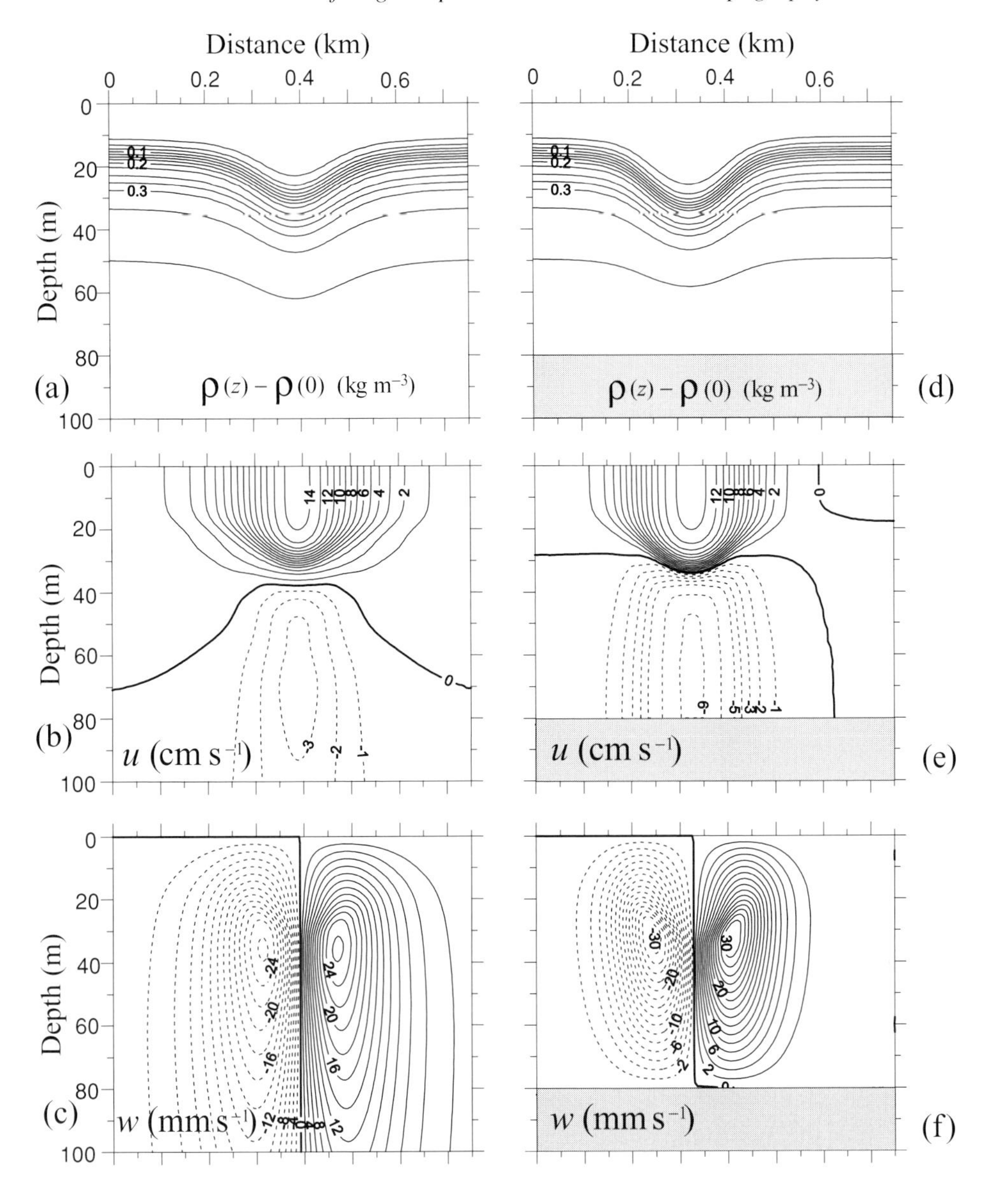

Figure 5.23. Parts (a)–(c) represent the initial K–dV solitary wave at the isobath of 200 m; (d)–(f) display the results of an adjustment of the incident wave to the conditions of the shallower basin with depth 80 m. Parts (a) and (d) show the density anomaly relative to the free surface (kg m^{-3}); (b) and (e) show the fields of the horizontal velocity u (cm s^{-1}); and (c) and (f) show those of the vertical velocity w (mm s^{-1}).

wave with an amplitude of $a_\xi = 15$ m in the deep part of the basin. Figures 5.23(d)–(f) represent the result of an adjustment of the solitary wave to the conditions of the sill of Mainau. No visible secondary SIWs or wave trains develop while the wave propagates from the deep part to the shallower part of the water. The incoming wave preserves its initial form as a solitary wave of deflection. The single evident qualitative change only concerns the nonessential transformation of the vertical structure of the horizontal and vertical velocities according to the vertical profile of the first baroclinic mode in shallow water. Quantitatively, the maxima of the u and w velocities become more pronounced as a consequence of the conservation of energy.

Note that a possible reflection, which can occur during the interaction of the SIWs with the bottom topography, does not take place either, because of the small inclination of the bottom topography (the bottom angle is less than 0.6°). Thus, backscattering of the wave is negligible, and, except for the losses due to friction, the wave conserves its energy during propagation.

So, it can be concluded that all characteristics of the medium and wave parameters change very slowly along profile 1. Therefore, at $a_\xi/(H - H_\xi) \ll 1$ adiabatic behavior of the wave prevails.

Scenario 2: Wave transformation at $a_\xi/(H - H_\xi) \sim 1$

Propagation of SIWs along section 2, which crosses the local sill A–A, leads to another scenario of the wave evolution; see Figures 5.20(b) and 5.21. This example of a wave transformation over the sill A–A is more typical and is shown in Figure 5.24. "Autumn stratification" was used as characterized by profile 2 in Figure 5.22, for which the density jump is located at about 20 m depth. For this case the maximum of the wave deflection of the incident wave lies at depth $z = -20$ m, i.e. $H_\xi = 20$ m. Note, however, that this value does not coincide with that for the K–dV soliton, which is equal to 35 m, due to reasons explained in the previous chapter. So, the distance between the horizon $z = -H_\xi$ and the top of the sill was about 30 m. Thus, the amplitude of the incoming wave, $a_\xi = 15$ m, is comparable to the distance $H - H_\xi$, and the parameter $a_\xi/(H - H_\xi) = 0.5$, implying that one should expect very strong effects of interaction. Figure 5.24 illustrates this fact but shows only the more interesting fragments of the final stage of evolution for $t \geq t_0$ (t_0 is the reference time) when the SIW approached 50 m depth. The time scale T_0 used in Figure 5.24, and henceforth employed, is λ/c_p. In the present case, $T_0 = 364$ s.

Conventionally, one can detect three stages of the SIW evolution:

(1) The first stage (prior to $t = t_0$) is characterized by a weak transformation of the propagating SIW and its adjustment to the ambient conditions over a slowly varying depth. No dispersive effects are visible. This stage is analogous to that already described.

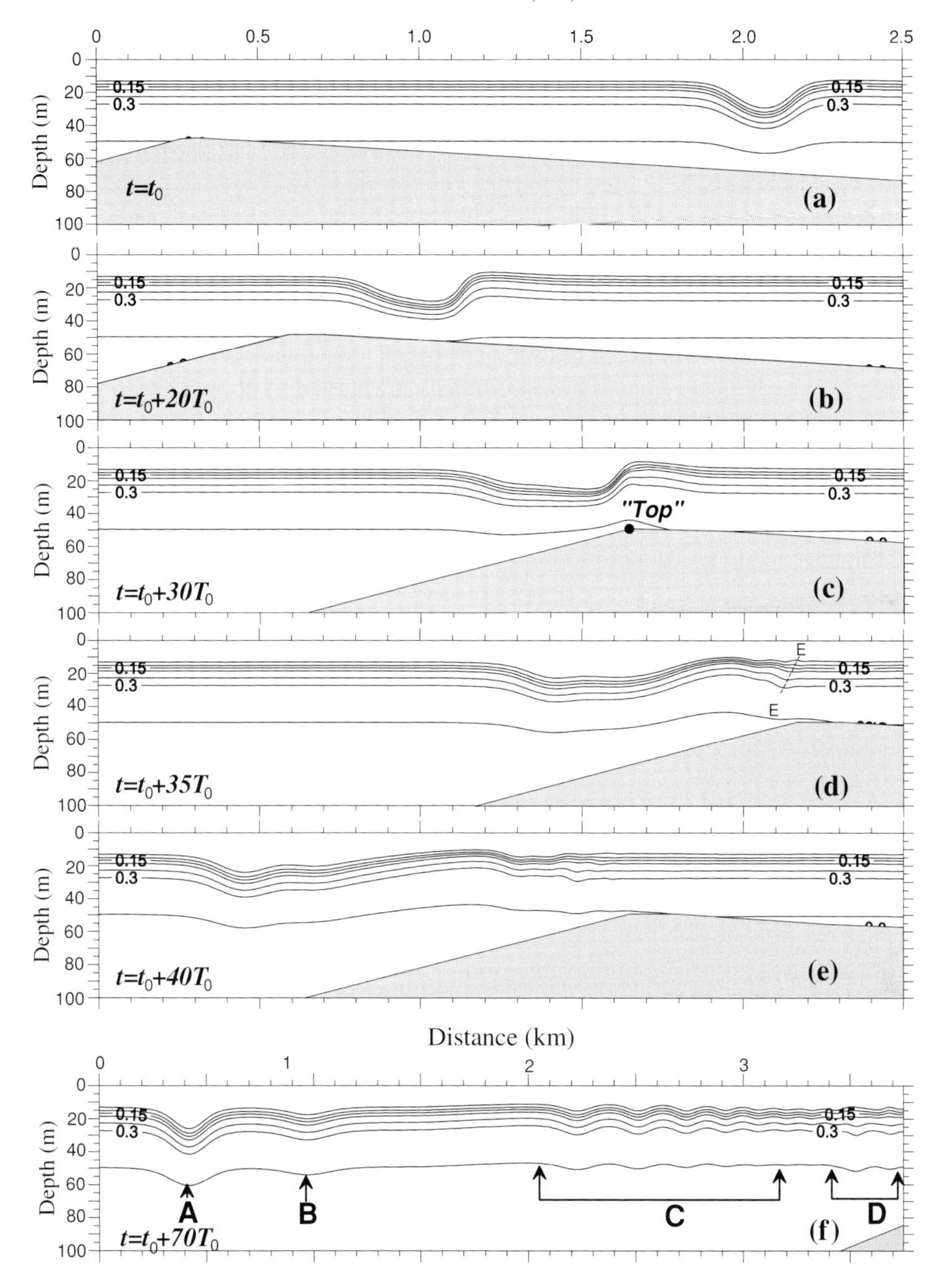

Figure 5.24. Evolution of the density field in autumn for stratification 2 in Figure 5.22 over the underwater ridge and along cross-section 2 in Figure 5.20. The amplitude of the incoming wave is $a_\xi = 15$ m. The density anomaly relative to the free surface is shown. The contour interval equals 0.05 kg m^{-3}. The leading solitary waves of depression are marked in (f) by A and B; the secondary attached dispersive wave tail is shown by segment C, and the second-mode internal waves are shown by segment D. Section E–E in (d) represents the initial stage of the generation of second-mode internal waves.

(2) During the time interval $t_0 < t < t_0 + 25T_0$, the SIW continues to propagate over a slightly inclined bottom, and its vertical scope becomes comparable to the total water depth. This leads to an increase in the nonlinear effects. First, this concerns the nonlinear phase speed, which strongly depends on the wave amplitude. The wave trough begins to propagate more slowly in comparison with the wave periphery. This effect leads to a steepening of the rear wave face; however, the frontal face becomes more and more gently sloping; see Figure 5.24(b). Very probably, the SIW would be destroyed further if the water depth continued to decrease, but beyond the point *"Top"* in Figure 5.24(c) the SIW leaves the shallow-water zone and propagates further into relatively deep water. At this point, the *dispersive* stage of the wave evolution commences.

(3) For $t > t_0 + 30T_0$, the rear steep face of the SIW, owing to dispersion, starts to transform and evolve into a secondary oscillating wave tail. The result of such a transformation is presented in Figures 5.24(f) and 5.25. One should note that the energy of the initial wave is transferred not only to the first baroclinic mode, but also to the higher modes. Fragment D in Figure 5.24(f) with counter-phase displacements of isopycnals represents second baroclinic mode waves.

Such an effect of energy transfer to a high baroclinic mode was found experimentally in ref. [105] and described theoretically in ref. [251]; it will be considered towards the end of this section. Due to the smaller phase speed, the second-mode wave packet (fragment D) separates effectively from the leading first-mode wave train (fragment C). The reason for the generation of the second and higher baroclinic modes is the sudden change of the bottom topography on the top of the sill (sharp edge). The initial stage of this second-mode generation is marked by the dashed line E–E in Figure 5.24(d).

A last remark on the *dispersive* stage of evolution: on the frontal side of the wave field, instead of the single incoming SIW, two newly born solitary waves of depression are formed (positions A and B in Figures 5.24 and 5.25). Their characteristics are very close to those of the K–dV solitons. Soliton B is weaker than A, which is why it is wider and propagates more slowly than soliton A.

In the above-described mechanism of internal wave evolution, wave reflection induced by bottom topography does not play any essential role, at least it is not discernible in Figures 5.24 and 5.25. This is understandable if one compares the scale of appreciable bottom changes, which is more than 10 km, with typical wavelengths, which are much smaller, $\lambda \sim 150$ m. The wave *behaves* locally.

The runs presented above were repeated for weaker incoming waves having amplitudes of 5 and 2 m, corresponding to nondimensional parameters $a_\xi/(H - H_\xi)$ equal to 0.2 and 0.13, respectively. All other conditions in those runs were kept the same as for the case discussed above. In these two cases the fission of an incoming wave into secondary waves over the top of the sill and behind it is much weaker. This is not surprising, because the parameter $a_\xi/(H - H_\xi)$ is smaller.

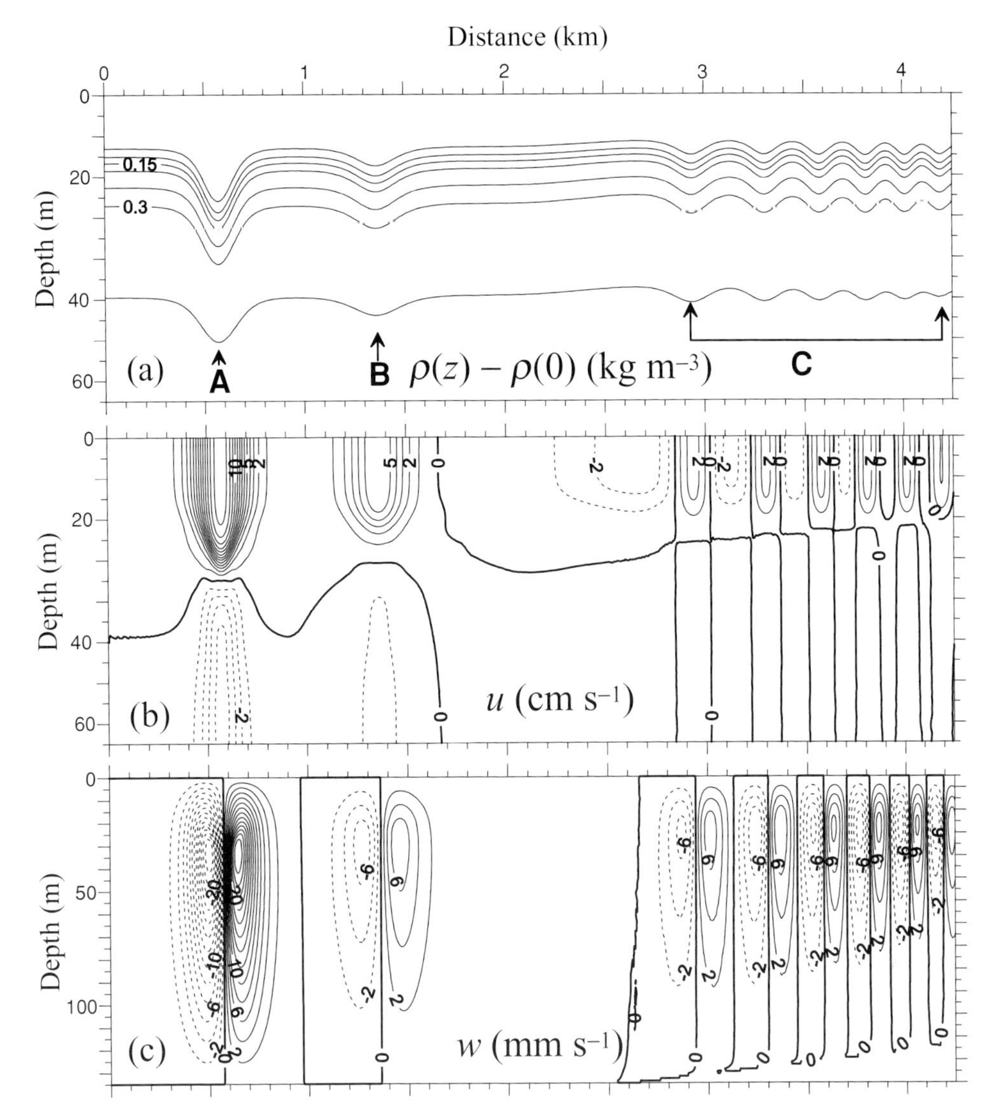

Figure 5.25. The resulting wave field at $t = 90T_0$ after the penetration of a 15 m SIW behind the underwater sill (profile 2 in Figure 5.20). (a) The density anomaly relative to the free surface (kg m^{-3}); (b) the field of horizontal velocity u (cm s^{-1}), and (c) that of the vertical velocity w (mm s^{-1}). A, B, and C indicate the same wave fragments as in Figure 5.24.

Under such conditions, the weak propagating SIW is not changed so dramatically over the top of the sill as it is for a strong wave, and it is not destroyed. As a result, the backscattered wave tail is also much weaker.

This fact of strong dependence of the efficiency of wave scattering upon the nonlinearity was checked for every wave listed above, for both summer and autumn stratification. It was found that for very weak waves, $a_\xi = 2$ m, the energy transfer from the incident wave to the wave tail does not exceed 1%.

Scenario 3: Wave breaking at $a_\xi/(H - H_\xi) > 1$

The last series of numerical runs was performed for section 3 of Figure 5.20(b); it is characterized by steep bottom topography and very shallow water of 5 m or less near Mainau Island. The bottom profile used in the model is depicted by the dashed line in Figure 5.21. Isopycnals run against the bottom, and a very strong effect of wave–bottom interaction must be expected in the shoaling process of the internal wave. An incoming wave can be completely destroyed after its overturning and breaking in a shallow-water area. An example of such destruction is displayed in Figure 5.26. The reference time $t = t_b$ is defined as the moment of wave breaking when the horizontal velocity at a point exceeds the local phase speed.

Due to nonlinear effects, the frontal face of the propagating wave flattens and, correspondingly, the rear face becomes steeper, Figure 5.26(a). Such a change of the wave profile occurs because of a smaller propagation velocity of a trough in comparison with the wave periphery; this was discussed in Scenario 2, when the evolution of SIWs propagating along cross-section 2 was discussed. The basic difference to those cases is that here the incident wave does not penetrate into a deep-water region after passing the top of the sill, but continues to propagate under the conditions of a narrowing wave guide. This leads to a progressive steepening of the rear wave face. At a definite stage of evolution, the rear wave front approaches a *vertical wall*, forms a baroclinic bore, and is subsequently destroyed.

Figure 5.26 illustrates the process of overturning of the steep rear front; it may be seen at that instant of time when the top of the propagating baroclinic bore outstrips the wave trough. In this situation, the heavier and denser water penetrates into the relatively light water layers and falls down to the wave trough, Figure 5.26(c). Thus, we conclude that wave breaking is *a kinematic instability of the wave profile*.

We will discuss this mechanism in more detail below. We want to emphasize that zones of internal wave breaking are the potential places of enhanced energy dissipation and intensive water mixing. Local coefficients of turbulent viscosity and diffusivity in such places increase dramatically, in fact by several orders of magnitude. This leads to a fast attenuation, dissipation, and overmixing of water layers and to the formation of a local zone with a new stable vertical fluid stratification. For instance, the horizontal pulsating stream described above and shown in Figure 5.26(e), at $t = t_b + 35T_0$ transforms into a more regular structure, almost without vertical density inversions (see A–A in Figure 5.26(f)).

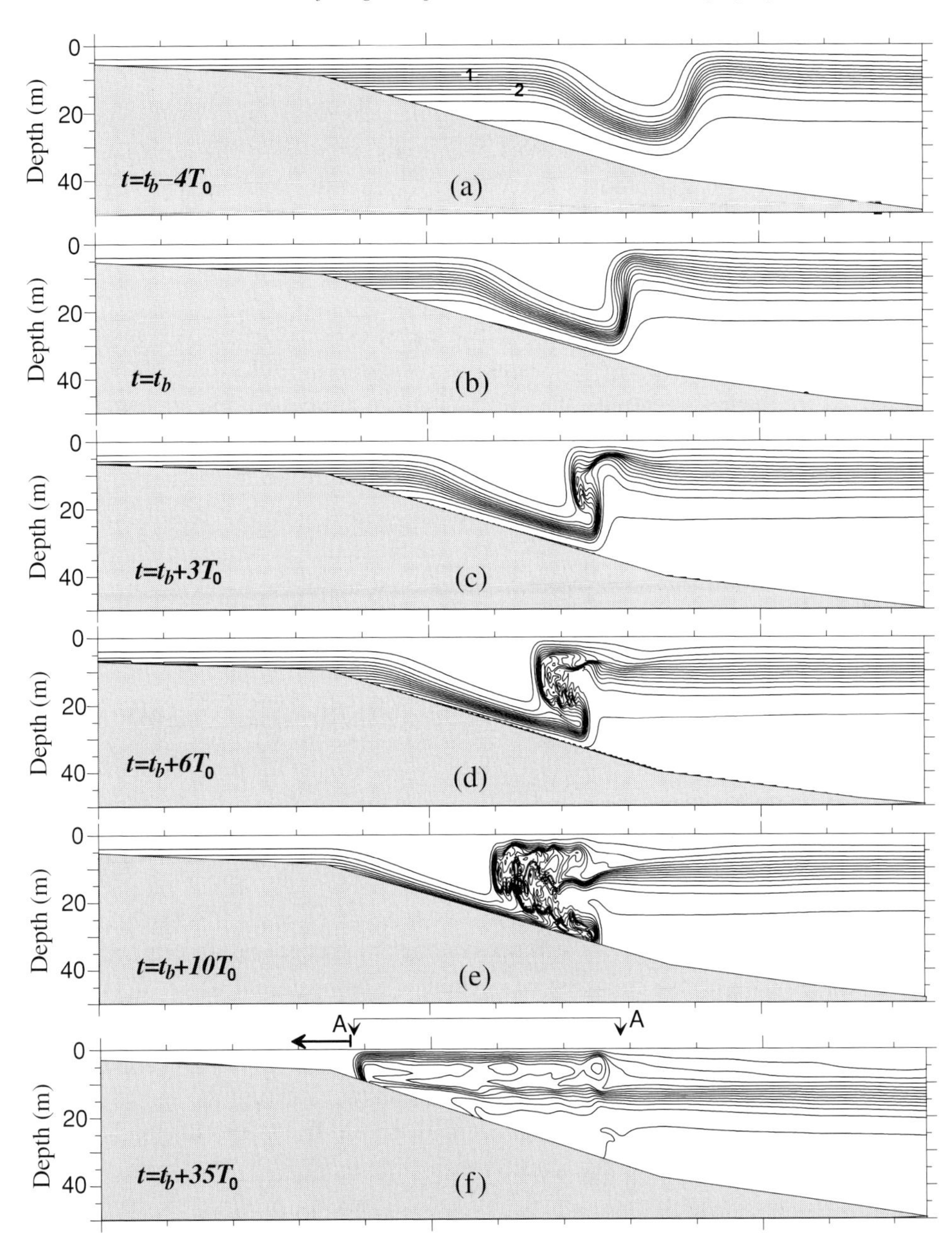

Figure 5.26. Evolution of the density field during internal wave breaking near the lake boundary along cross-section 3 in Figure 5.20 for a fluid stratification as depicted by line 1 in Figure 5.22. The amplitude of the incoming wave is $a_\xi = 15$ m. The density anomaly relative to the free surface is shown; the contour intervals equal 0.2 kg m^{-3}; and the time scale T_0 equals 83 s. Segment A–A represents a mixing zone with a newly formed fluid stratification after wave breaking.

Comparison of Figures 5.26(e) and (f) obviously shows the reconstruction of the hydrophysical fields during the corresponding 2075 s ($= 25T_0$): during this interval the density field became more regular, the maximum horizontal velocity decreased substantially, namely 1.5–2 times, and the vertical pulsations almost completely disappeared as the vertical circulating motion survived on the frontal side of the moving intrusion and on its rear side. The average velocity of propagation of the density intrusion between $t = t_b + 25T_0$ and $t = t_b + 35T_0$ was 18 cm s^{-1}.

The above results concern the evolution and destruction of a relatively large wave. Similar effects of overturning and breaking also occurred for all other considered waves. As for the basic features, the mechanisms of their evolution are very similar and exhibit the same characteristic stages: nonlinear transformation, formation of a baroclinic bore, its overturning and breaking with the formation of density intrusion, and its dissipation. Deviation from such a scenario is only observed for small-amplitude waves. We shall consider these differences in more detail in Section 5.4.2, which will be devoted to a detailed study of the breaking mechanism of internal waves.

5.4.2 Strong wave–topography interaction: breaking criterion

The objective of our modeling efforts here is to investigate the process of evolution of a solitary internal wave in a slope-shelf region during its strong overturning and breaking. Certainly such events take place in many regions of the World Ocean where strong SIWs are usually observed, for instance in the Andaman Sea [180], [188], the Sulu Sea [5], [142], at the Mascarene Ridge [120], [208], among others. Up to the present time, there is still no satisfactory theoretical description of the observational data on the breaking of oceanic solitary internal waves obtained in the Sulu Sea [30] and the Pechora Sea [216].

We follow ref. [252], which considers strong wave–topography interaction. The model assumes a plane SIW of depression with amplitude a_ξ, propagating orthogonally from the deep part, III, with depth H_3, to the linear continental slope, II, and further to the shelf, I; see Figure 5.27(a). The vertical structure of the incoming wave is defined by typical density profiles, as depicted in Figure 5.27(b), which have been measured in the Andaman Sea and the Sulu Sea [137], [138].

As we anticipate relatively strong vertical motions during wave breaking and quite intensive water mixing at the location of wave breaking, we include in our governing equations (5.30) the non-Boussinesq terms, as in ref. [157], and the dissipative terms with Richardson number dependent parameterization for the vertical coefficients of turbulent mixing in accordance with ref. [185]. Thus, in their final

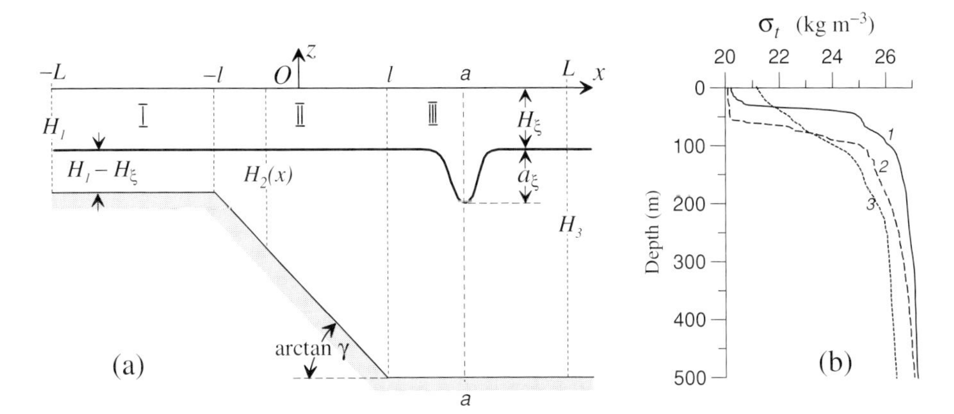

Figure 5.27. (a) Schematic diagram of the computational area; (b) profiles of conventional density used in the model measured in the Andaman Sea (1, 2) and in the Sulu Sea (3).

form the governing equations read

$$\left.\begin{aligned}(\nabla^2\psi)_t + J(\nabla^2\psi, \psi) &= g\rho_x/\rho + B(x, z, t) + D(x, z, t),\\ \rho_t + J(\rho, \psi) &= (K^{\mathrm{H}}\rho_x)_x + (K^{\mathrm{V}}\rho_z)_z.\end{aligned}\right\} \tag{5.61}$$

Here, $\rho(x, z, t)$ is the water density, and $B(x, z, t)$ and $D(x, z, t)$ represent the non-Boussinesq and dissipative terms, respectively, given by

$$\begin{aligned}B(x, z, t) &= \{-\rho_z\,[\psi_{xt} + J(\psi_z, \psi)] - \rho_x\,[\psi_{xt} + J(\psi_x, \psi)]\\ &\quad + \rho_z[(A^{\mathrm{H}}\psi_{xz})_x + (A^{\mathrm{V}}\psi_{zz})_z] + \rho_x[(A^{\mathrm{H}}\psi_{xx})_x + (A^{\mathrm{V}}\psi_{xz})_z]\}/\rho,\\ D(x, z, t) &= (A^{\mathrm{H}}\psi_{xz})_{xz} + (A^{\mathrm{H}}\psi_{xx})_{xx} + (A^{\mathrm{V}}\psi_{zz})_{zz} + (A^{\mathrm{V}}\psi_{xz})_{xz}.\end{aligned}$$

The coefficients of vertical turbulent mixing, A^{V} and K^{V}, are determined by the Richardson number dependent turbulent parameterizations,

$$\left.\begin{aligned}A^{\mathrm{V}} &= \frac{a_0}{(1 + \alpha_0 \mathrm{Ri}(x, z, t))^p} + A_{\mathrm{b}}^{\mathrm{V}},\\ K^{\mathrm{V}} &= \frac{A^{\mathrm{V}}}{(1 + \alpha_0 \mathrm{Ri}(x, z, t))} + K_{\mathrm{b}}^{\mathrm{V}},\end{aligned}\right\} \tag{5.62}$$

where the local Richardson number is defined by

$$\mathrm{Ri}(x, z, t) = \frac{N^2(x, z, t)}{u_z^2(x, z, t)}. \tag{5.63}$$

Furthermore, $A_{\mathrm{b}}^{\mathrm{V}}$ and $K_{\mathrm{b}}^{\mathrm{V}}$ are the coefficients of diffusivity, describing background turbulence, and a_0, α_0, and p are adjustable parameters; $N(x, z, t)$ is the local buoyancy frequency. These parameterizations for the vertical turbulent kinematic

Table 5.3. *Model parameters used in the numerical experiments on wave breaking.*

Δt (s)	Δx (m)	Δz (m)	a_0 (m^2 s^{-1})	$A_b^V = K_b^V$ (m^2 s^{-1})	p	$A^H = K^H$ (m^2 s^{-1})
0.2–2	1–5	0.25–1	(0.5–2) $\times 10^{-3}$	10^{-5}	1; 2	10^{-5}–10^{-1}

viscosity and diffusivity increase the coefficients A^V and K^V in regions with small values of Ri. In regions with vertical density inversions, the Richardson number is taken to be zero.

The calculations were conducted with the parameters listed in Table 5.3. It was found that the numerical scheme was quite stable, even during the overturning of a wave front. The given selection of the parameters is in accordance with a long propagation distance of SIWs without essential attenuation (500 wavelengths and more). This is in agreement with the *in situ* observations discussed in Chapter 4, which show that SIWs can propagate several hundred kilometers from their source of generation.

The background density distribution in the numerical runs was set by choosing the profiles typical for the Andaman and Sulu Seas; see Figure 5.27(b). The water depth H_3 in the deep part III of the basin was taken to be 1 km; in the shelf zone I, H_1 was constant and, in different runs, taken to have values between 200 and 100 m. We did not choose any specific bottom profiles, because of the wide variety of bottom topographies occurring in these regions. Instead, we defined the transition zone II as a linear slope region with inclination $dH/dx = \gamma$ and performed numerical runs in a wide range of inclinations to make the results more applicable to different sites of the coastline. The slope angle arctan γ varied from 0.52° to 21.8°.

The approach used here to obtain the initial fields for the incident wave is the same as that described in Section 5.3.1. The amplitudes a_ξ, equal to the maximum displacement of the isopycnals of the incident internal waves of permanent form, were 19.2, 37.9, 57.6, 81.4, and 84.4 m.

Kinematics of wave breaking

Before describing the wave breaking criterion, let us investigate a kinematic mechanism of overturning and strong breaking of internal waves. All breaking events, though different in detail, had similar features and peculiarities, which will be described in the following.

Figure 5.28 displays the fields of density and horizontal, u, and vertical, w, velocities at three different stages of wave evolution: before overturning (Figures 5.28(a)–(c)); just at the beginning of overturning (Figures 5.28(d)–(f)); and after overturning (Figures 5.28(g)–(i)). It should be mentioned here that a

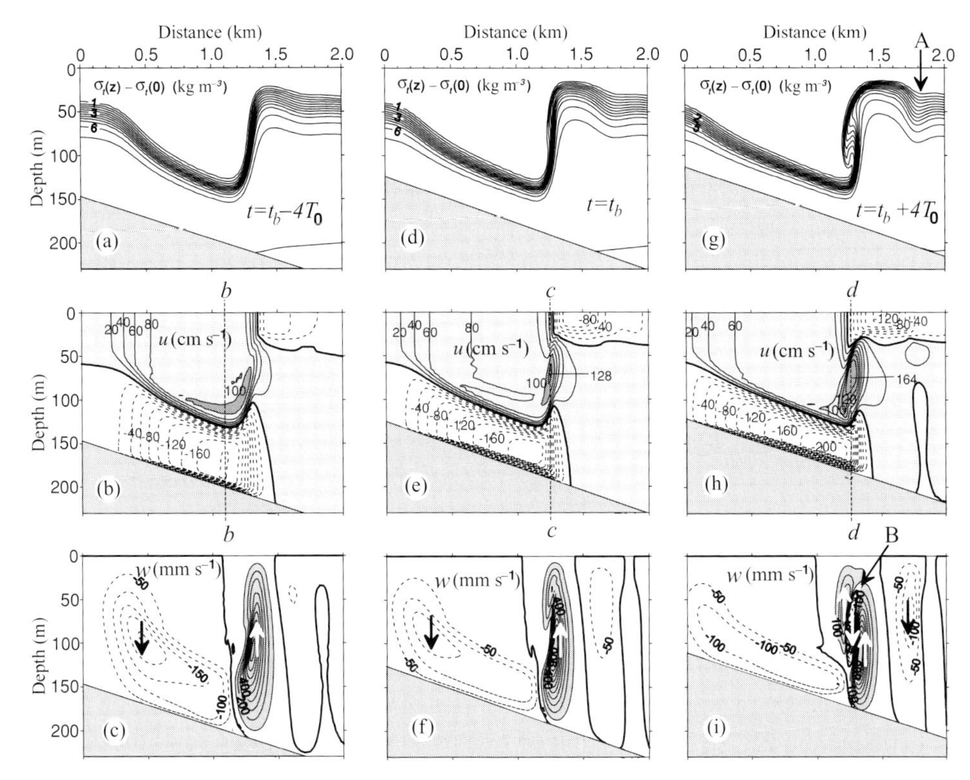

Figure 5.28. Evolution of the conventional density σ_t ((a), (d), (g)), horizontal velocity u ((b), (e), (h)), and vertical velocity w ((c), (f), (i)) during the breaking of an solitary internal wave. Parts (a)–(c) represent the wave field before breaking; (d)–(f) occur just at the beginning of the overturning of a rear wave face; and (g)–(i) depict the situation after overturning. The time scale $T_0 = 75.5$ s. Arrows in (c), (f), and (i) show the downward and upward water fluxes; position A in (g) represents the secondary generated wave; position B in (i) shows the downward water flux at the site of wave breaking.

solitary wave with an amplitude of 84.4 m, propagating from the deep part of the basin with $H_3 = 1000$ m over the slope with a relatively moderate inclination, $\arctan \gamma = 2.9^{\circ}$, penetrates into the shallow zone to the isobath of 260 m without any discernible change in its symmetry. Note that at this early stage dispersive effects are not yet developed, and a dispersive wave tail has not arisen at the downstream side of the wave. The solitary wave serves essentially as the wave profile until the 250 m isobath is reached.

Starting from the isobath at approximately 250 m, the frontal face of the wave becomes more gently sloping and, correspondingly, its trailing edge becomes steeper. This behavior of wave shoaling, obtained here by direct numerical simulations, was previously described in several experimental papers; see, e.g., refs. [91], [117],

[154]. This change in the wave profile occurs because of a smaller propagation velocity of a trough in comparison with the wave periphery. Nevertheless, this peculiarity of the shoaling of strong SIWs deserves further comment.

Generally, according to the dependence of the local phase velocity on the fluid stratification, a deeper pycnocline leads to a larger internal wave speed in a typical oceanic situation, when the interface is closer to the free surface than to the bottom. For instance, in a two-layer system, according to formula (5.16) the wave trough must propagate faster than the wave crest. Only when the pycnocline is located closer to the bottom than to the free surface does the situation reverse. Such scenarios of internal wave evolution of different forms and origins – waves of depressions, long baroclinic tidal waves, etc. – were observed in the ocean and have been described in a series of theoretical papers; see, e.g., refs. [75], [129], [130], [243].

This remark does not concern internal solitons. As has been shown above, in the latest stages of evolution the wave field disintegrates into a series of SIWs. In such waves nonlinearity and dispersion compensate each other; these waves propagate as waves of permanent form. The trough and periphery of these waves move with equal velocity, in spite of their different distances to the free surface. This result is a consequence of weakly nonlinear theories. The idea of the existence of SIWs of permanent form beyond the limit of weakly nonlinear theories was also used in Section 5.3.

As expected, our study found that large-amplitude SIWs, propagating over a slowly variable depth, do not reveal any discernible dispersive effects in the deep part of the ocean. Over variable depth such waves adjust during their evolution to the ambient conditions without signs of disintegration or formation of a secondary wave tail. Only when the wave has reached the approximate depth of 250 m does the above described steepening of the rear wave face set in. Beyond this site (at $H_2 < 250$ m), the SIW depresses the pycnocline below the middle of the water depth: the isopycnal $\sigma_t = \sigma_t(0) + 3$ kg m^{-3}, which lies in the undisturbed fluid at a depth of about 50 m falls by an additional 84.4 m in the wave center and reaches the horizon of about 134 m; see Figure 5.28(a).

Figures 5.28(d) and (g) display the isopycnal lines at the moment of overturning of the steep rear wave front; they show time slices when the top of the propagating baroclinic bore just reaches overturning or outstrips the wave trough. In this situation, the heavier and denser water penetrates into the relatively light water layers and falls to the wave trough, Figure 5.28(g). Thus, we conclude that wave breaking is due to the *kinematic instability* of the wave profile.

In the following, we will discuss in more detail the analysis of this fact, by comparing the orbital velocities of the fluid particles and their propagation (phase) speed. Before addressing these points, let us analyze the qualitative changes in the patterns of the horizontal and vertical velocities which take place during an overturning event.

The striking feature of these patterns is the following. Before breaking, the maximum of the horizontal velocity shifts from the free surface, where it was located in the initial wave, into the deep layers, i.e. to the wave trough, see Figure 5.28(b). It then shifts to the lateral boundary of the steep slope, Figure 5.28(e), where the value of the maximum velocity increases up to the wave breaking. The overturning event is clearly seen from scrutiny of the evolution of the w-field. Comparison of Figures 5.28(c), (f), and (i) shows the appearance of the downward water fluxes after the wave overturning, at position B in Figure 5.28(i), in a zone where upward fluxes of water otherwise occur.

The kinematic nature of the breaking of internal waves described above becomes more understandable when horizontal velocity profiles are compared with the local phase speed of the wave. Figure 5.29 shows snapshots of $u(z)$-profiles, corresponding to the vertical cross-sections a–a in the wave center in the deep part of the

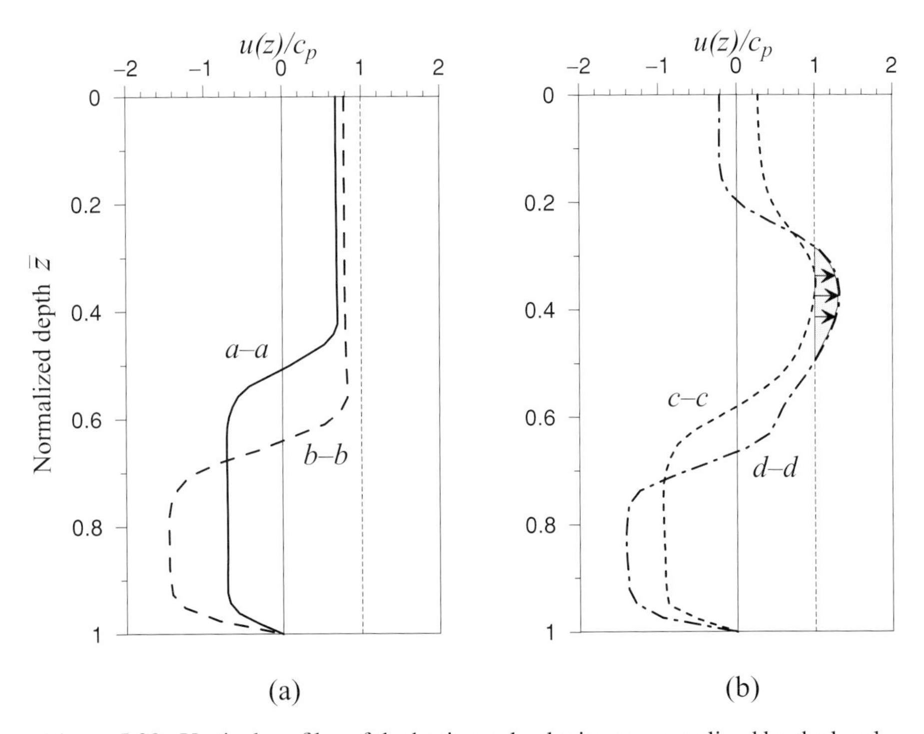

Figure 5.29. Vertical profiles of the horizontal velocity, u, normalized by the local phase speed, c_p. Profiles are computed for cross-section a–a, related to the wave center far from the slope, and for cross-sections b–b, c–c, and d–d, depicted in Figure 5.28. Part (a) corresponds to the vertical structure of the u velocity before wave breaking, and (b) corresponds to that during the overturning of the rear wave face.

ocean of zone III (Figure 5.27), and $b-b$, $c-c$, and $d-d$ (Figures 5.28(b), (e), and (h)). Vertical dashed thin lines in Figure 5.29 mark the local phase speed, which in fact diminishes with the shoaling of the wave. Before breaking, a local maximum of both profiles $a-a$ and $b-b$ exists at depths $\bar{z} = 0.44$ and 0.57, respectively. Nevertheless, the magnitude of the velocity at this site is everywhere less than the local wave speed. The next two profiles, $c-c$, and $d-d$, reveal wave breaking. Now the magnitude of the horizontal velocity in the peaks exceeds the local phase speed.

So, as follows from the numerical runs, *a kinematic instability is obviously responsible for the mechanism of strong wave breaking rather than a shearing instability*, as was suggested by Kao *et al.* [117]. Internal wave breaking occurs on the steep rear slope of the wave at the location where on-shore fluid velocities exceed the phase speed of the wave.

Such a mechanism was assumed in ref. [262] to explain the destruction of periodic internal waves over inclined bottom waves. For laboratory conditions, the process described above was reported in refs. [91] and [92], and more recently in refs. [85] and [154], where solitary internal waves were studied. Despite this fact, the likelihood of attenuation and breaking of internal waves under the action of a shear instability should not be neglected as a possibility as it was found in ref. [22] and was also shown in our numerical experiments. Figure 5.30 displays the fields of the Richardson number $\mathrm{Ri}(x, z)$ computed for the initial incident wave in zone I (Figure 5.30(a)) and just before wave breaking (Figure 5.30(b)), corresponding to the situation displayed in Figure 5.28(d). The zone with $\mathrm{Ri} < 0.25$ is rather small in the incident wave, only in the center of the solitary wave, and the minimum value of the Richardson number here does not drop below 0.2.

The situation is quite different at the instant of wave breaking. In the region of breaking, when the vertical fluid stratification is still stable, the Richardson number falls to the level 0.1, and the zone with $\mathrm{Ri} < 0.25$ occupies a rather extended area. Thus, the generation of shear instability and the increasing of turbulent mixing are very likely here, although, as was discussed and concluded above, the basic input for wave breaking is provided by the kinematic instability. Nevertheless, for a complete analysis, the existence of shear instability should also be taken into account.

Figure 5.31 represents the ultimate fate of a solitary wave after breaking. As seen from the evolution of the density field, the water does not mix instantaneously at the location of the overturned rear face. Observe, however, the formation of the "mushroom structures" at the beginning of the breaking of the baroclinic bore, Figures 5.31(b) and (c). The dense water that is penetrating into the light layers behaves as an upstream propagating jet. It undergoes multiple reflections from the near surface and near bottom layer, Figures 5.31(c)–(e). At the latest stage

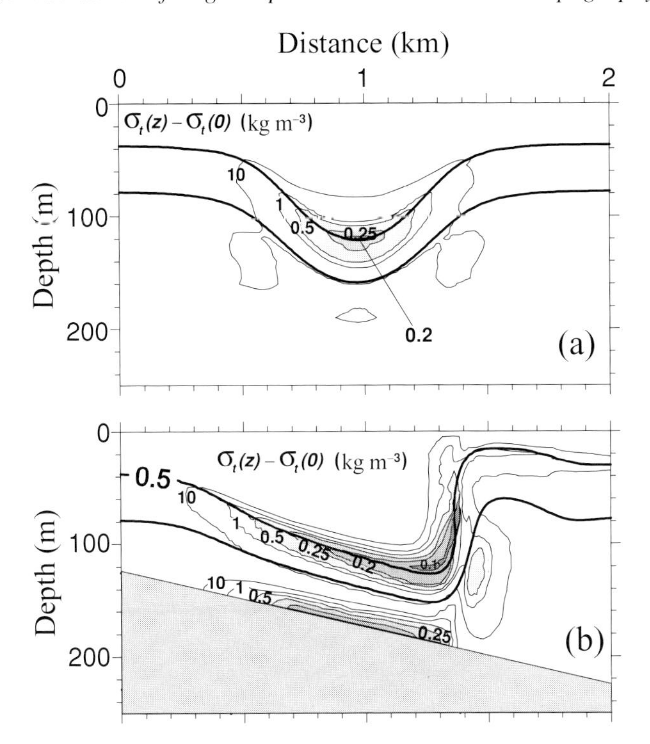

Figure 5.30. Fields of the Richardson number calculated for the incident wave (a) in the deep part of basin III and (b) over the slope area II, immediately before wave breaking. The thick solid lines represent the isopycnals with values 0.5 and 6.0. The thin lines are isolines of the Richardson number.

of the evolution, the propagating jet gradually loses its coherent structure and transforms into a horizontal turbulent pulsating density current. In Figure 5.32 this jet is graphically displayed in greater detail. It is characterized by the upslope directed core, which is located in the intermediate layers and which is exposed to vertical oscillations.

The peculiarities of the internal wave dynamics in the slope-shelf area considered above can serve as an explanation of some experimental evidence of oceanic SIWs breaking over an inclined bottom topography. For instance, in the Sulu Sea, the shoreward flux (internal swash) was fixed near the shelf edge off the coast of Palawan Island; this was why a sudden shift occurred of the CTD probe to the coast by the arrival of the upslope flow produced by the shoaling of the large solitary wave [30]. Besides, the inverse structure of the density, as exhibited in Figure 5.32(a), in the area of the upslope flux below the core was measured on the Pechora Sea Shelf after internal wave breaking had occurred there [216].

Note that the above example of overturning and strong breaking is one of the possible scenarios – probably the crucial one – of the evolution of large-amplitude

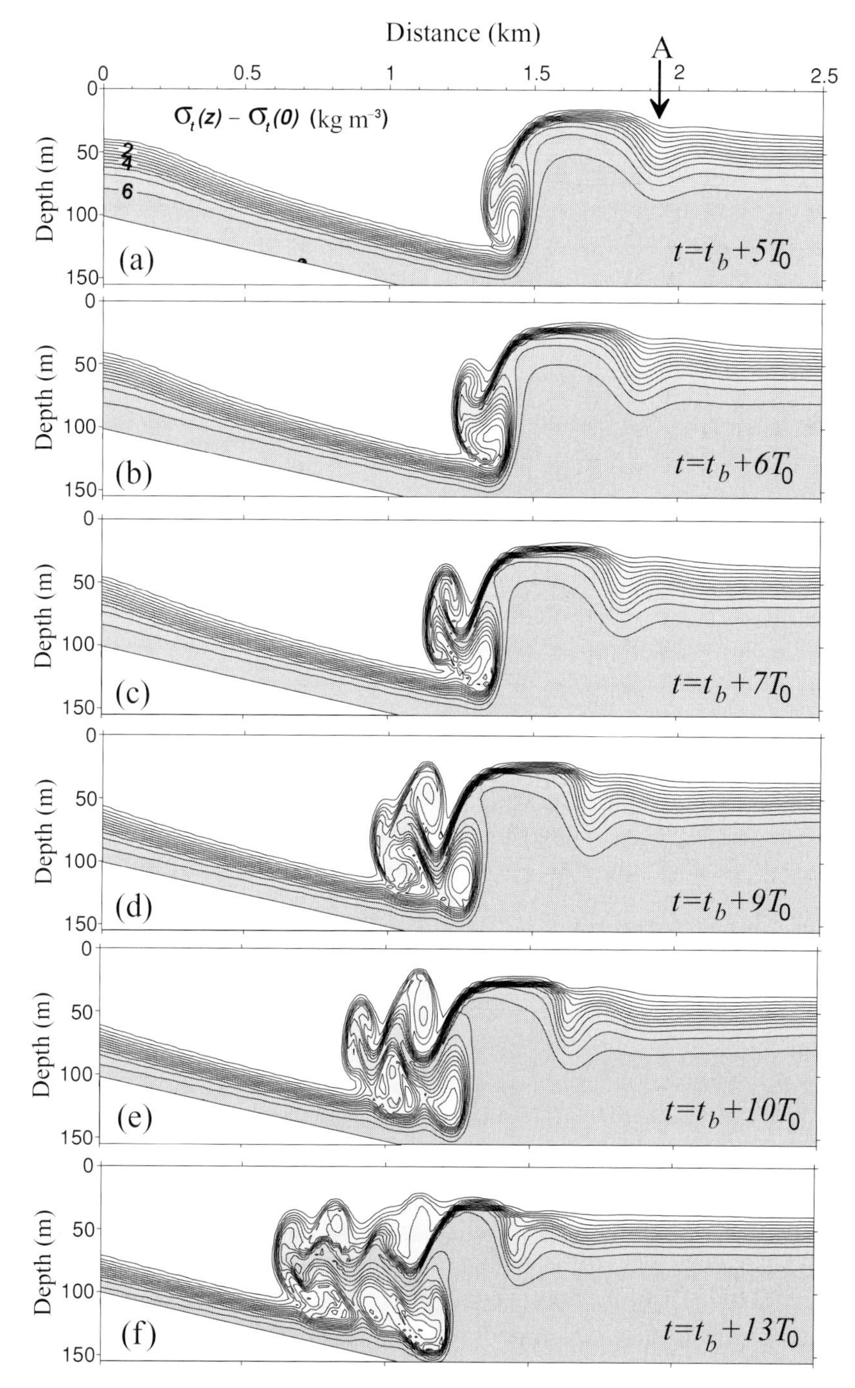

Figure 5.31. Evolution of the density field during wave breaking. The contour interval is 0.5 kg m^{-3} and the time scale $T_0 = 75.5$ s.

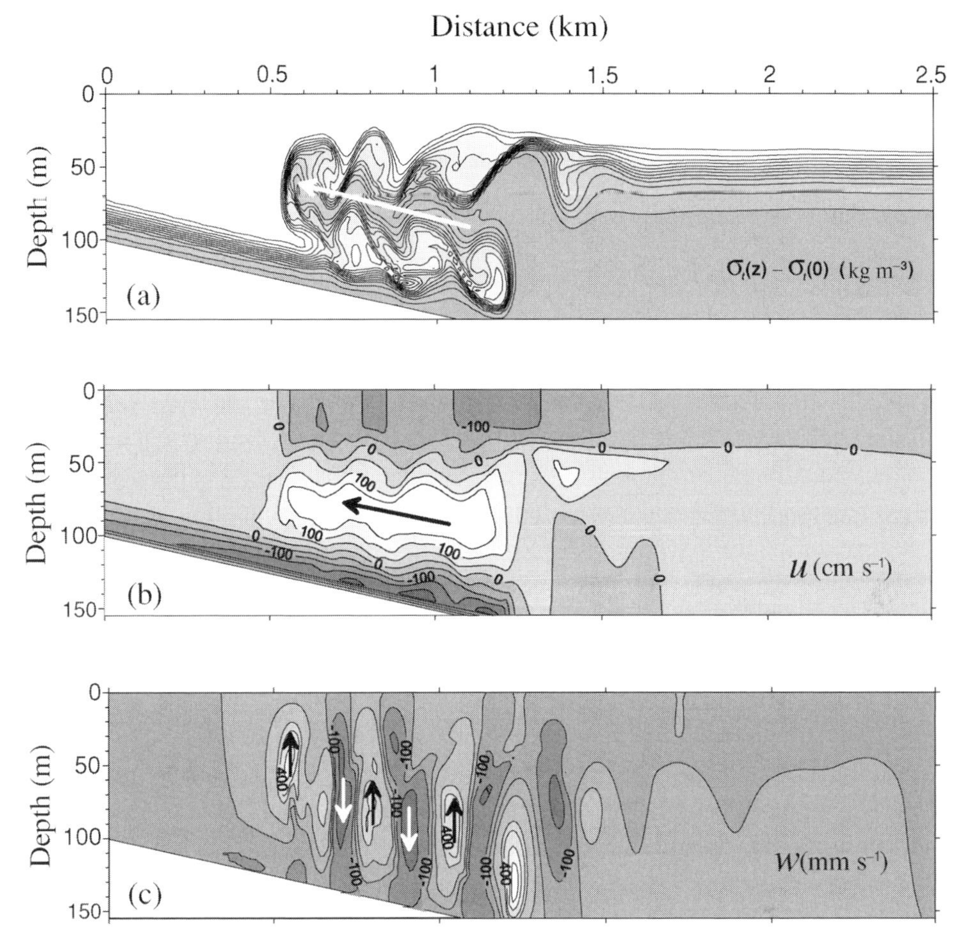

Figure 5.32. Fields of (a) the conventional density σ_t, (b) the horizontal, u, and (c) the vertical, w, velocity 1057 s after breaking from the rear wave face. The contour interval for the density is 0.5 kg m^{-3}; those for horizontal and vertical velocities are shown in the graphs. White and black arrows in (a) and (b) show the direction of propagation of the density intrusion. White and black arrows in (c) indicate down- and upwelling, respectively.

SIWs over an inclined bottom. Of course, intense SIWs can also penetrate onto a shelf without destruction. To know the conditions of internal wave breaking is therefore vital; these will be discussed below. When a SIW evolves in a shelf-slope area without overturning, and when its parameters satisfy the conditions of weak nonlinearity, the evolution process can be investigated with the K–dV equation, as discussed above. Since the nonlinear and dispersion coefficients of the K–dV equation are very sensitive, not only to the variations of the bottom topography, but also to the vertical stratification and to shear, the temporal and spatial changes of the latter – expressed as temporal and spatial variabilities of stratification and background currents – can lead to large variations of the coefficients of the K–dV equation (the quadratic nonlinearity can even change its sign). A change of sign leads to several interesting effects in the wave evolution, because, e.g., the resulting wave field significantly depends on the sign of the cubic nonlinear term and on the amplitude of the incoming wave. For more details, see ref. [84].

In this section we focus only on the investigation of the mechanism of strong SIW breaking and the conditions of its occurrence. However, other interesting effects were also found. For instance, the incoming wave in the transition area may split into transmitted and reflected waves, and this fission is more visible when $\arctan\gamma > 10^\circ$. The characteristics of the wave field in the shallow-water zone and the back-radiated waves via their structure, amplitudes, and modal composition depend strongly on the bottom angle, the parameter $H_1 - H_\xi$, the position of the pycnocline, and the wave amplitude. At large bottom inclination, it is possible to generate not only first- but also second-mode secondary transmitted and/or reflected baroclinic waves, solitons, and soliton trains. More details on this point are considered in Chapter 6.

Breaking criterion

Scrutiny of all the numerical runs showed that a breaking criterion could be determined. It was found that, besides the slope angle, another fundamental parameter controls the process of whether a solitary wave passes into the shallow shelf zone I or will break somewhere over the slope region. This parameter is the ratio of the wave amplitude a_ξ to $H_b - H_\xi$, i.e. the distance from the undisturbed isopycnal of the maximum amplitude, located at the depth H_ξ, to the bottom (see Figure 5.27(a) and inset in Figure 5.33).

Very often, a two-layer approximation of water stratification is used for the analysis of oceanic solitary waves. In such a case, the position of the maximum amplitude coincides with the interface, and the value $H_1 - H_\xi$ is, in fact, the depth of the lower layer on the shelf. In our case, for a typical oceanic fluid stratification, the situation is more complicated, and the maximum of the isopycnal depression lies below the seasonal pycnocline.

To estimate the expediency of using a two-layered approximation for the breaking criterion, note that the maxima of the vertical displacement eigenfunction for the presented density profiles 1 and 2 with density jump at 50 and 80 m depth (see Figure 5.27) are located at 79 and 180 m depth, respectively. So, the depth of the maximum displacement in the two-layer model and that of the continuously stratified model differ by approximately a factor of two, if we define the position of the buoyancy frequency maximum as an interface. As a consequence, all estimations obtained with the use of the parameter H_ξ in the two models will

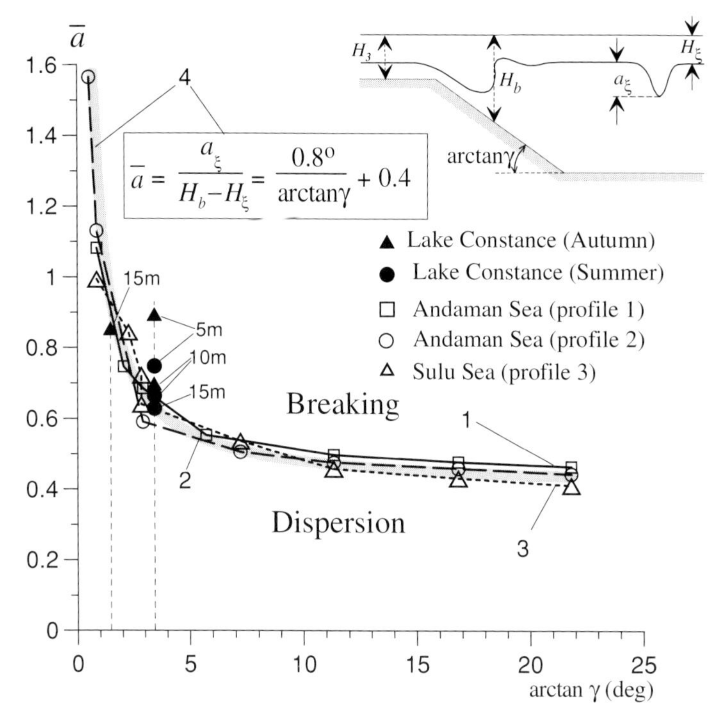

Figure 5.33. Breaking events obtained in the model. Curves 1–3 correspond to the density profiles depicted in Figure 5.27(b). The thick shaded line 4 is the best fit for lines 1–3. It separates the fields of parameters for which internal wave breaking takes place (above the curve) and of those for which the wave passes onto the shelf as a dispersive wave train (below the curve). The inset explains the parameters used for the breaking criterion.

differ by about 100%. The situation is even worse for profile 3 in Figure 5.27. In this case, the pycnocline is very wide. It is not evident where the interface in a two-layer model should be placed if one tries to simplify the fluid stratification depicted by line 3. Probably a multilayered model is more appropriate in this case.

Thus, we define the value H_ξ by the undisturbed position of the isopycnal line which has its maximum depression, a_ξ, at the center of the wave (see the inset in Figure 5.33). Note that H_ξ coincides with the depth H_ξ^{lin} of the eigenfunction maximum defined by the linear boundary value problem only in the limit of infinitesimal waves. For strongly nonlinear solitary internal waves, the maximum of the wave amplitude, as found in Section 5.3, is shifted away from the depth $z = -H_\xi^{\text{lin}}$, and this shift increases with the growth of the wave amplitude.

The breaking criterion is shown in Figure 5.33, where $a_\xi/(H_\text{b} - H_\xi)$ is plotted against the slope angle $\arctan\gamma$. Only results with breaking events are presented here. The parameter H_b is the water depth at the point of wave breaking (see inset in Figure 5.33). The location of wave breaking, or the break point, is defined as the position where the orbital velocities begin to surpass the phase speed.

One should note the excellent coincidence of the three curves obtained for the different density profiles. The best fit for these curves is curve 4, with the analytical representation

$$\bar{a} = \frac{a_\xi}{H_b - H_\xi} = \frac{0.8^\circ}{\arctan \gamma} + 0.4. \tag{5.64}$$

This curve separates the field of parameters $\bar{a}$, γ, where breaking takes place (above curve 4) from the region where the wave passes onto the shelf without breaking if the point $(\arctan \gamma, a_\xi/(H_3 - H_\xi))$ lies below curve 4.

The formulated breaking criterion for continuously stratified fluids generalizes the results obtained in refs. [91] and [92] for a two-layer system. Over the range of parameters examined in these references for laboratory conditions, breaking occurs when the undisturbed lower-layer depth was about two to three times the wave amplitude a_ξ. It was found that for $a_\xi/(H_b - H_\xi) < 0.3$ the incident wave moved onto the shelf without instabilities, and for $a_\xi/(H_b - H_\xi) > 0.4$ a strong overturning occurred above the slope. The results reported in refs. [91] and [92] showed a slight dependence on the bottom angle.

In this connection, one should also mention a weak dependence of the wave breaking on the bottom angle for a continuously stratified fluid in the range $\arctan \gamma > 5^\circ$, when the wavelength of the incident wave is comparable to the length of the transition zone. At the same time, wave breaking strongly depends on the bottom angle when $\arctan \gamma < 5^\circ$, i.e. over a gently sloping bottom. This result has not been previously cited, and we will discuss it in more detail.

The reason for such a strong dependence consists in the dispersive effect working together with the nonlinearity. In the incident wave these two effects compensate each other, and the initial wave propagates in the basin of constant depth as a solitary wave of permanent form. Over the inclined bottom this balance is violated, and the solitary wave begins to change its shape. Over a short transition zone II ($\arctan \gamma > 5^\circ$), the nonlinear effects predominate, and the incident wave is destroyed before the secondary waves can be radiated. For instance, one can identify only a weak secondary wave in Figures 5.28 and 5.31 behind the rear face of the overturning solitary wave (position A) as a manifestation of dispersion.

The situation is quite different when a solitary wave propagates over a gently sloping bottom. An example illustrating this situation is presented in Figure 5.34. Because of a long propagation over a slightly inclined bottom, a secondary wave tail may evolve effectively because of dispersion. Position B in Figure 5.34(b) shows the initial stage of the wave tail development. Finally, a well developed wave train is clearly visible in Figure 5.34(d) (position range C). A significant part of the energy

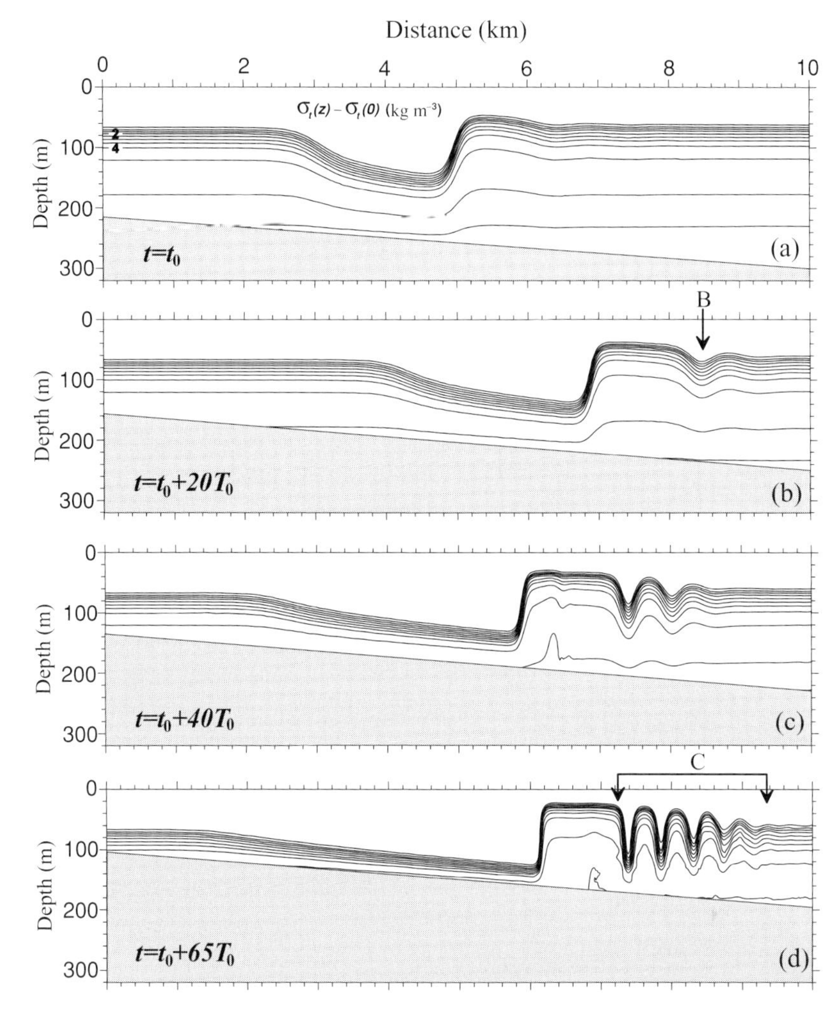

Figure 5.34. Evolution of the density field over a gently sloping bottom ($\arctan\gamma = 0.52°$). The contour interval is 0.5 kg m^{-3} and the time scale $T_0 =$ 155 s. The secondary dispersive wave tail is shown by segment C in (d).

of the incident wave is transferred to the dispersive wave tail; this explains why larger waves can penetrate longer distances without breaking over a less inclined bottom topography. This effect is the more pronounced the more gently sloping the bottom topography is.

These results concern the case of a strong wave–topography interaction. If the wave amplitudes are relatively small, the breaking event is not so pronounced; it would lead in this case to a definite correction to the above formulated criterion. Figure 5.35 shows the evolution of a 2 m amplitude SIW in the slope-shelf area of Lake Constance for summer stratification. Qualitatively, the processes of

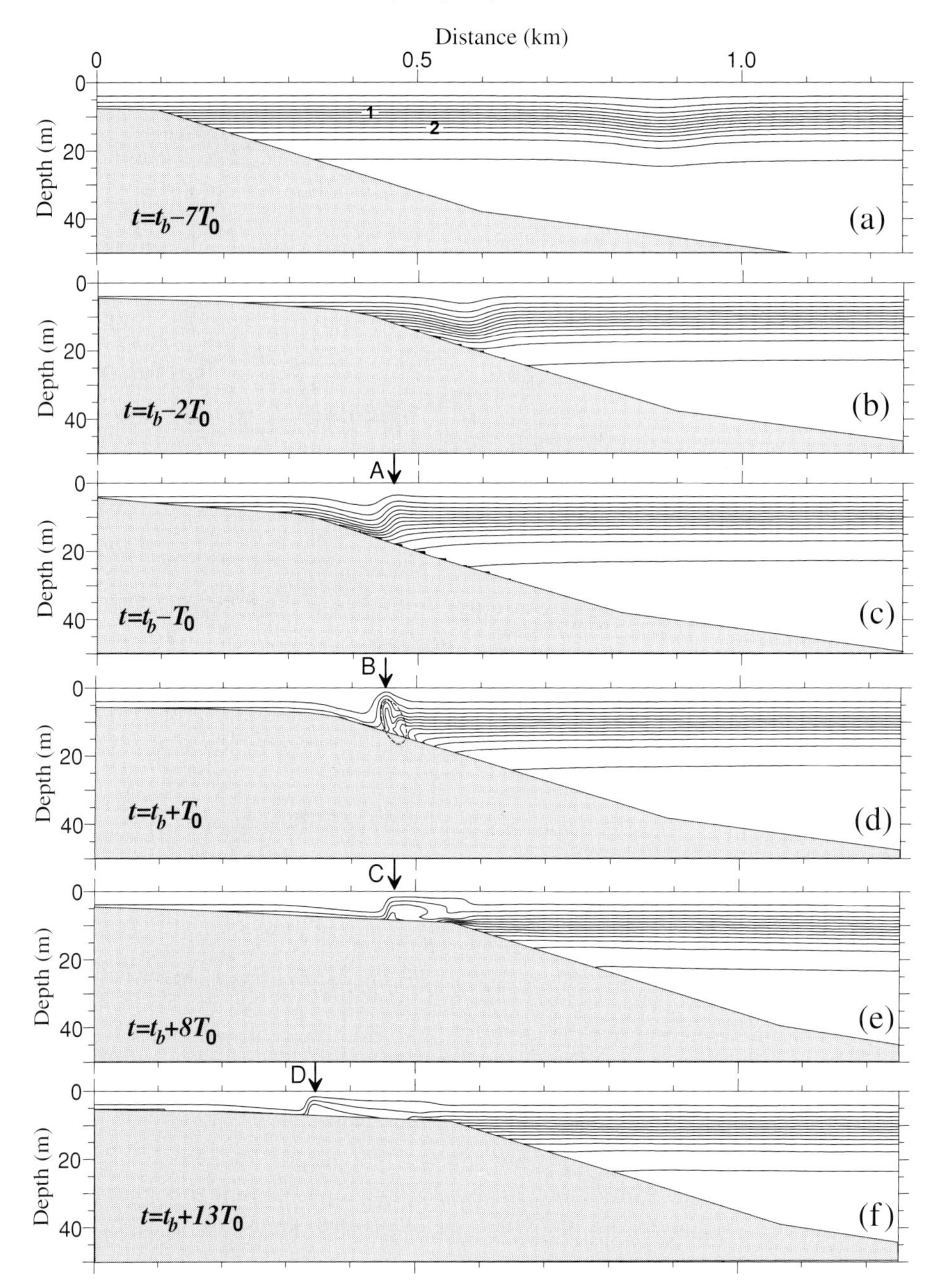

Figure 5.35. Interaction of a small-amplitude internal wave ($a_\xi = 2$ m) with the bottom topography. All designations are the same as in Figure 5.26, except the time scale; here $T_0 = 330$ s. Arrows marked by letters A–D show the formation of an upslope propagating bolus.

transformation of the small-amplitude wave at the initial stages look very similar to those described for strong waves. At the beginning of the wave transformation, a SIW penetrates into a shallow-water zone without any visible changes of its form up to the depths where the pycnocline crosses the bottom surface, Figures 5.35(a), (b). The essential nonlinear transformations arise at $t = t_b - T_0$, when the vertical scales of the isopycnal depressions and the pycnocline thickness are comparable to the total water depth, Figure 5.35(c). Position A identifies the local isopycnal elevation, formed due to the nonlinearity of the wave process.

Position B shows the place where the internal wave breaks. Note, however, that the breaking process of the relatively weak wave is not so pronounced as that for large-amplitude waves. A relatively weak nonlinearity leads to the result that a small-amplitude wave overturns not over the whole depth, but only near the bottom. The local zone of overturning is marked by an ellipse in Figure 5.35(d). The near-surface isolines do not reveal a tendency to overturn.

This last circumstance leads to a slightly different scenario of wave breaking, which is valid for a weak nonlinearity: the internal wave overturns and is destroyed only near the bottom, where a bottom boundary layer (BBL) is formed; the turbulent pulsating spot of overmixed water, which arises after overturning, disappears rapidly due to the high level of turbulent viscosity and diffusivity in the BBL. Beyond this, the wave motion does not have stochastic character, but is rather deterministic.

At the same time, the near-surface layers of the SIW, which were not exposed to overturning, form a wave of elevation, which propagates upstream. Fragments of such a wave are shown in Figures 5.35(e) and (f) by the letters C and D. So, in case of weakly nonlinear waves, instead of a turbulent pulsating jet, waves of elevation (or *boluses*) are produced in the breaking area [253]. This conclusion is in agreement with the results of numerous laboratory experiments on the breaking of internal waves; see, e.g., refs. [91] and [92].

The last remark concerning the difference between the processes of wave breaking for strong and weak waves implies that there are consequences for the breaking criterion. The breaking events obtained in the present series of numerical runs for relatively weak waves (conditions relating to Lake Constance) are presented in Figure 5.33 by the black triangles and circles. Even though the breaking criterion was constructed with lake applications in mind, one should note its good agreement for results obtained with similar data found for oceanic conditions of relatively large waves ($a_\xi > 5$ m). Breaking events for small-amplitude waves ($a_\xi \leq 5$ m) lie above the breaking curve. This result can be linked to the different manifestations of the breaking mechanism for strong and weak waves. Some remarks about this difference were made above. Of course, a definite discrepancy can follow also from the fact that the breaking criterion (5.64) was found for a flat inclined linear bottom topography, whereas the bottom profile 3 in Figure 5.21 is not uniform and has a whole range of characteristic angles of inclination. For instance, in autumn a SIW with amplitude 15 m is destroyed further away from the coastline than in summer (bottom inclination angle of 1.5° instead of 3.4°; see Figure 5.33). Nevertheless, one can state that the breaking criterion found for oceanic conditions can be applied for lakes if internal waves are relatively strong ($a_\xi > 5$ m).

5.4.3 Generation of high baroclinic modes by wave–topography interaction

The goal in this section is to study the dynamics of the interaction of SIWs with a sill, and so answer the question about the possibility of the generation of second-mode solitons during such interactions. This was formulated as a hypothesis [105] on the basis of laboratory experiments, and was studied theoretically in ref. [251].

Experimental setup and measuring technique

The experimental arrangement and measuring techniques have been reported in refs. [52] and [266]. Here, we shall briefly describe the experimental setup and typical experimental data obtained in ref. [105].

Experiments were conducted in a glass-walled wave tank 10 m long, 33 cm wide, and 35 cm high. A schematic of the experimental arrangement with explanations of the basic parameters is presented in Figure 5.36. A salt-stratified system was constructed by filling the tank with fresh water of density 10^3 kg m^{-3} in the upper layer and then carefully pouring salt water, with an approximate density of 1022.5 kg m^{-3}, below it. The result is a two-layered configuration with a steep density transition through a diffuse interface (Figure 5.37).

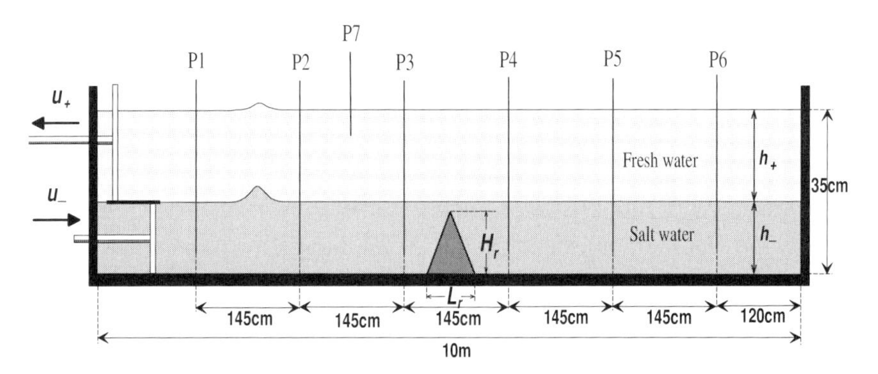

Figure 5.36. Sketch of the experimental arrangement (frontal view). On the left are two pistons, separated by a thin plate at the interface level, which simultaneously move in opposite directions with velocities u_+ and u_-, displacing the same volume of water. Six electrical resistivity gauges P1–P6 record the interface elevation. A seventh gauge P7 records the free surface motion. The triangular sill with height H_r and width L_r is placed in the middle of the tank.

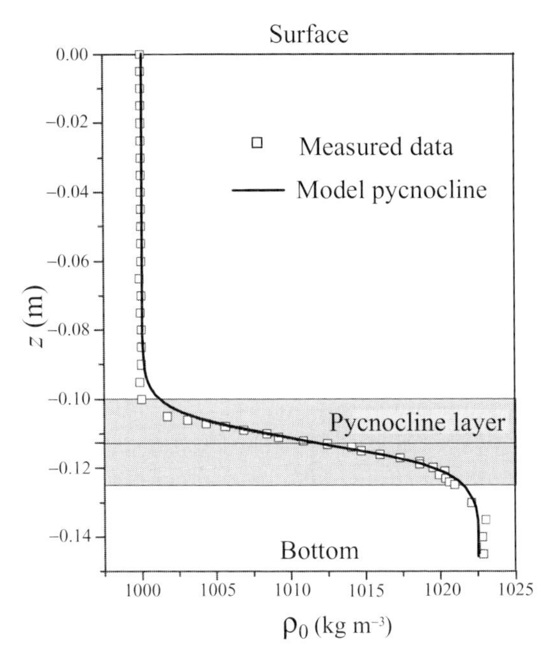

Figure 5.37. Typical density profile in the channel. The measured data points (□) were fitted with the smooth model pycnocline (5.17).

Internal water waves, excited by the wave generator at left, travel down the channel to the right. The wave generator consists of two pistons at the heights of the two water layers and covering the whole width of the channel; they are moved in opposite directions in such a way that the transported water volume is the same in the two layers ($u_+h_+ = -u_-h_-$). The main idea in generating solitary waves is to produce a positive elevation of soliton-like shape that will then develop into a solitary wave during propagation. To achieve pure baroclinic waves free surface elevations should be kept as small as possible.

The internal wave propagation and its interaction with the sills that are built in later are measured by six gauges, P1–P6, along the channel. An additional surface gauge, P7, is installed to ensure that there is no significant surface elevation.

The experiments

The experiments investigated the interaction of solitary waves with a sill, built on the ground of the channel, Figure 5.36. When the lower layer is partially blocked by a sill, an incoming solitary wave will be split into a reflected part and a transmitted part. Depending on the degree of blocking, $B = H_r/h_-$, defined as the ratio of the sill height H_r to the height of the lower layer h_-, the forward and backward moving split waves either keep a solitary character or are changed into oscillating wave trains. As a special case, the excitation of a second transmitted solitary wave following the first at lower speed was observed. This phenomenon is the main subject of the ensuing description of the experiments and theoretical modeling.

There are two main parameters characterizing the system: first the soliton amplitude, and secondly the degree of blocking. The soliton amplitude is connected to the Froude number $\mathrm{Fr} = u_{\mathrm{max}}/c_{\mathrm{p}}$, which is defined as the maximum piston velocity of the wave generator u_{max} normalized by the linear phase speed c_{p}. The value of Fr was changed between $\mathrm{Fr} = 0.1$ and $\mathrm{Fr} = 0.8$. The best results for soliton excitation were found with $\mathrm{Fr} = 0.2$. This value generated an elevation amplitude of approximately 5 mm. Higher Froude numbers produce oscillating wave trains behind the solitary wave, especially when the wave interacts with the sill. The degree of blocking, B, was varied from 0.5 to 1. Lower values of B do not have an observable effect on the transmitted wave, whilst higher values will break the transmitted soliton down to a simple oscillating wave train. The best results, i.e. the splitting of the incoming soliton into two parts, while keeping the solitary character, were achieved for $0.7 < B < 0.9$.

An additional variable, the steepness of the sill (or its width L_{r}), typifying the geometry of the sill, is important but does not concern the higher mode structure. It may be seen from the experiments that L_{r} seems to have only a small influence on the interaction with the waves. The transmitted waves are very similar to those transmitted by a short sill with the same value of B, except for their smaller amplitude. Other parameters, such as the water density and temperature, were kept approximately constant and are not considered any further here.

Typical experimental data

Figure 5.38 shows a typical data sheet obtained from the experiments. The elevation curves, measured by the gauges, are plotted versus the time at the positions of the gauges. For each curve the same elevation scale, indicated in the upper right corner, is used, while the vertical position of the curve corresponds to the gauge position. The initial peak was approximately 5 mm, decaying to about 1 mm while propagating all the way down to the end of the channel. The sketch of the channel to the left of Figure 5.38 explains the position of the sill and the gauges along the channel length. The dotted lines with arrows mark the path of the first-mode soliton; its gradient determines the velocity, which is approximately 7 cm s^{-1}. One sees easily the splitting of the wave at the sill in the middle of the channel and the multiple reflections at both channel ends. The solid lines with arrows show the propagation of the second-mode wave generated by the sill. Here the speed is only 2 cm s^{-1}, even though the amplitude is almost equal to the transmitted first-mode peak. One can also indicate in Figure 5.38 a reflected second-mode wave; however, owing to the smallness of the signal its existence in the experiment is questionable.

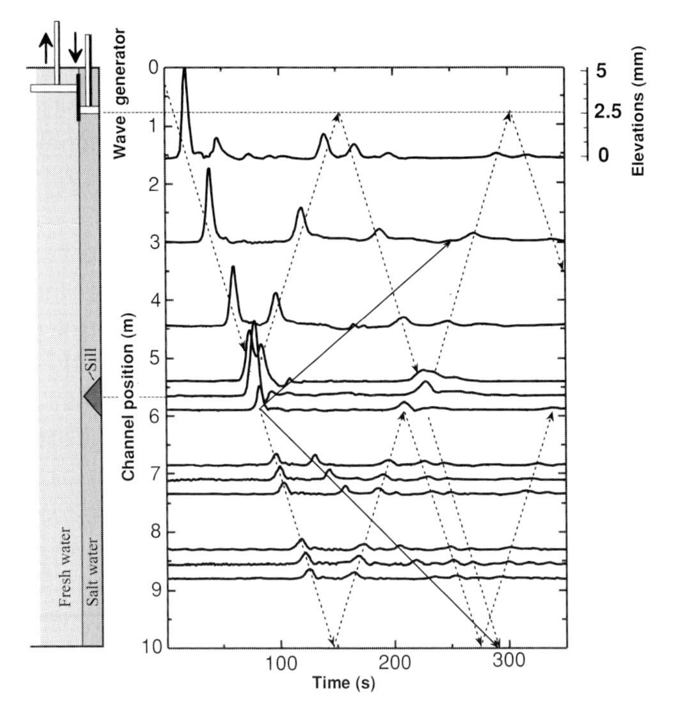

Figure 5.38. Typical experimental data. Time series of the interface deflection at the six gauge positions P1–P6 (By repeating the experiment under identical conditions at gauges P4, P5, and P6, three different positions are shown.) The wave channel with the exact location of the sill is shown on the left. The time axis of the individual gauges is at the position of the vertically drawn channel on the left. Dotted lines with arrows mark the propagation of the first mode, and solid lines with arrows denote that of the second, higher mode. Parameters are $\mathrm{Fr} = 0.19$, $B = 0.92$.

Comparison of the elevation profiles of the measured first- and second-mode peaks with the shape of the typical K–dV soliton, which is given by the sech^2 profile, showed that the theoretical profiles match the experimental ones very well except in the tails, where disturbances occur [105]. The most significant difference between the two observed wave modes is their traveling speed. Theoretical estimates show that the values of the wave speed of the first two baroclinic modes are $c_{\mathrm{p}}^{(1)} = 7.1\ \mathrm{cm\ s^{-1}}$ and $c_{\mathrm{p}}^{(2)} = 2.1\ \mathrm{cm\ s^{-1}}$. So $c_{\mathrm{p}}^{(2)}$ can be estimated to be about one-third of $c_{\mathrm{p}}^{(1)}$. The values of the wave speed from the experiment are $c_{\mathrm{p}}^{(1)} = 7.0\ \mathrm{cm\ s^{-1}}$ and $c_{\mathrm{p}}^{(2)} = 2.0\ \mathrm{cm\ s^{-1}}$. The agreement between experimental and theoretical results is quite good.

Thus, the main conclusion from the experiments described above is that the single first-baroclinic-mode solitary wave, riding on the interface and interacting with a sill, was split by the obstacle into pairs of reflected and transmitted soliton-type waves representing the first-mode wave of the layered system with diffusive interface, plus the second-mode waves due to the finite thickness of the interface. Good agreement of the horizontal profiles of the reflected and transmitted soliton-like waves and their phase speeds of propagation with the theoretical ones permits us to draw such inferences.

Unfortunately, not all characteristics of the generated waves could be measured in the laboratory experiment. For instance, because of the use of electrical gauges instead of optical ones, the vertical and horizontal structures of the reflected and transmitted waves (velocity, density) could not be identified. In fact, only the analysis of the spatial structure of the solitary waves and comparison with exact analytical solutions – horizontal profiles with the sech^2 function, vertical structure with profiles of the standard boundary value problem – together with the data on the phase speed can give us an affirmative answer to the question of whether the observed solitary waves in the experiment are internal solitons of first or second mode or not.

The more successful laboratory experiments with the generation of secondary reflected and transmitted waves of second mode were carried out for incident solitary waves with relatively small amplitudes, $\text{Fr} \sim 0.2$. Higher Froude numbers, as mentioned above, produce more oscillating wave trains behind the solitary waves, especially when interacting with the sill. Thus, the basic question which remains as a consequence of the laboratory modeling, and which can be important for the interpretation of oceanic *in situ* measurements, is as follows: what is the range of applicability of the conclusion, derived as a hypothesis, about the possibility of generation of second-mode solitons during the interaction of first-mode solitons with bottom obstacles? The theoretical model will allow us to obtain more reliable and more substantial conclusions over a wide range of the controlling parameters.

Results of the numerical modeling

The numerical model and its initialization used in the present investigation were the same as in 5.4.1 and 5.4.2. The basic differences were the spatial and temporal scales of the model, which were close to the conditions of the laboratory modeling.

The mechanism of interaction of the solitary internal wave with an obstacle is discussed in detail on the basis of one numerical experiment called the basic case run, after which we will discuss the sensitivity of the model results to the height of the sill and to the amplitude of the incident wave. All input parameters of the model (stratification, configuration of the sill, amplitude of the incident wave) in the basic case run are very close to those in the laboratory experiment.

To reproduce the laboratory experiment, the amplitude of the incident wave was taken to be 5 mm, corresponding to the laboratory Froude number 0.13. In the numerical runs, the Froude number was defined as the ratio of the maximum horizontal orbital velocity of the incident wave, u, localized at the bottom, to the linear phase speed, $c_p^{(1)}$. The conformity of the numerical runs with the laboratory experiments was maintained by choosing the same amplitude of the incident wave, which in the laboratory experiments was measured very accurately. The height of the sill, H_r, was equal to 3.2 cm (blocking parameter $B = 0.92$), and its length L_r was 60 cm, as in the laboratory experiments.

Figure 5.39 shows the process of interaction of the solitary wave with the triangular sill. Snapshots of the isopycnal lines are presented for four consecutive times. Only the more representative time slices are shown. The initial wave motion is from right to left. The beginning of the interaction occurs at $t = 5T_0$ (Figure 5.39(a)), the passage of the wave behind the sill occurs at $t = 6T_0$ (Figure 5.39(b)), the formation of the first and second transmitted and reflected waves occurs at $t = 9T_0$ (Figure 5.39(c)), and the final stage of evolution occurs at $t = 17T_0$ (Figure 5.39(d)) with fully established transmitted and reflected solitary waves. The directions of propagation of all waves are indicated in Figure 5.39(d) by arrows.

The first transmitted and first reflected waves look qualitatively like the first-mode solitary waves with co-phase displacements of isopycnals through the entire depth from the surface to the bottom. At the same time, the more intensive second transmitted and reflected waves are characterized by the counter-phase displacements of the isopycnals below and above the center of the pycnocline. This counter-phase displacement qualitatively coincides with the behavior of the solution of the standard boundary value problem for internal waves. In this case, the first and second eigenfunctions have at depth, i.e. below the pycnocline, only one or two extrema. Thus, in the numerical run the incident solitary wave is split over the sill into two pairs of reflected and transmitted solitons of the first and second modes. This conclusion is drawn from purely qualitative considerations of Figure 5.39, but only careful analysis of the horizontal and vertical structure of such waves, their kinematic characteristics (phase speed, length scale), and comparison with the analytical solution (K–dV soliton in the present case) can provide the answer to the question of whether these waves are really internal solitons of the first and second modes.

To check this hypothesis, let us analyze the detailed structure of the waves and compare them with the K–dV solitons of the same amplitude. This comparison is performed for the time $t = 35T_0$. Figure 5.40 shows the structure of the first transmitted wave. Along with the isopycnal lines, the vertical profiles of the horizontal and vertical velocities, u and w, are presented. The latter are constructed at those places where the maximum of u or w occurs. For comparison, the corresponding profiles of the K–dV soliton, having the same amplitude, are shown by dashed lines.

It is clear that the analytical K–dV soliton describes the first transmitted wave well. The nonessential discrepancy in the analytical and numerical profiles is linked

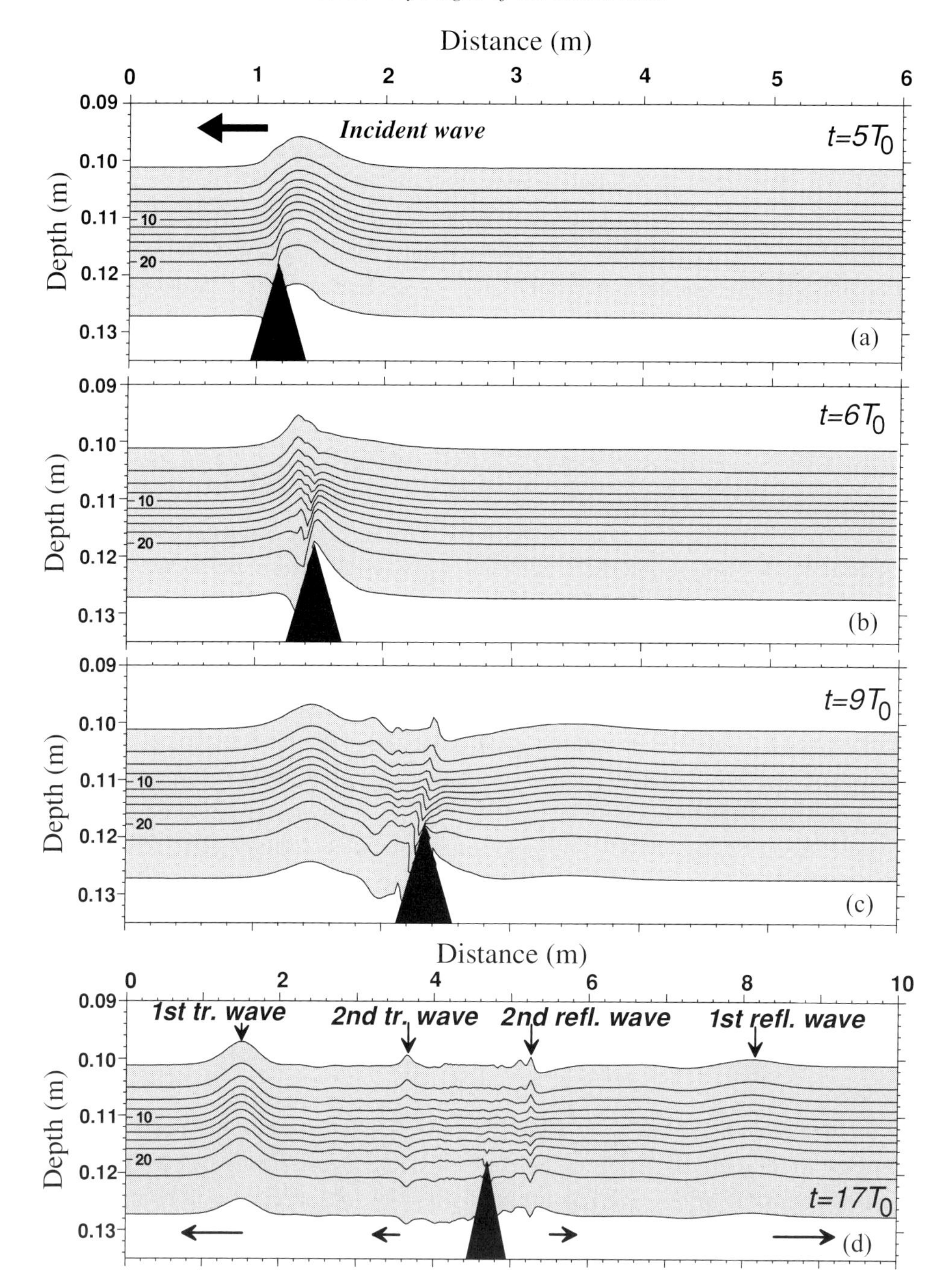

Figure 5.39. Evolution of the density field (density anomaly relative to the free surface (kg m^{-3})) during the interaction of a first-mode solitary wave with the sill. First and second reflected and transmitted waves are presented in (d) at $t = 17T_0$, $T_0 = \lambda_1/c_{\mathrm{p}}^{(1)} = 4.2$ s.

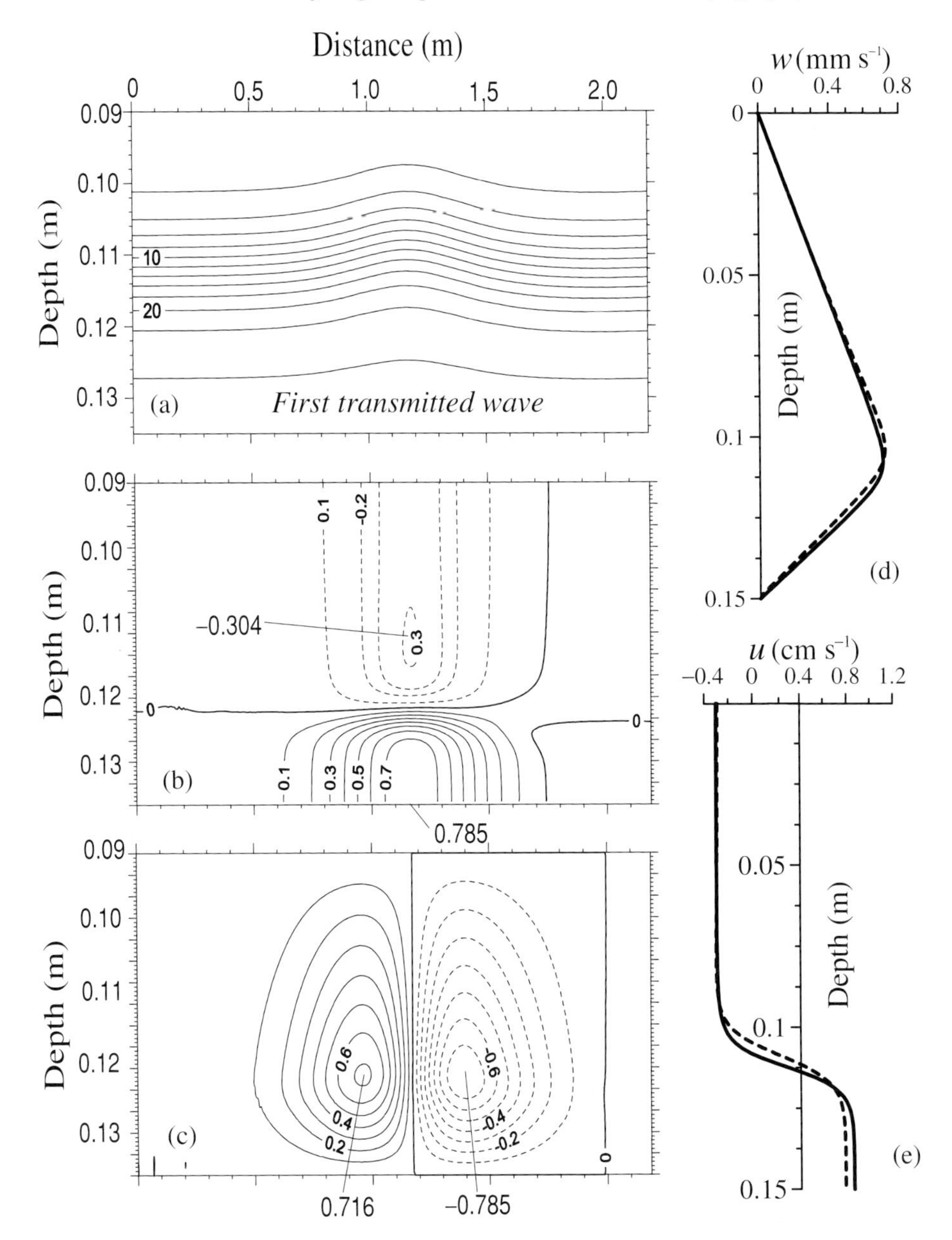

Figure 5.40. (a) Density anomaly field (kg m^{-3}), (b) horizontal (cm s^{-1}) and (c) vertical (mm s^{-1}) velocity fields of the first transmitted wave. The vertical profiles of (d) the vertical and (e) the horizontal velocity are represented by dashed lines. For comparison, the vertical profiles of the K–dV soliton of the same amplitude are shown by solid lines.

to the difference between the real finite-amplitude solitary wave and the analytical soliton defined for weakly nonlinear theories as seen in Figure 5.40. For the wave with amplitude 3.86 mm (Fr = 0.102) this difference is not very large. Moreover, the phase speed of its propagation equals $0.996c_{\mathrm{p}}^{(1)}$. So, one can conclude that the characteristics of the first real transmitted solitary wave are very close to the analytical ones; moreover, this wave can be considered almost as a K–dV soliton.

The same conclusion can be reached for the first reflected wave. It has an amplitude of 1.28 mm (Fr = 0.032) and is presented in Figure 5.41. A somewhat less convincing coincidence of this wave with the K–dV soliton can be explained by the relatively short distance of propagation from the sill (approximately 20 units of wavelength). This wave has a three times smaller amplitude than the first transmitted wave, and consequently it is considerably wider; its wavelength equals 51.6 cm as compared with 32.5 cm for the first transmitted wave, but its phase speed differs from the linear speed (speed of propagation of a wave tail) by only 0.5% as compared with 2.5% for the first transmitted wave. So, the first reflected wave must travel an additional distance of several tens of wavelengths for the separation from the tail. The latter is clearly seen in Figure 5.41. Nevertheless, one can conclude that somewhere far from the sill its characteristics will be very close to the theoretical K–dV soliton.

Let us consider now the structure of the second transmitted wave, which was treated in the laboratory experiment as a second-mode soliton. Its structure is shown in Figure 5.42 and is compared with the second-mode analytical K–dV soliton of the same amplitude, $a_\xi = 1.47$ mm (solid lines denote u and w profiles). Agreement between the numerical and analytical curves is fair. The discrepancy can be explained by the fact that the second transmitted wave propagates not into an undisturbed medium, but into the background of the wave tail remaining behind the first transmitted first-mode soliton.

The second reflected wave, Figure 5.43, shows completely different characteristics. In fact, this signal cannot be considered as a single solitary wave but must be viewed as a wave train. It consists of three consecutive waves arranged in sequence according to the magnitude of their amplitudes. In the front part of the train the waves possess the structure of the second baroclinic mode with a counter-phase displacement of the isopycnals above and below the pycnocline center, but in the tail one can also find manifestations of the third and fourth baroclinic modes (see A–A in Figure 5.43(b)). Moreover, comparison of such waves with the second-mode K–dV solitons of the same amplitude shows that the analytical wave is several times wider and, as a consequence, its vertical velocity is almost three times smaller. The last feature of the numerical waves is that they "live" on the pycnocline. Their horizontal velocity attenuates exponentially in the directions away from the pycnocline to the free surface and to the bottom. Thus, our analysis shows that the second

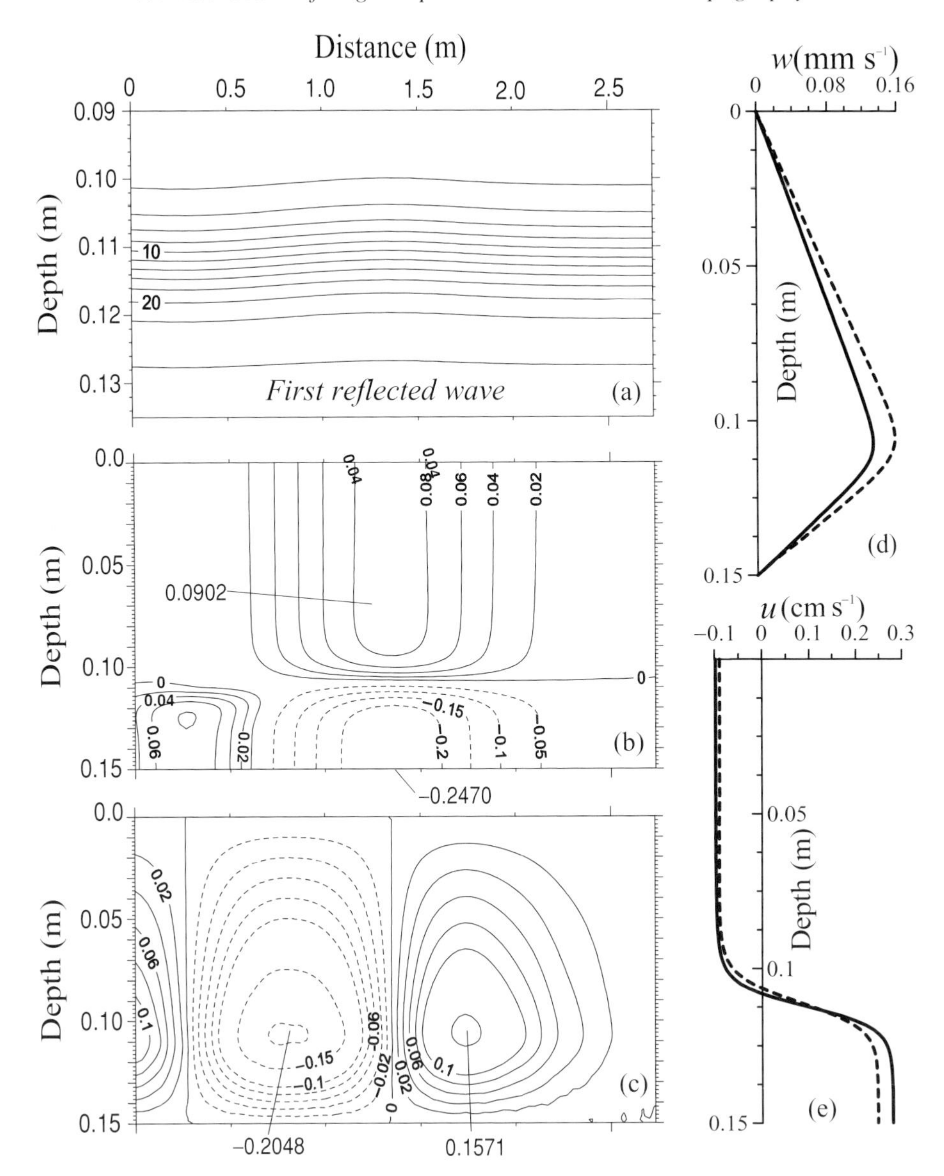

Figure 5.41. The first reflected wave. Notations are the same as in Figure 5.40.

reflected wave signal cannot be treated as a solitary wave of the second baroclinic mode.

Now let us evaluate how sensitive the characteristics of the reflected and transmitted waves are to the height of the sill. In the experimental work [266] the main question under study was to delineate the portions of an incoming wave that are reflected, transmitted, and dissipated. It was found that for a two-layer fluid the

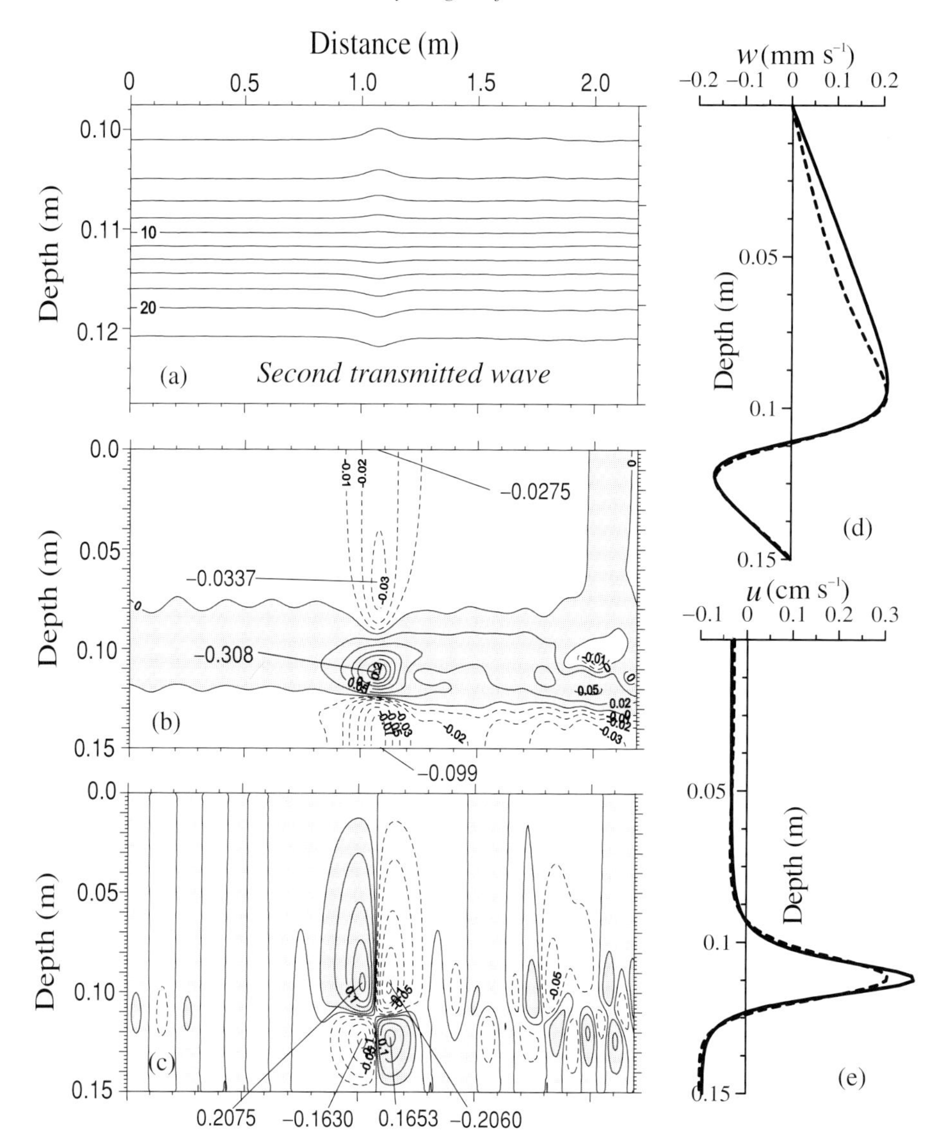

Figure 5.42. The second transmitted wave. Notation is the same as in Figure 5.40.

blocking parameter B is the basic parameter controlling the transfer of the energy across an obstacle. One would expect that in our case of a smooth pycnocline this parameter not only controls the amount of transmitted and reflected energy but also the redistribution, i.e. the splitting of the energy between first- and higher order baroclinic modes. So, contrary to the analysis in ref. [266], we formulate in this section a wider question, and we try to understand what might be the influence of the blocking parameter on the quantitative characteristics (amplitudes) and on the

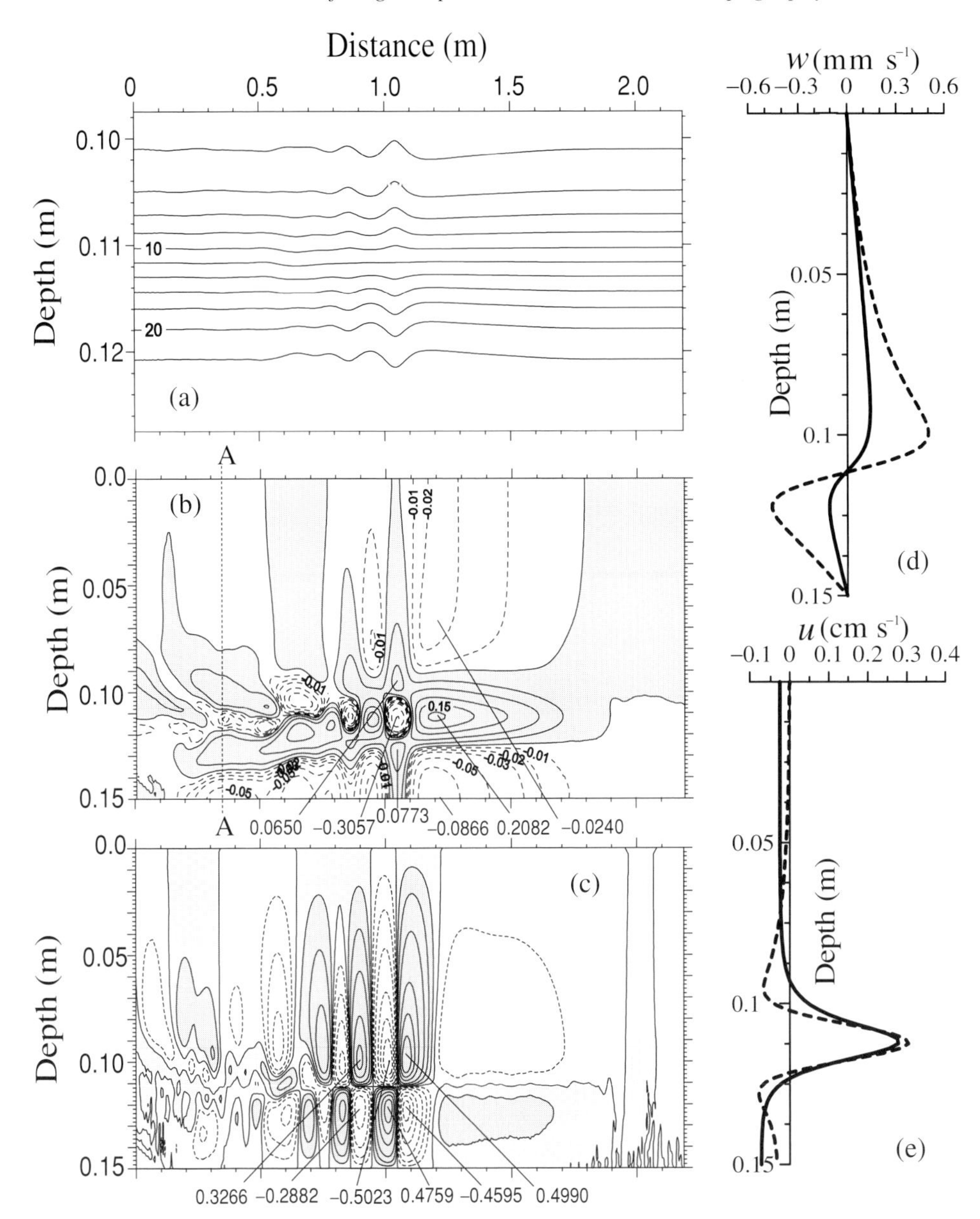

Figure 5.43. The second reflected wave. Notation is the same as in Figure 5.40.

qualitative structure (comparison with K–dV solitons) of the reflected and transmitted waves.

Figure 5.44 displays the dependence of the amplitudes of the reflected and transmitted first and second waves on the sill height. These are defined as the maximum displacements of the isopycnals from the state of equilibrium. This figure indicates a decrease of the amplitude of the first transmitted wave – the incoming wave loses

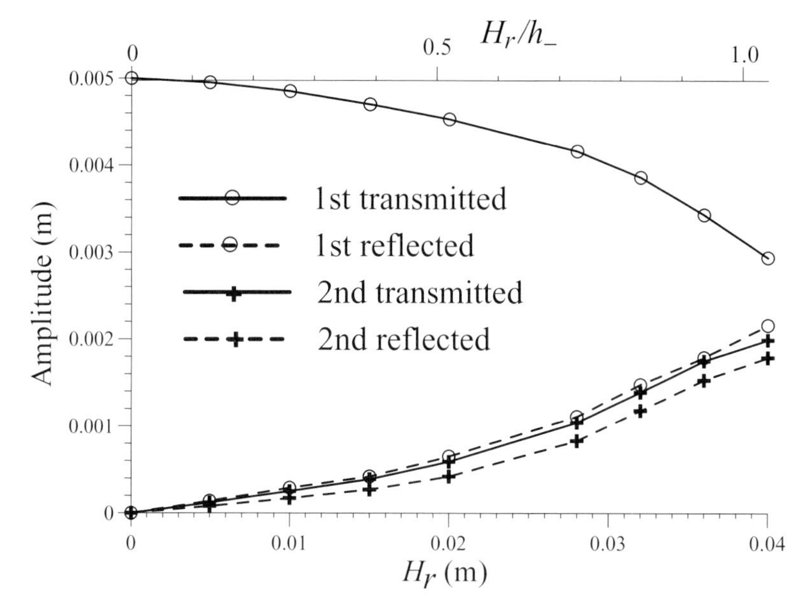

Figure 5.44. Amplitudes of the transmitted and reflected waves plotted against the degree of blocking, $B = H_r/h_-$.

part of its energy – and an increase of the amplitudes of both the reflected and the second transmitted wave with an increase in B. This demonstrates a growing effective energy transfer with increasing blocking parameter from the incoming first-mode solitary waves to the secondary scattered waves. As in the experimental work [266], the scattering of the energy of the incoming wave may be ignored, however, when $B < 0.6$.

The next interesting feature is that all secondary generated waves (both first reflected and second transmitted) have almost equal amplitudes in the entire range of B values considered. This is probably connected with the fact that these waves are generated by a single wave disturbance which arises over the sill during the interaction of the incident wave with it (see, for instance, Figure 5.39(b)). After some time it splits, due to the dispersion, into first and second baroclinic wave disturbances.

The reflected and transmitted waves for the basic case run were compared with the analytical solution (5.52). Let us analyze how the structure of the first reflected and both transmitted waves depends on the height of the obstacle. As in Section 4.1, we compare all solitary waves with the K–dV solitons having the same amplitude. Figure 5.45 summarizes this comparison. All curves are normalized to the maximum value of the appropriate K–dV soliton. Figure 5.45 suggests that first and second transmitted waves fit the analytical solution (5.52) over a wide range of the

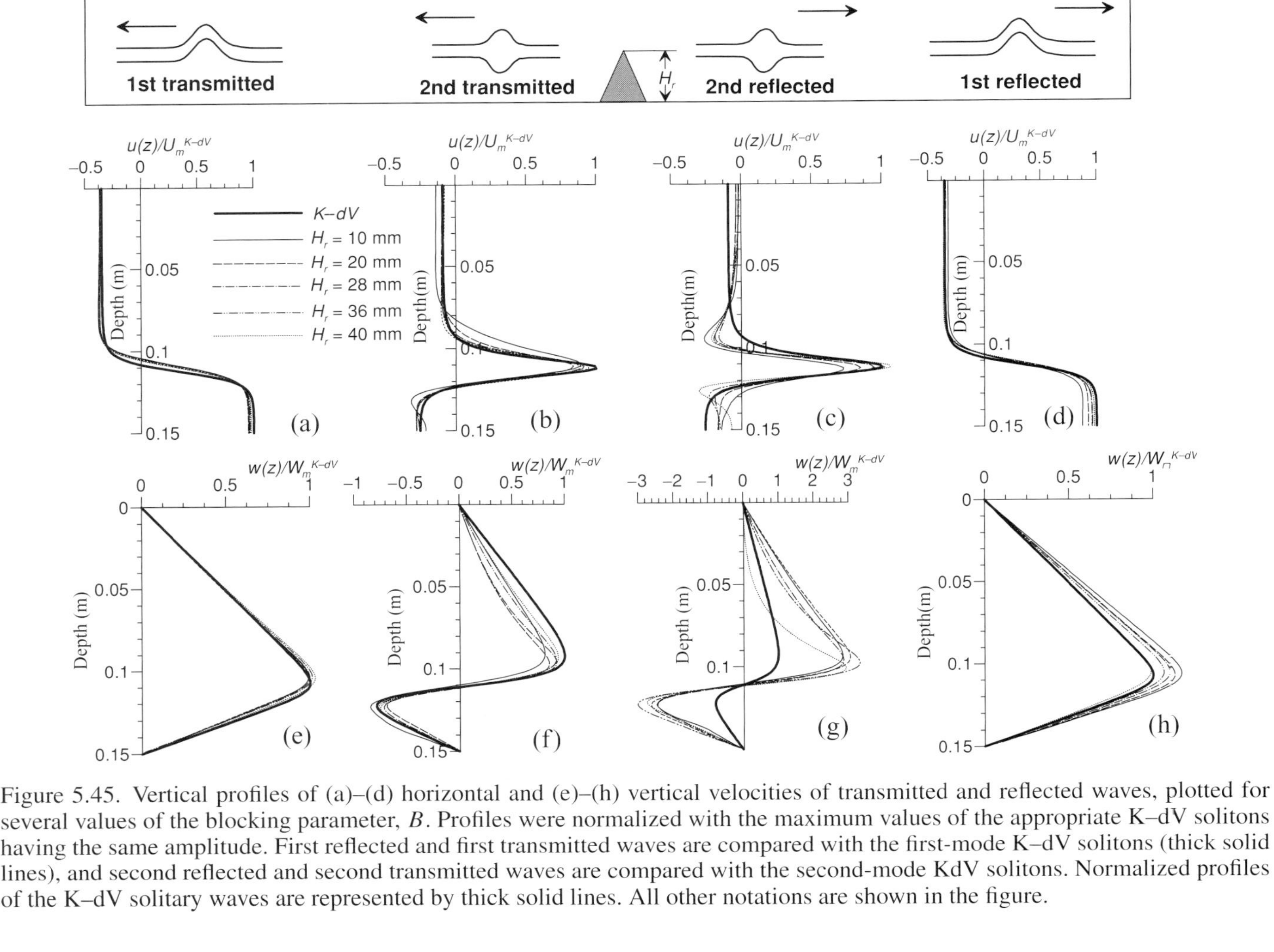

Figure 5.45. Vertical profiles of (a)–(d) horizontal and (e)–(h) vertical velocities of transmitted and reflected waves, plotted for several values of the blocking parameter, B. Profiles were normalized with the maximum values of the appropriate K–dV solitons having the same amplitude. First reflected and first transmitted waves are compared with the first-mode K–dV solitons (thick solid lines), and second reflected and second transmitted waves are compared with the second-mode KdV solitons. Normalized profiles of the K–dV solitary waves are represented by thick solid lines. All other notations are shown in the figure.

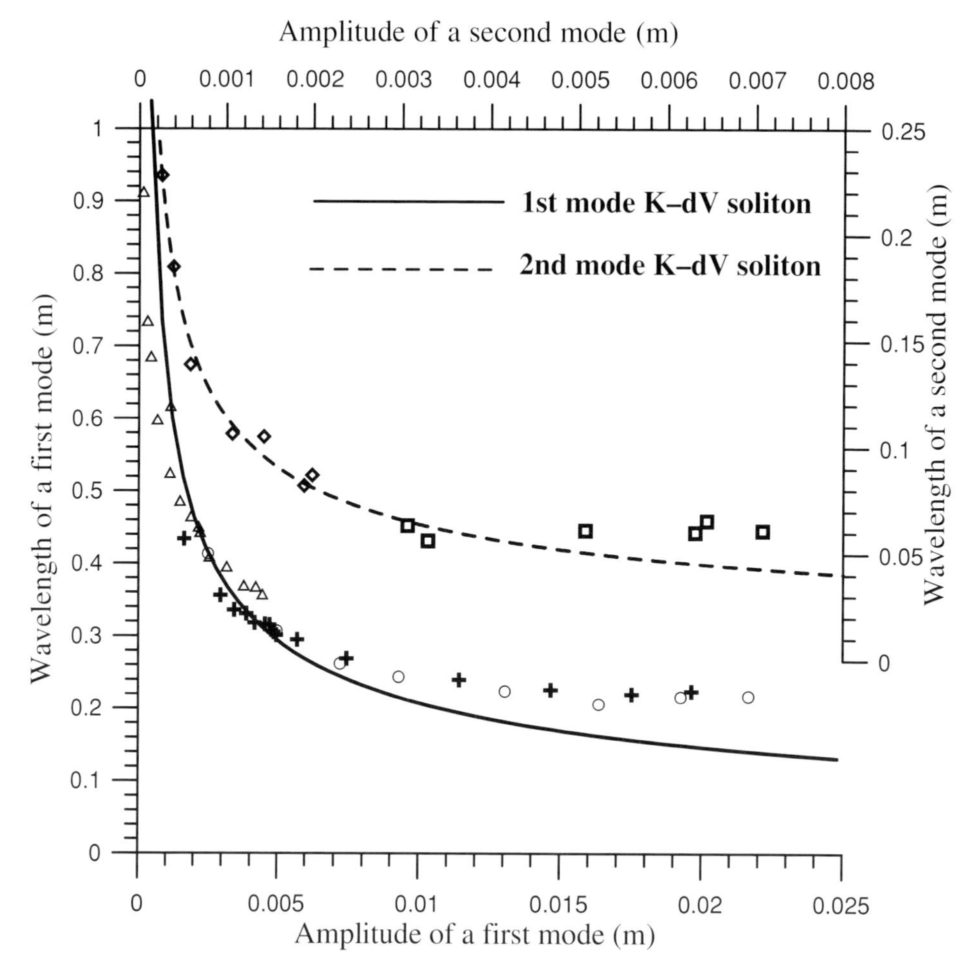

Figure 5.46. Wavelengths of solitary waves plotted against wave amplitude. Solid and dashed lines denote the theoretical curves for the first- and second-mode K–dV solitary waves, respectively. The data points were calculated with the help of formula (5.60) for numerical incoming, reflected, and transmitted solitary waves: (○) incoming wave; (+) first transmitted wave; (△) first reflected wave; (◇) second transmitted wave; (□) second reflected wave.

parameter B $(0 < B < 1)$ quite well. Large discrepancies between numerical and analytical curves for the second transmitted waves can be explained by the fact that these waves propagate not in the undisturbed fluid but on the background of the weak first-mode wave train and thus remain in the tail of the first transmitted wave. The existence of such a dispersive wave train was also indicated in ref. [266].

The behavior of the vertical profiles of the first and second reflected waves corroborates the conclusion, formulated above, that the first reflected wave can be interpreted as a first-mode K–dV soliton-type wave, but that the second one belongs to a second-mode dispersive wave train, of which the characteristics are very far from a second-mode K–dV soliton (these waves are considerably shorter than the K–dV soliton).

As a rule, the maximum of the vertical velocity of the computed first reflected wave is slightly larger than that of the analytical solution. This is probably because the first reflected waves are shorter than those of the analytical solitons. To check this idea, the integral wavelength defined by formula (5.60) was estimated for every wave. Figure 5.46 shows the wavelengths of all computed reflected and transmitted first and second solitary waves compared with theoretical values defined by formulas (5.53). Reflected first-mode solitary waves are represented by triangles. Evidently, weak reflected waves having amplitudes less than 0.2 cm are shorter than the analytical K–dV solitons. This can probably be explained by the relatively small distance of propagation from the source of generation compared with its wavelength.

In summary, the blocking parameter and the height of the sill control the energy loss of the incoming wave and its transfer to reflected and transmitted secondary waves, but they have only a weak influence on the modal and spatial structure of the wave field.

6

Generation mechanism for different background conditions

The linear wave theory described in Chapters 2 and 3 is useful for interpreting small-amplitude wave phenomena occurring near oceanic bottom features or near density fronts. The nonlinear approach considered in Chapter 4 shows how large the difference can be between the linear and nonlinear waves when the external generating forcing is sufficiently strong. Several examples of the generating processes were studied for both subcritical (Fr $<$ 1) and supercritical (Fr $>$ 1) cases for typical oceanic conditions. However, when considering the large variety of oceanic background stratifications, bottom profiles, or values of external forcings, it is often not possible to apply the results described in Chapter 4 directly to the specific realistic conditions encountered in the field.

In this chapter, we concentrate on phenomena that are new when compared with those described in Chapter 4. These phenomena involve the influence of the vertical and horizontal fluid stratification on the nonlinear mechanism of internal wave generation, as well as the effects related to the rotation of the Earth (especially at high-latitude seas). Nonlinear baroclinic tides over steep bottom topography are also considered. Note that all topics in Chapter 4 were motivated by observational results obtained during field measurements at different sites of the World Ocean. This will also be the approach taken here; thus, all theoretical reasonings will be accompanied by illustrations and interpretations of real *in situ* data.

6.1 Effects related to the rotation of the Earth

According to the linear theory presented in Section 2.6, the group velocity of the internal tidal waves together with their amplitudes approaches zero, and the phase speed tends to infinity, when the latitude φ tends to its critical value φ_c (for the M_2-tide $\varphi_c = 74.5^\circ$; see Figures 2.21 and 2.22). The wave equation (1.42), which at low latitudes ($N < \sigma < f$) is hyperbolic, changes its type to elliptic if $\sigma > f$, and baroclinic oscillations cannot be obtained as freely propagating internal waves, but rather represent exponentially damped solutions that are driven by waves existing at $\varphi < \varphi_c$.

Thus, the theory prohibits the existence of periodic baroclinic tides above the critical latitude. However, in ref. [273] observations of the baroclinic tide M_2 were mentioned for the first time. The registration of semidiurnal internal tides at high latitudes was also reported in refs. [146], [186], and [196]. The results of these measurements during the Barents Sea Polar Front (BSPF) experiment and the results of the theoretical modeling described in ref. [259] are summarized in the following.

6.1.1 Barents Sea Polar Front experiment

The BSPF experiment took place over the southern flank of the Spitsbergen Bank, about 60 km east of Bear Island, during August 1992 (Figure 6.1(a)). The experiment was a combined physical oceanographic and acoustic tomographic study, with CTD and ADCP surveys, oceanographic moorings, and an acoustic tomography measurement. The aim of the experiment was to study the Barents Sea Polar Front features and their dynamics and the influence on the regional oceanography. Repeated shipboard surveys were conducted within a $70 \times 80\,\text{km}^2$ box, shown as the black frame in Figure 6.1(b). Velocity observations were performed during the surveys from a shipboard ADCP probe. Moorings with temperature sensors and current meters at depths of 20, 50, and 80 m were deployed at three corners of the survey area.

The current meter and temperature records showed that the M_2-tidal signal predominated at all deployed moorings. The baroclinic component of the measured signal was well pronounced. This fact is illustrated in Figure 6.2, which shows frequency spectra from current meter measurements. Whilst the 20 m spectra represent the sum of the barotropic and baroclinic oscillations, and the 50 m spectra present rather well the barotropic component, the spectra of the difference between currents at 20 m and 50 m display the baroclinic oscillations. This is due to the specific structure of the buoyancy frequency profiles measured in the area of the BSPF experiment, because the horizontal velocity of the first baroclinic tidal mode changed direction at a depth of approximately 50 m.

The basic question suggested by the analysis of the experimental data is as follows. How could the barotropic tide generate propagating internal waves in the vicinity of 74.5°, if the theory prohibits the existence of such a phenomenon here?

The next interesting question concerns the possibility of the tidal generation of short internal waves given by the data of the moorings (Figure 6.3(a)), which show evidence of oscillatory motions, perhaps soliton-like waves. The semidiurnal signal analyzed above is evident, most clearly in the u-velocity component (directed to the

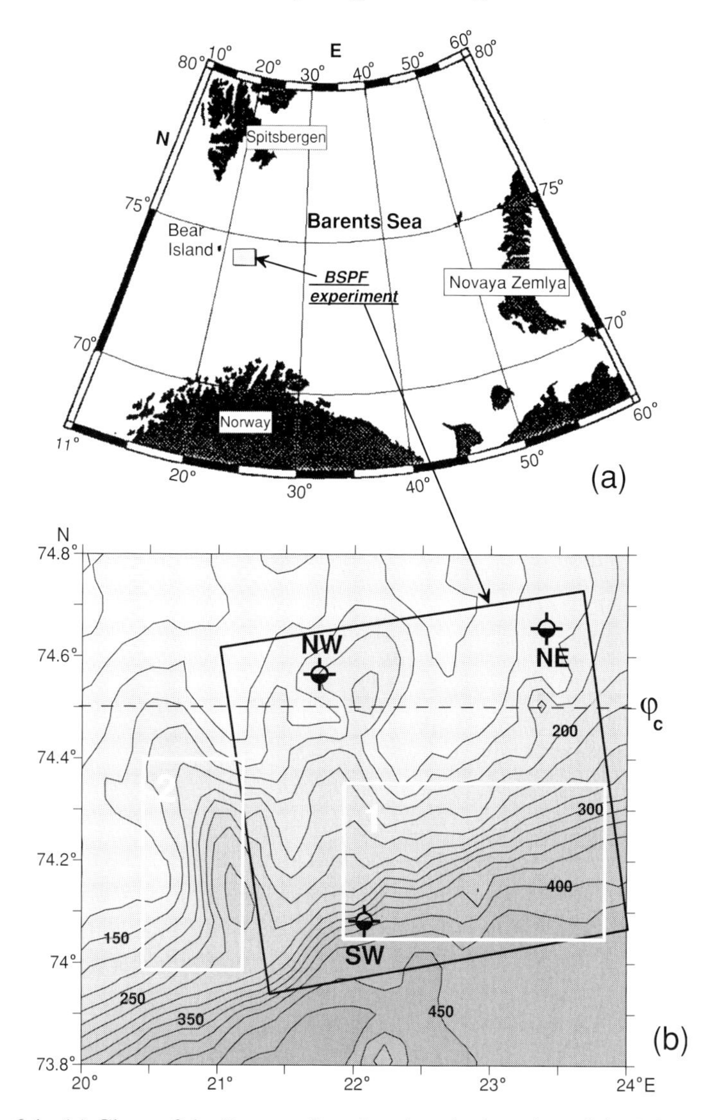

Figure 6.1. (a) Chart of the Barents Sea showing the location of the BSPF experiment. (b) Local bathymetry with black rectangle showing the sampling area of the BSPF experiment. The positions of the northeast (NE), northwest (NW), and southwest (SW) moorings are marked. Two white rectangles, 1 and 2, identify the bathymetries used in numerical calculations.

east). However, at higher frequency, strong oscillations are seen in both temperature and velocity records which appear to be manifestations of internal soliton-like waves. For example, just after midday on August 19, a 2 °C temperature spike is seen (Figure 6.3(b), asterisk), indicating a depression of the thermocline. At the same time, a large spike is seen in the v-component of the velocity (directed to

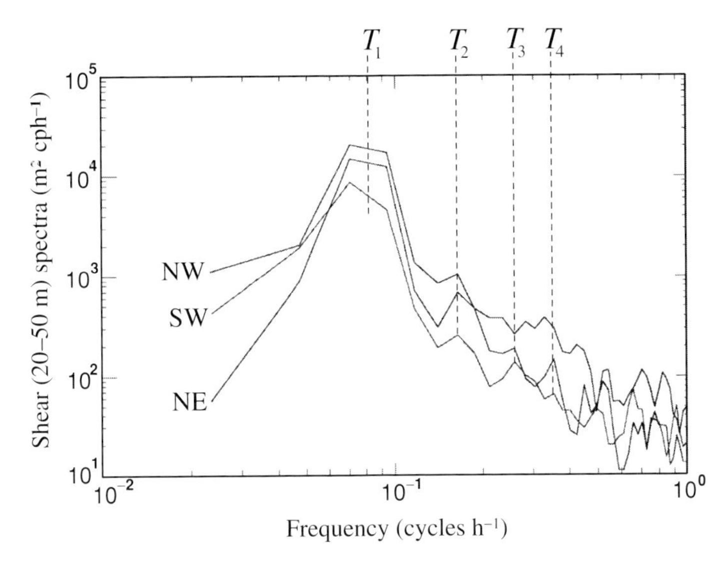

Figure 6.2. Spectra of the velocity difference between the 20 m and 50 m depth from the NW, SW, and NE moorings. Periods T_1, T_2, T_3, and T_4 are equal to 12.4, 6.2, 4.1, and 3.1 h, respectively.

the north) and a smaller spike in the u-component. Several vertical profiles of the temperature in the vicinity of the SW mooring recorded by CTD measurements are presented in Figure 6.3(c). A rough estimation of the pycnocline depression amplitude made for the large spike (marked by the asterisk) on the basis of the 20 m depth temperature series (Figure 6.3(b)) and vertical temperature profiles (Figure 6.3(c)) gives a value of approximately 25 m.

Since the linear model does not predict any conspicuous tidal activity in the vicinity of the critical latitudes, but the BSPF experiment shows it, one can conclude that the nonlinearity of the wave process is responsible for this generation.

Qualitatively, the specific input of the nonlinearity to the generation process can be estimated from the data of the barotropic tides and bottom topography using the parameter of nonlinearity, ε_1, see (4.1). During the BSPF experiment, the magnitude of the M_2 barotropic tidal flux, Ψ_0, over the shelf break of the tested area was in the range 20–60 $m^2\,s^{-1}$ [186]. According to ref. [146], a typical barotropic tidal excursion can be estimated to span 5 km. Thus, two basic sites of internal tide generation in the BSPF experiment region can be identified: the first is marked by rectangle 1 in Figure 6.1(b) and is typical for this area with generating conditions $\varepsilon_1 = 0.06$; region 2 is characterized by a steeper bottom topography and stronger cross-topography tidal currents. For this site, $\varepsilon_1 = 0.4$ (the shallow-water tidal excursion is used for the estimation), and one can expect an effective generation of baroclinic tides.

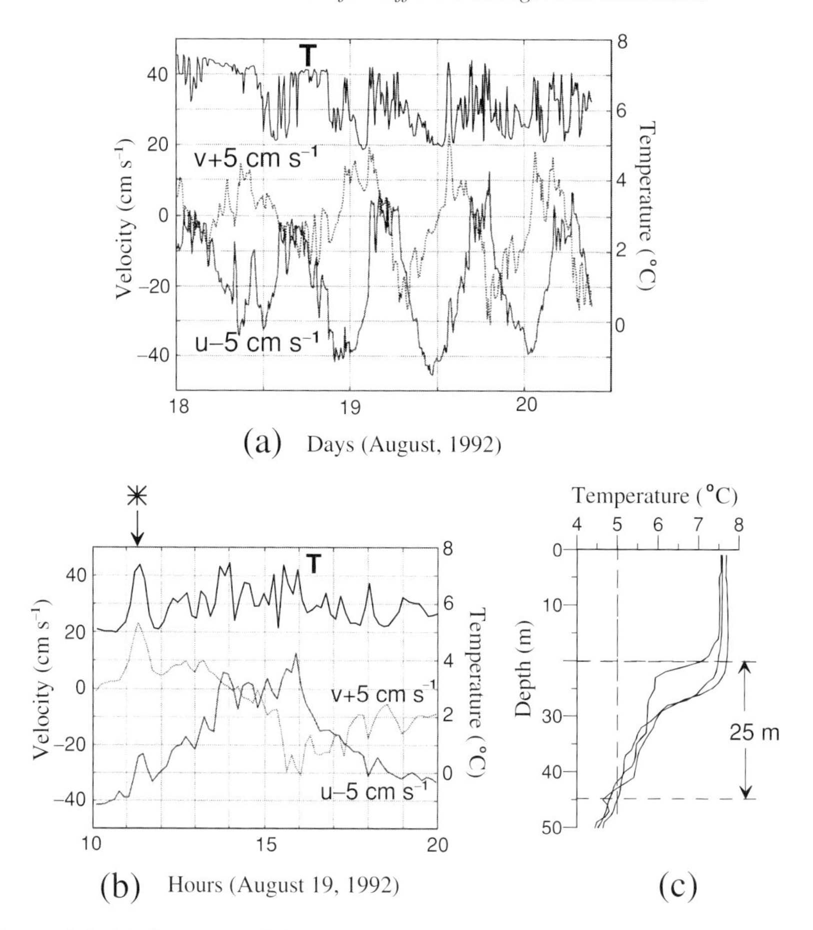

Figure 6.3. (a) Segment of the temperature and velocity records from the 20 m depth on the southwest (SW) mooring; see Figure 6.1. (b) Part of these records for August 19, 1992, showing evidence of soliton-like internal waves. The most pronounced spike is marked by an asterisk. (c) Temperature profiles near the SW mooring.

In the following we consider "typical" as well as "extreme" conditions of internal wave generation in the BSPF experimental area. In the modeling we use a "smoothed" background fluid stratification obtained during BSPF.

6.1.2 Baroclinic tides

First, let us analyze the predictions of the nonlinear model for the "typical" case. The results of the numerical modeling for region 1 are presented in Figure 6.4. These runs were performed for $\Psi_0 = 45\,\mathrm{m}^2\,\mathrm{s}^{-1}$ [186], and the latitude $\varphi = 70^\circ$ was used instead of $\varphi = 74^\circ$, where the selected bottom is located. In fact, the local inertial

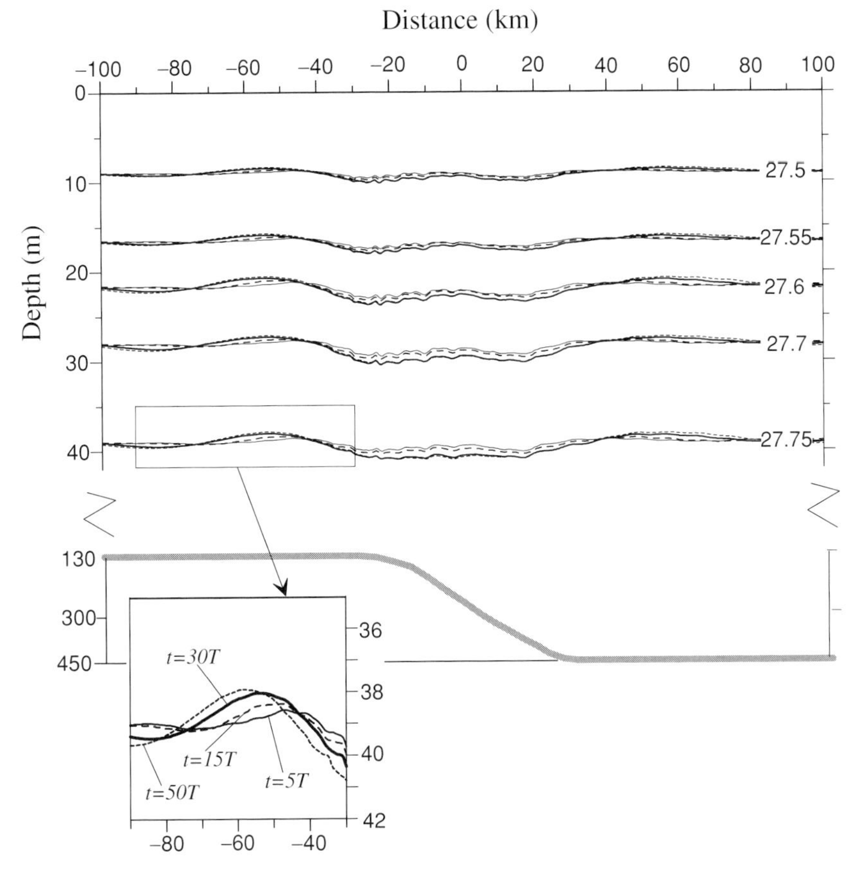

Figure 6.4. Cross-section of conventional density σ_t (kg m^{-3}) obtained numerically for four slices of time, corresponding to 5 (thin line), 15 (dashed line), 30 (thick line), and 50 (dotted line) tidal cycles. These calculations were performed for the bottom topography marked by rectangle 1 in Figure 6.1(b). The inset is a detailed demonstration of the density behavior.

frequency, f, can be changed by vorticity, V_x, of the background currents to an effective frequency, f_{eff}, where $f_{\text{eff}}^2 = f(f + V_x)$ [125], and free internal inertial waves can exist at anomalously low frequencies σ such that $f_{\text{eff}} < \sigma < f$. This correction to the lower bound of a frequency band arises whenever one takes into account the background vorticity. We introduced such a correction in Chapter 3 (see (3.17)).

Positive background vorticity in the vicinity of the critical latitude may close the wave guide for M_2, whereas negative vorticity will promote a poleward shift of the effective critical latitude. In fact, the BSPF was exposed to a negative vorticity (up to 5.5×10^{-6} s^{-1}) caused by the mean currents. That is why the effective inertial frequency was shifted towards the pole by 4° [121]. This shift was introduced in our model to link the basic characteristics of the baroclinic tides more closely to those of the experiment.

As the intensity of the tidal forcing in region 1 was relatively small ($\varepsilon_1 = 0.06$), one should expect also a weak baroclinic response of the system to the external impact. Figure 6.4 illustrates this rather clearly. The maximum deflection of the isopycnals beyond the bottom topography from their undisturbed position is only about 1 m, and waves do not reveal any tendency to nonlinear disintegration and fission into a sequence of solitary waves. So, the generated waves can be considered to be linear, and for this reason they must reveal characteristics of linear internal tidal waves. Realistically, rough estimates from Figure 6.4 give values for the tidal lengths of approximately 60 km in the shallow region and almost 100 km in the deep region. Thus, the parameter kU_0/σ introduced in Section 4.3 is of the order of 0.1, and the generated waves can be related to the baroclinic tides.

Estimation of the value of the group speed obtained from the time series of Figure 6.4 on the basis of the analysis of energy propagation (see the inset) during 45 tidal cycles yields the value $c_g = 0.015\ \mathrm{m\,s^{-1}}$. This is in accordance with the conclusions from the linear theory presented in Chapter 2: long internal tidal waves with small group speed are to be expected. It is unlikely that the large-temperature oscillations measured by the thermistor at the SW mooring (Figure 6.3) were connected with the waves generated in area 1. Even the 4° latitude shift introduced in the model due to the background vorticity could not significantly intensify the generation process as the excited waves remained very weak.

6.1.3 Short internal waves

Quite a different situation was encountered when the calculations were performed for region 2 in Figure 6.1(b), for which the appropriate parameter ε_1 was estimated to be 0.4. So, a conspicuous nonlinear response is now possible, and no latitude shift due to the background vorticity was introduced. That is why the calculations were performed for $\varphi = 74°$. A time series showing the evolution of the density field during the two first tidal cycles is presented in Figure 6.5. Similar series, however, obtained for low latitude, $\varphi = 10°$, are shown in Figure 4.19. Comparison of these two figures is very useful in understanding the generation mechanism at high latitudes.

Figure 6.5 shows that by the instant $t = 1T$ the propagating waves A_1 and B_1 (which look like solitary waves) are formed in a similar way to those taking place at low latitude (Figure 4.19). The parameter kU_0/σ, estimated for the generated short waves, takes values 14 and 6 for the shallow and deep zones, respectively, if one takes their length scales as 1.5 km. Thus, the generated waves can be considered as intermediate between unsteady lee waves and mixed lee waves; compare with Table 4.1.

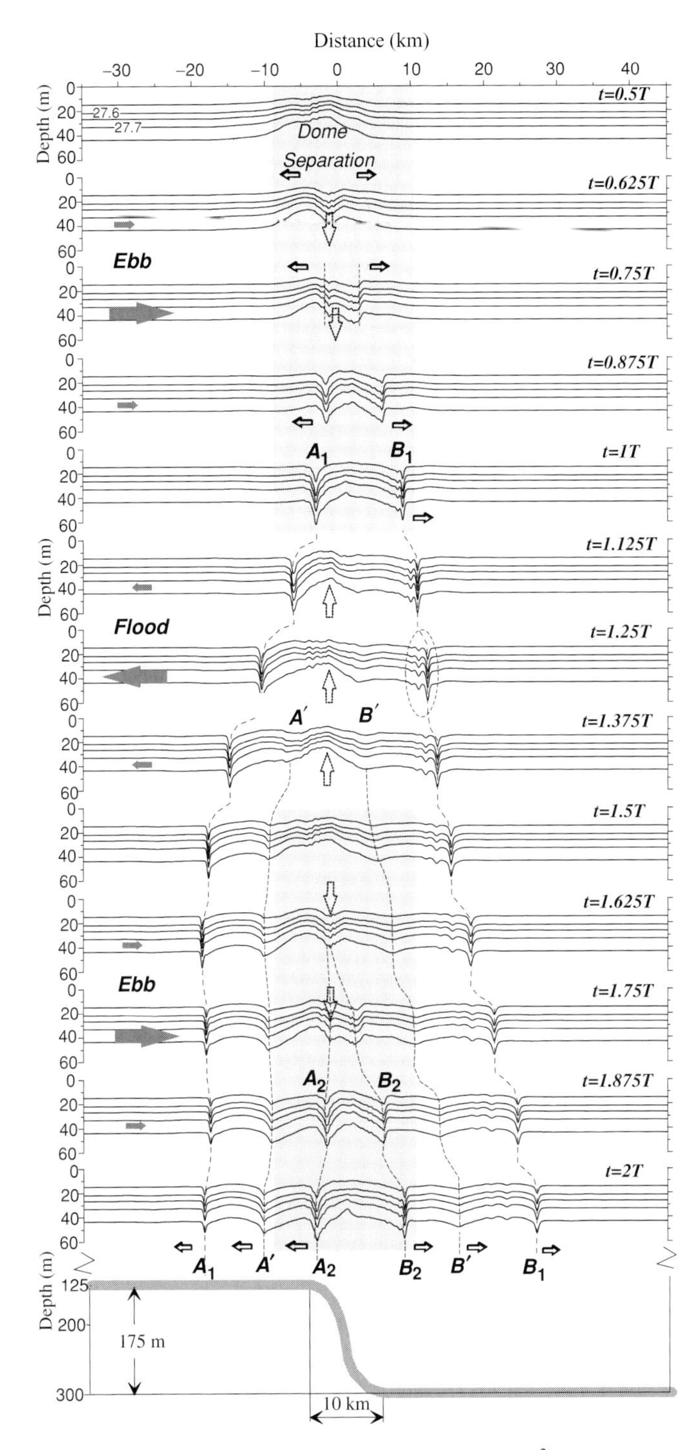

Figure 6.5. Evolution of the conventional density σ_t (kg m^{-3}) field during two tidal cycles near the topography identified by rectangle 2 in Figure 6.1(b). The value and direction of the barotropic tide are shown by arrows with different dimensions. The series of waves moving in opposite directions are marked by A_1, A', A_2 and B_1, B', B_2. Waves marked A' and B' correspond to double harmonic waves.

During the second tidal cycle (between $t = 1T$ and $t = 2T$), the process recurs, but instead of finding two waves, A_2 and B_2, as in Figure 4.19, two additional waves of depression, A′ and B′, are generated between $t = 1.25T$ and $t = 1.5T$, and it is only later that the waves A_2 and B_2 are formed. The positions of both pairs are traced in Figure 6.5 by dashed lines.

This unusual generation of short "double waves" is probably related to the manifestation of the nonlinearity in the presence of strong Coriolis effects: the wave energy is then transferred directly not to the shorter scales (e.g. to wave trains, as observed at lower latitudes and shown in Figure 4.19), but to the higher harmonics. Such a generation of a "double wave" by tidal flow over topography was also observed in a laboratory experiment [61]. We will return to the problem of generation of multiple harmonics below.

Comparison of the isopycnal shapes within the shaded areas in Figure 6.5 (presenting the density field with one tidal cycle delay) reveals that they look very similar. This leads to the conclusion that the initial nonperiodic waves excited over the slope left the source of generation very quickly, and that a periodic motion was established in the central segment of the calculation area during two to three tidal periods. However, from what was said above, this conclusion is in contradiction with the fact that (i) baroclinic tides near the critical latitude must have a very low group speed, and (ii) the initial transient processes will continue for a very long time, i.e. an indefinite time in the limit, if one approaches the critical latitude. Such behavior was discussed above, and is shown in Figure 6.4 for region 1.

In fact, there is no contradiction. Contrary to the results obtained for region 1, where an almost linear baroclinic tide was generated, in region 2 unsteady lee waves are excited. They have relatively large amplitudes and short spatial scales, which are comparable with the scales of the bottom topography. The speed of their propagation equals the velocity of the tidal flux, which is why they leave the place of generation very quickly; moreover, all transient processes only continue for several tidal cycles.

Figure 6.6 confirms this conclusion. The coincidence of the wave profiles at $t = 9T$ and $t = 12T$ is fairly good. An appreciable discrepancy between them can be explained by the fact that, apart from the lee waves, baroclinic tides are present in the total wave field. However, input of baroclinic tides to the wave field seems to be negligible.

Now let us discuss the phenomenon of generation of higher harmonics found in the analysis of Figure 6.5. The nonlinearity usually leads to an energy transfer from the background wave to the harmonics with multiple frequencies [175]. The double-frequency wave is usually excited first. To check this, we repeated calculations without the Jacobian in the governing system (4.3) (linear case) with double forcing frequency ($T_2 = 6.2\,\mathrm{h}$). The resulting isopycnal profile is shown by the dashed line in Figure 6.6. A qualitative comparison of the linear ($T_2 = 6.2\,\mathrm{h}$) and nonlinear

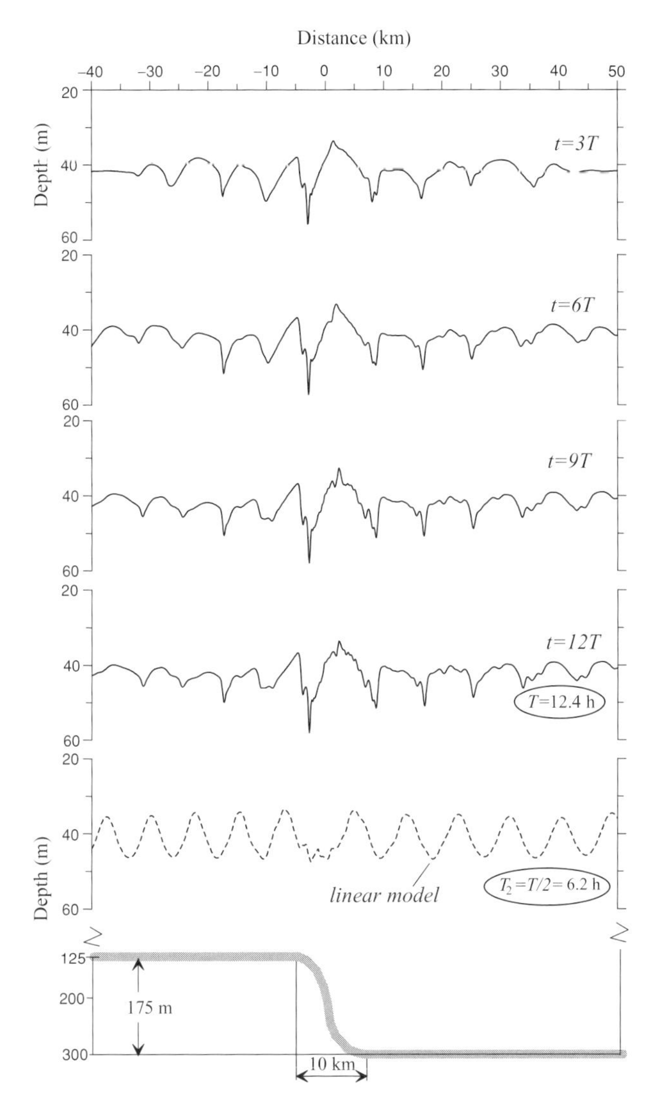

Figure 6.6. Isopycnal contours for $\sigma_t = 27.75\,\mathrm{kg\,m^{-3}}$ after 3, 6, 9, and 12 tidal cycles (solid lines); results are obtained from the linear model ($T_2 = 6.2\,\mathrm{h}$, dashed line).

($T = 12.4\,\mathrm{h}$) cases clearly confirms the generation of short internal waves with double frequency in the nonlinear case.

Nonlinear transfer of energy takes place not only at the double harmonic with period 6.2 h, but also at other multiple harmonics. Time series between $t = 4T$ and $8T$ of the baroclinic horizontal velocity, calculated without the barotropic tidal flux for the shallow and deep parts, respectively, are presented in Figures 6.7(a) and (c). These curves show that not only the first double harmonic, but also higher harmonics

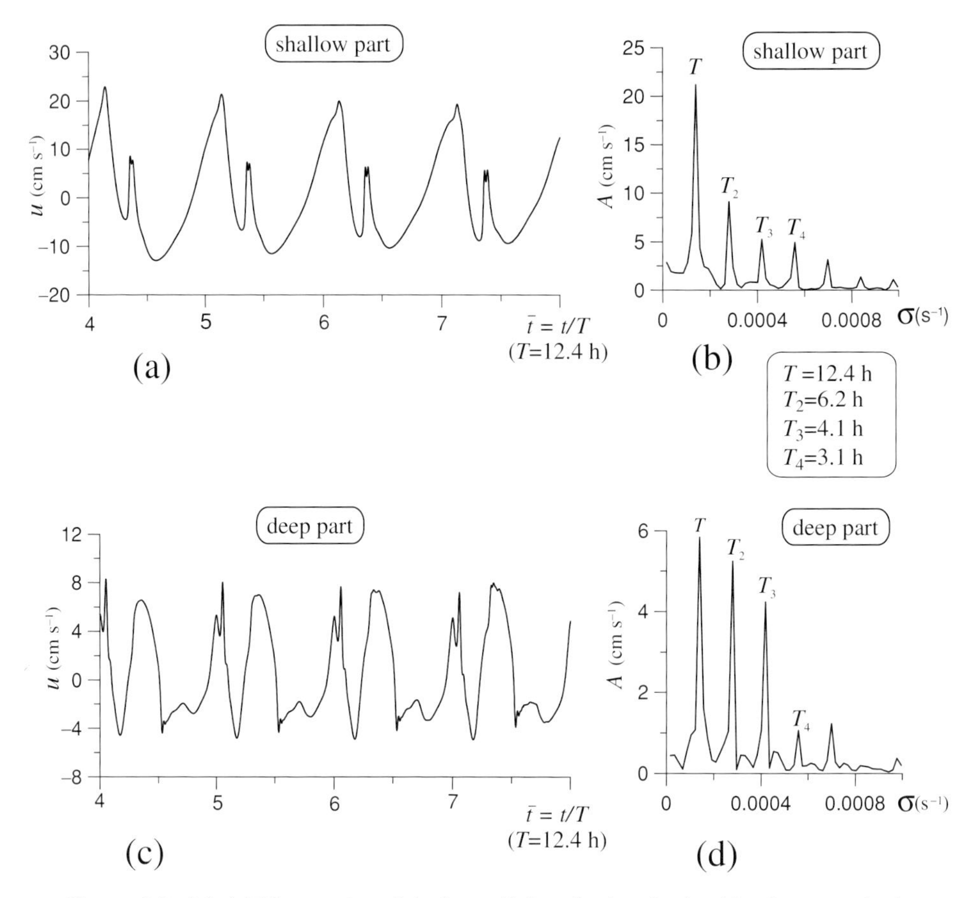

Figure 6.7. (a), (c) Time series of the baroclinic velocity obtained by the numerical model at the free surface in shallow- and deep-water zones, respectively. (b), (d) Spectra of the time series of the horizontal velocity amplitudes at deep and shallow parts.

have been generated in the baroclinic wave field. To illustrate this, the spectra of the horizontal velocity amplitudes were calculated from the 12-period time series of the baroclinic velocity as described by ref. [123] (see Figures 6.7(b) and (d)). Both spectra show the presence of not only waves with the tidal frequency σ, but also spikes at multiple frequencies with periods $T/2$, $T/3$, and $T/4$, corresponding to 2σ, 3σ, and 4σ, and with falling intensity. Thus, it is obvious that an energy transfer occurs from the tidal harmonics to the multiple frequencies due to the nonlinearity. It is interesting that the experimental spectra in Figure 6.2 also reveal the presence of multiple-frequency harmonics. The maximum of the baroclinic wave energy with double frequency is seen quite well.

Consider next the form of the excited waves beyond their place of generation. As was concluded in refs. [73] and [75], the tidally generated waves under conditions of strong rotation appear as a coherent cnoidal-type wave, not as a disintegrated set of solitary waves. This result, proved for weakly nonlinear waves, may, however, be violated for a stronger nonlinearity. This is probably the case for the episode displayed in Figure 6.5. The fragment within the ellipse at $t = 1.25T$ shows that even close to the critical latitude the tidally generated waves can take the form of packets of solitary internal waves. Note, however, that the mechanism of their generation is not so pronounced as it is without rotation.

The spatial structure of wave B_1 becomes evident in a close-up view of the episode displayed in Figure 6.5, presented in Figures 6.8(a), (c), and (e). For comparison purposes, the equivalent fields for the K–dV solitary internal wave with the same amplitude are shown in Figures 6.8(b), (d), and (f). It is clearly seen that the agreement between the two waves is quite satisfactory. Thus, without any doubt, the characteristics of the waves excited in gently sloping topography (region 1) and steep topography (region 2) are quite distinct from one another. The ensuing investigation, performed with different values of the Coriolis parameter, will be helpful in understanding the reasons for such differences.

6.1.4 Dependence on the rotation of the Earth

Figure 6.9 displays the density field for identical, tidally driven waves for three values of the latitude, $\varphi = 10^\circ$, 30°, and 70°, in the linear (dashed lines) and nonlinear (solid lines) cases calculated for region 2. Comparison of these two classes of solutions allows us to estimate the effects that the nonlinearity imparts to the generated fields. The linear solution, obtained from the numerical model, behaves qualitatively, as was predicted by the linear analytical solution presented in Chapter 2. First, it is periodical in space, as one would expect in the linear theory; see formulas (2.38) and (2.39). Next, the dependence of the amplitude of the linear waves upon the latitude, obtained numerically, also corroborates the found analytical decrease of the generation efficiency when the critical latitude is approached (compare the dashed line in Figure 6.9 for the linear case). For instance, the amplitudes of the waves in the deep part of the basin were approximately 13, 11, and 2 m for the latitudes $\varphi = 10^\circ$, 30°, and 70°, respectively. Furthermore, scrutiny of Figure 6.9 and the computational results discloses an increase of the wavelengths of the linear waves with growing rotation by almost a factor of 5 from the equator to $\varphi = 70^\circ$. This is in agreement with the linear dispersion relation (1.78); see also (2.80).

A qualitative comparison of the wave profiles in the linear and nonlinear cases at low and relatively moderate latitudes (Figures 6.9(a) and (b)) nevertheless discloses in part surprisingly good correlations. The distance between two successive

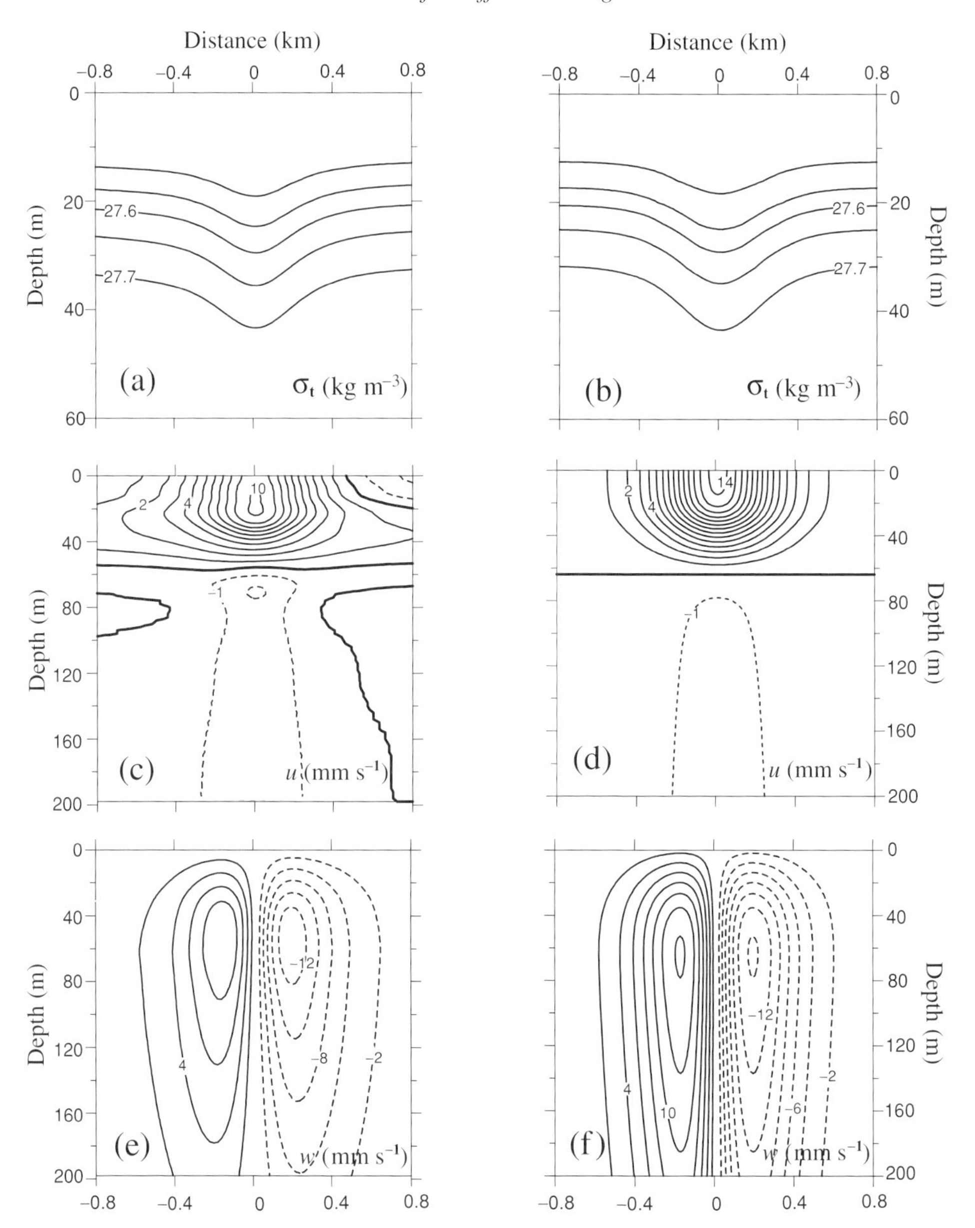

Figure 6.8. (a) Close-up view of wave B_1 from Figure 6.5, together with (c) its horizontal and (e) its vertical velocity. (b) Density profile and (d) horizontal and (f) vertical velocities for the K–dV soliton.

nonlinear wave trains coincides with the wavelength of the linear baroclinic tidal waves. The basic feature that distinguishes the profiles of the nonlinear waves from the linear ones is their asymmetry and disintegration into packets of short localized internal oscillations. In fact, these wave trains are developed because of the strong

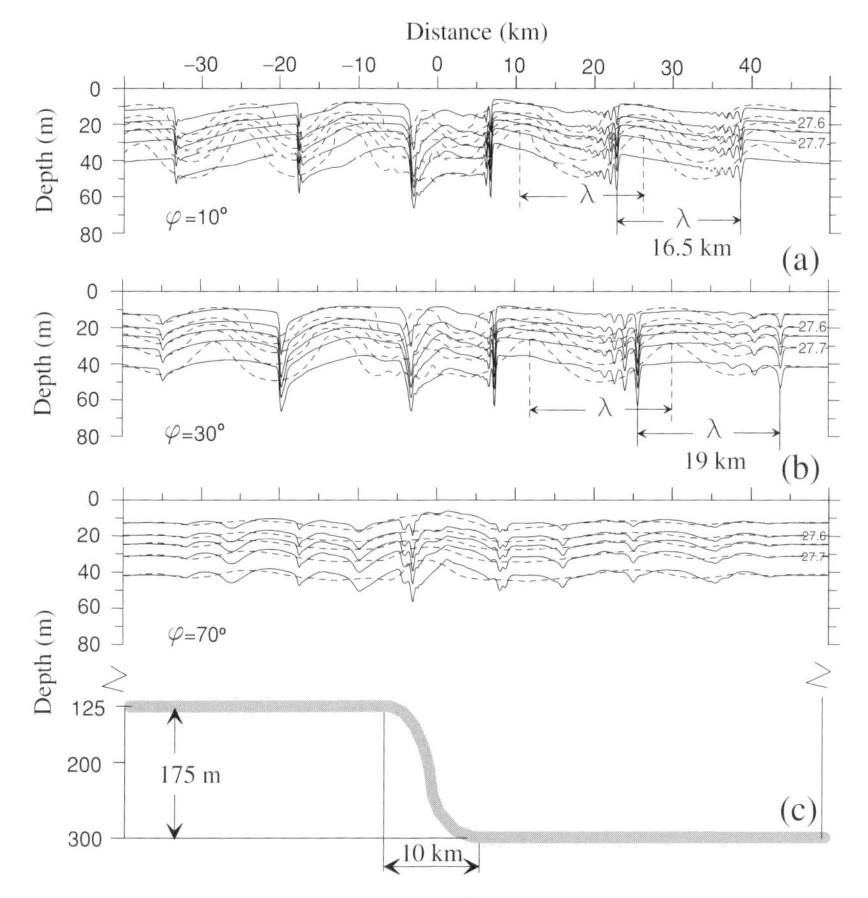

Figure 6.9. Conventional density σ_t (kg m^{-3}) sections obtained for tidally driven waves at three latitudes $\varphi = 10°$, $30°$, $70°$ after three tidal periods (solid lines). Results from the linear model over the same time periods are presented by dashed lines.

nonlinearity and the nonhydrostatic dispersion taking place at the rear steep face of a long tidal wave, as discussed in Chapter 4.

Now let us investigate nonlinear waves which are generated if region 2 is located above the critical latitude. Figure 6.10 displays the density field calculated for four values of the latitude, $\varphi = 75°$, $80°$, $85°$, and $89°$. The isolines in all cases coincide nearly perfectly; this demonstrates that the generation mechanism is independent of the rotation. The suppression of the generation of internal tides with increasing rotation leads to a decrease of the specific input of internal tides into the total wave field and an increase of the input of unsteady lee waves. Since lee waves originate from the hydraulic jump at the lee side of an obstacle, they are short, with wavelengths comparable to the length scales of the bottom topographic feature,

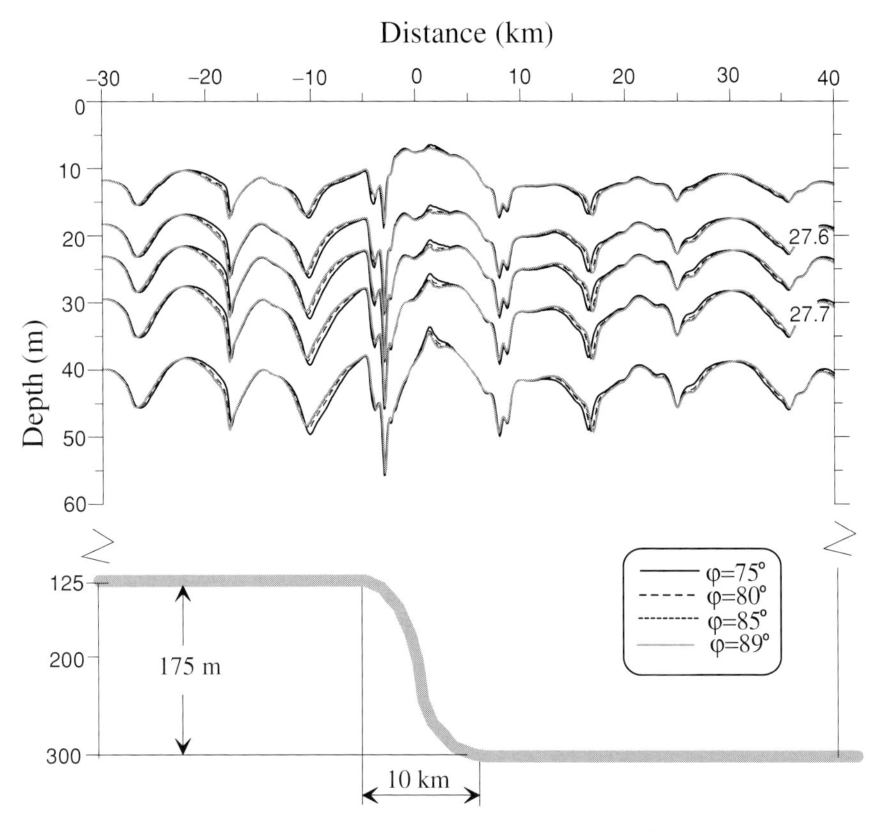

Figure 6.10. Cross-section of conventional density σ_t (kg m^{-3}) for tidally induced internal waves calculated after three tidal cycles for the latitudes $\varphi = 75°$, $80°$, $85°$, and $89°$.

and their characteristics are defined by the shape of the bottom topography, the intensity of the tidal flux, and stratification, and they do not depend on the rotation. This is why at high latitudes the tidally generated internal waves exhibit properties inherent in lee waves. The generation of waves with multiple harmonics is very probably the manifestation of a strong transfer of the energy from the barotropic tidal flux to short waves.

6.2 Influence of the fluid stratification

The fluid stratification of the World Ocean varies in a wide range: from an almost homogeneous fluid in the polar zones of deep winter convection to well stratified equatorial waters. This can have a substantial influence on the mechanism of the topographic generation of baroclinic tides. In this section we investigate how strongly the generation mechanism depends on the local peculiarities of the buoyancy frequency profile. We shall use results from the linear theory, which can be useful also in the nonlinear case for the interpretation of data obtained in numerical runs and field experiments.

According to the results found by the linear theory of Section 2.2, see formula (2.21), the amplitudes of the generated internal waves depend on the value of the parameter ε_0, which is the ratio of the height of the bottom topography, $h_{\max}$, to h_0 given in the *ad hoc* variables (x, z_1), according to (2.2):

$$\varepsilon_0 = \frac{\int_{H_{\max}-H_0}^{-H_0} (N(z)^2 - \sigma^2)^{1/2}\, dz}{\int_0^{-H_0} (N(z)^2 - \sigma^2)^{1/2}\, dz}. \tag{6.1}$$

For simple estimations we can use the fact that, usually, $N(z) \gg \sigma$ for baroclinic tides, and thus (6.1) takes the approximate form

$$\varepsilon_0 \cong \frac{\int_{H_{\max}-H_0}^{-H_0} N(z)\, dz}{\int_0^{-H_0} N(z)\, dz}. \tag{6.2}$$

It is seen that the "topographic" parameter ε_0 depends not only on the topographic height $H_{\max}$, but also considerably upon the fluid stratification. The physical meaning of (6.2) becomes clear with the aid of Figure 6.11, where the underwater ridge is shown together with two profiles of the buoyancy frequency, $N_1(z)$ and $N_2(z)$. Even though they possess different forms, these two profiles have the same integral characteristics, i.e.

$$S = \int_0^{-H_0} N_1(z)\, dz = \int_0^{-H_0} N_2(z)\, dz.$$

This means that the value of ε_0 in (6.2) for the two profiles, $N_1(z)$ and $N_2(z)$, differs only in the magnitudes of the numerators S_1 and S_2, which read

$$S_1 = \int_{H_{\max}-H_0}^{-H_0} N_1(z)\, dz, \quad S_2 = \int_{H_{\max}-H_0}^{-H_0} N_2(z)\, dz.$$

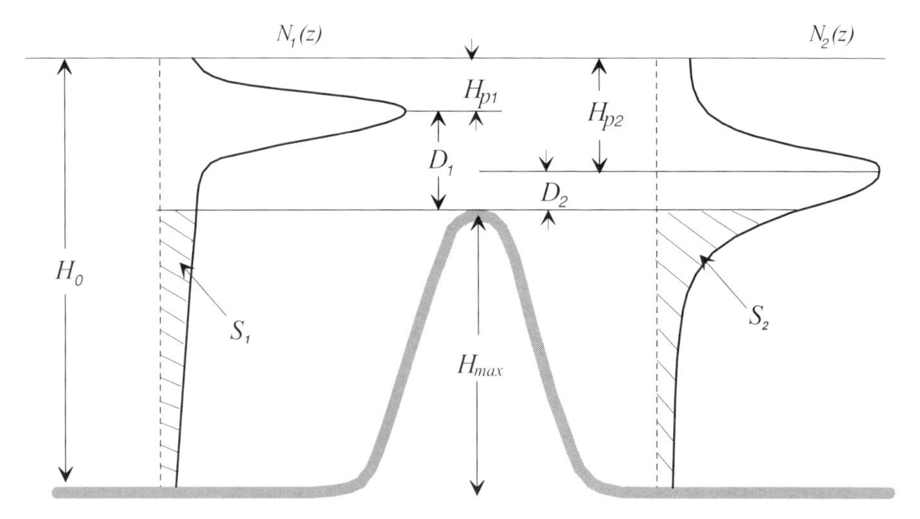

Figure 6.11. Sketch to explain (6.2) for two buoyancy frequency profiles.

The depth H_{p_2} of the pycnocline maximum for the buoyancy frequency $N_2(z)$ is greater than the corresponding value H_{p_1} for $N_1(z)$. So, it is clear that $S_2 > S_1$; see Figure 6.11. Thus, the value of ε_0 and, consequently, the amplitudes of the generated waves for the stratification $N_2(z)$ will be greater than for $N_1(z)$.

The results given above lead to the following conclusion: the intensity of the generated wave fields strongly depends on the distance between the pycnocline and the top of a topographic obstruction, i.e. D_1 and D_2 in Figure 6.11. So, the closer the pycnocline position is to the bottom, the larger will be the amplitudes of the generated waves (keeping all other conditions the same).

The above is a qualitative estimation to see how large the generated waves may become according to the position of the pycnocline. Apart from this, other parameters, i.e. the pycnocline width, the bottom profile extent, and its inclination, are crucial for the generation mechanism. To answer the question of which parameters in the linear theory control the topographic generation of internal waves, we follow the idea of internal tidal forcing introduced in ref. [9]. According to Baines, the determination of internal tides that are caused by the interaction of barotropic tides with a localized bottom topographic obstruction splits into two problems: first, the solution of the barotropic tidal problem, i.e. finding the functions u_{b}, v_{b}, w_{b}, and P_{b} of an "equivalent" unstratified ocean; and, secondly, determination of the additional motion in the stratified ocean, i.e. finding the functions u_{s}, v_{s}, w_{s}, ρ_{s}, and P_{s}, required to satisfy the dynamical equations and the radiation conditions.

According to this procedure, we must solve linear equations,
for the barotropic tide,

$$\left.\begin{aligned} (u_\mathrm{b})_t - f v_\mathrm{b} &= -(P_\mathrm{b})_x/\bar{\rho}_0, \\ (v_\mathrm{b})_t + f u_\mathrm{b} &= 0, \\ (w_\mathrm{b})_t &= -(P_\mathrm{b})_z/\bar{\rho}_0, \\ (u_\mathrm{b})_x + (w_\mathrm{b})_z &= 0, \end{aligned}\right\} \tag{6.3}$$

and, *for the baroclinic tide*,

$$\left.\begin{aligned} (u_\mathrm{s})_t - f v_\mathrm{s} &= -(P_\mathrm{s})_x/\bar{\rho}_0, \\ (v_\mathrm{s})_t + f u_\mathrm{s} &= 0, \\ (w_\mathrm{s})_t &= -(P_\mathrm{s})_z/\rho_0 - g(\rho_\mathrm{s})/\bar{\rho}_0, \\ (\rho_\mathrm{s})_t + w_\mathrm{s}\rho_{0z} &= -w_\mathrm{b}\rho_{0z}, \\ (u_\mathrm{s})_x + (w_\mathrm{s})_z &= 0. \end{aligned}\right\} \tag{6.4}$$

The fourth equation in (6.4) contains as a coupling the vertical velocity of the barotropic tide w_b; its appearance is a consequence of the fact that the bottom streamline of the tidal motion follows the bottom topography.

To find w_b, the continuity equation of system (6.3) is integrated over the depth $(-H, z)$ using the kinematic boundary conditions at the bottom, (1.36),

$$w_\mathrm{b}(x, z, t) = -\frac{\partial}{\partial x}(H u_\mathrm{b}) - z\frac{\partial u_\mathrm{b}}{\partial x}. \tag{6.5}$$

Because $H u_\mathrm{b} = \Psi_0 \exp(\imath\sigma t)$, we obtain from (6.5)

$$w_\mathrm{b}(x, z, t) = -\Psi_0 z \left(\frac{1}{H}\right)_x \exp(\imath\sigma t).$$

Accounting for the periodicity of the processes, an exponential factor $\exp(\imath\sigma t)$ is extracted from all variables. If this is done and (6.5) is used, the fourth equation in (6.4) can be rewritten as

$$g\rho_\mathrm{s} + \imath\bar{\rho}_0 w_\mathrm{s}\frac{N^2(z)}{\sigma} = \imath\bar{\rho}_0\Psi_0 z\frac{N^2(z)}{\sigma}\left(\frac{1}{H}\right)_x. \tag{6.6}$$

This makes it explicit that the external barotropic tidal forcing is present in this wave equation as a driving term on its right-hand side: isopycnals are displaced vertically by an oscillating barotropic current moving back and forth over the topographic obstruction. The amplitude of the restoring force or the tidal forcing is

$$F = \bar{\rho}_0\Psi_0 z\frac{N^2(z)}{\sigma}\left(\frac{1}{H}\right)_x. \tag{6.7}$$

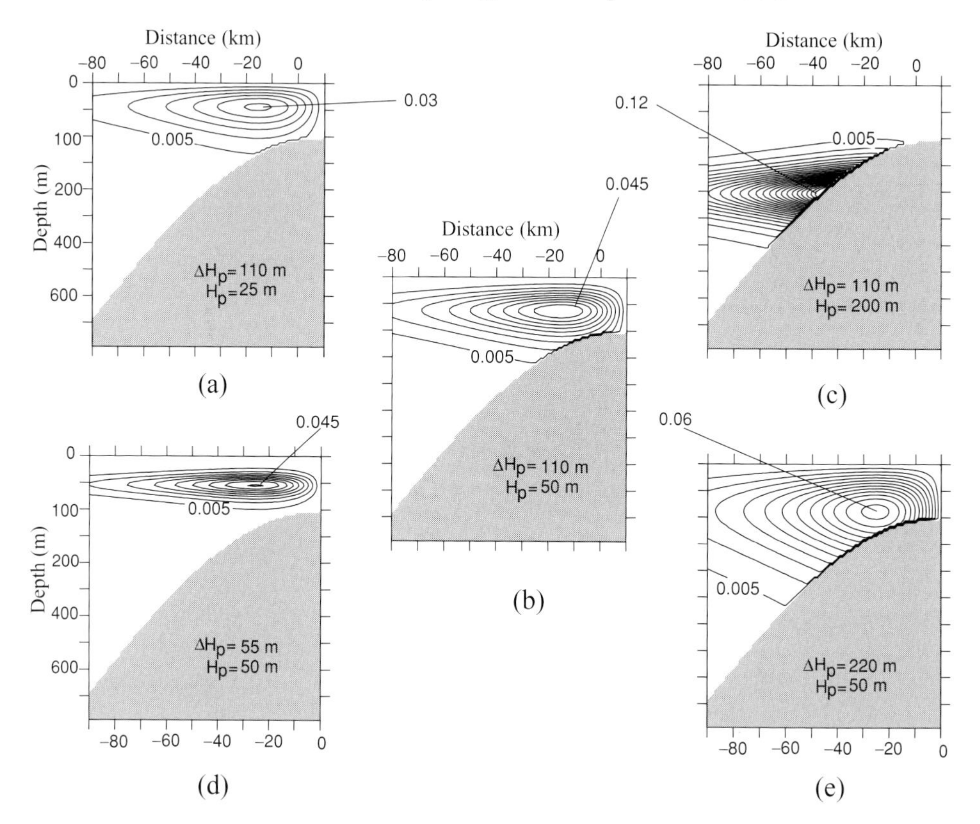

Figure 6.12. Vertical sections of the density of the tidal forcing F (N m^{-3}) calculated for the fluid stratification (1.70) with different parameters H_p and ΔH_p. The value N_p used in all cases was 0.03 s^{-1}.

Figure 6.12 shows the distributions of the tidal forcing over the slope domain calculated for different profiles of the buoyancy frequency. The greatest value of the tidal forcing is shown in each of these five cases. The difference among the first three cases ((a)–(c)) is only in the position of the pycnocline depth: $H_\mathrm{p} = 25, 50, 200$ m, respectively. The dependence of the value of the amplitudes of generated internal waves upon the value of H_p found earlier from estimates of the parameter ε_0 is also clearly seen from Figures 6.12(a) and (b): the greater H_p is, the larger will be the value of the generating forcing. It increases dramatically when the pycnocline resides below the top of the bottom topographic feature (Figure 6.12(c)). The change in the width of the pycnocline also results in different distributions of the density of the generating forcing (Figures 6.12(d), (b), and (e)).

Taking into account the concept of the internal generating forcing, let us investigate the influence of the different fluid stratifications on the nonlinear topographic generation of baroclinic tides.

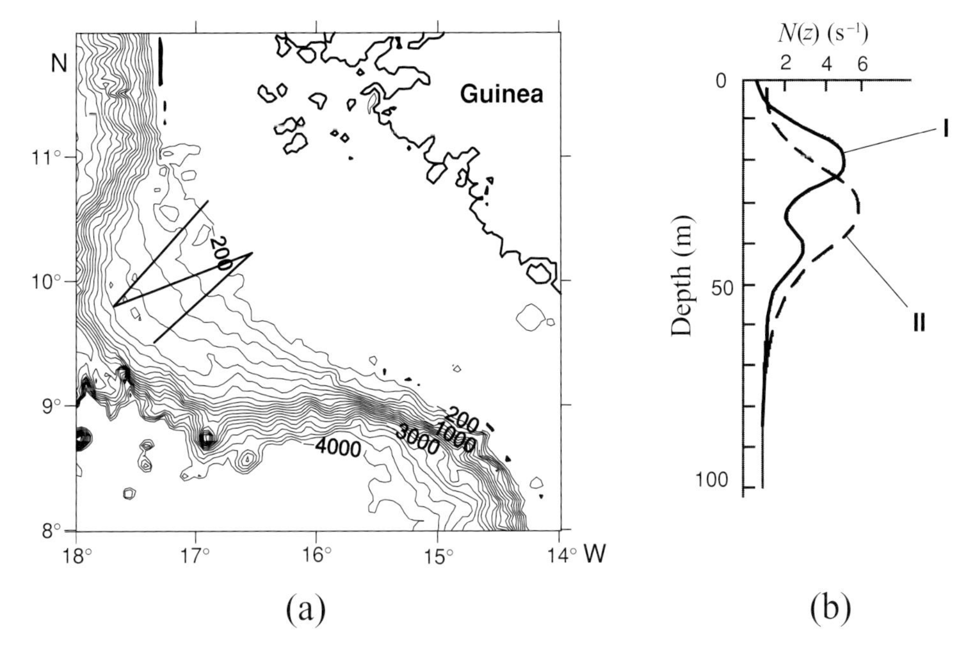

Figure 6.13. (a) Bottom topography in the shelf zone of Guinea with indication of ship tracks. (b) Two buoyancy frequency profiles measured in the test area. Profile I is typical for the deep-water zone, whilst profile II is characteristic of the shelf zone.

6.2.1 Variation of the vertical position of the pycnocline

The influence of the vertical position of the pycnocline on the baroclinic wave process was studied in the shelf zone of Guinea (8–12° N, 14–18° W) (Figure 6.13(a)) [244]. A wave experiment [79] was performed there in December, 1988; its objective was to measure tidally generated intensive internal waves. An example of such a wave, recorded at the water depth of 500 m, is presented in Figure 6.14(a). During its shoreward propagation it transformed into the wave train of Figure 6.14(b), and was also recorded at the 300, 250, and 200 m isobaths, Figures 6.14(b)–(d).

The generation of the measured waves and their evolution should obviously be considered in the context of the nonlinear theory presented in Chapter 4. For the modeling we considered the shallow-water zone with a depth of 30 m and the deep-water zone with a depth of 4.5 km; the transition zone with bottom topography $H(x)$ was taken as typical for the tested area. Two types of the fluid stratification were measured during the experiment. Their buoyancy frequency profiles are shown in Figure 6.13(b). Profile I (solid line) is characterized by two maxima at depths $\sim$20 m and $\sim$40–45 m; it was observed at all deep-water stations. The second type of stratification (II, dashed line) was measured at the shelf; it was characterized by one pycnocline located at a depth of 30–35 m.

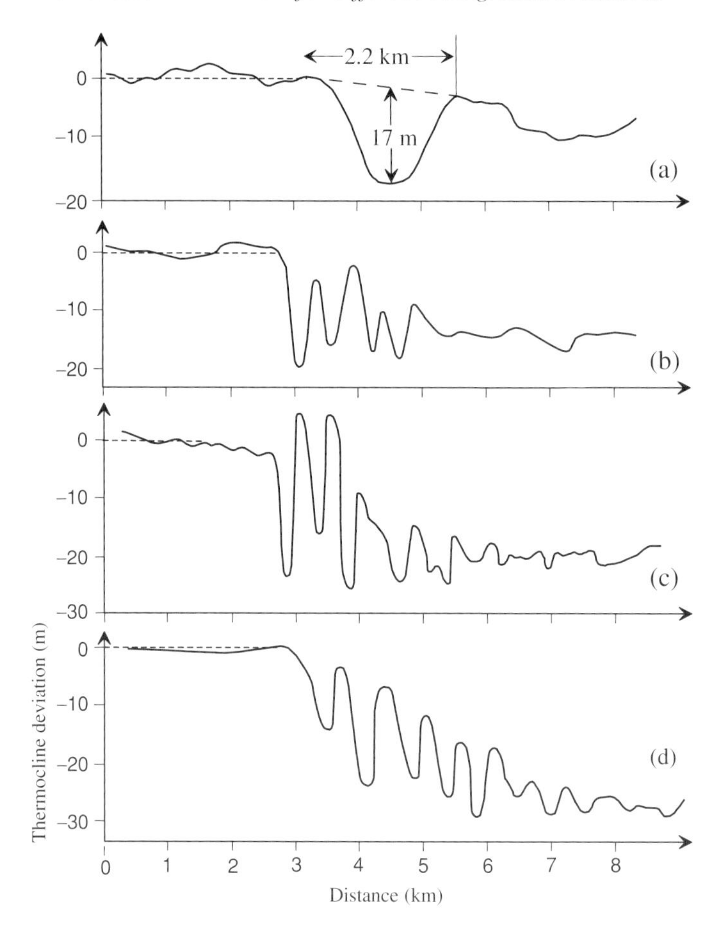

Figure 6.14. Thermocline profile found experimentally along isobath tracks at (a) 400 m, (b) 300 m, (c) 250 m, and (d) 200 m.

Strong horizontal density gradients (frontal zones) associated with steady oceanic currents were absent in the studied area. Thus, it was tempting to assume that such a dramatic change of the buoyancy frequency from the deep part to the shelf could be explained by the action of internal waves propagating shoreward and being destroyed somewhere on the shelf. Incidentally, an example of strong thermohaline reconstruction after wave breaking is also presented in Figure 5.26. As a result of this wave breaking, the $N(z)$-profile changes its shape: the second maximum vanishes and the first is slightly shifted down to the bottom. Variations of the depth of the pycnocline at the shelf were 10–20 m; these values are comparable to the wave amplitudes; see Figure 6.14.

The density field, calculated for the Guinean shelf for stratification I by the numerical model described in Chapter 4, is shown in Figure 6.15, which is an

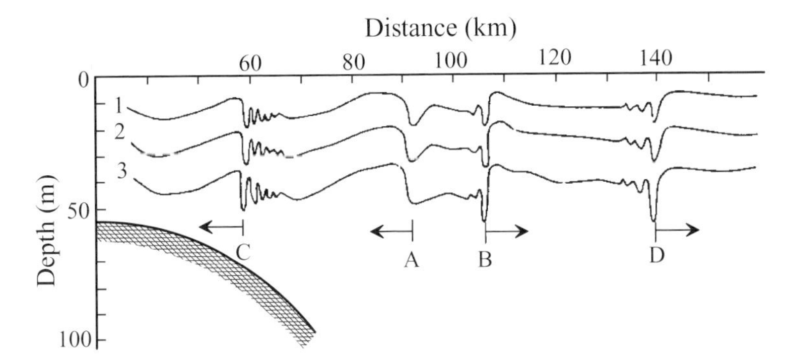

Figure 6.15. Snapshot of density anomaly $(\sigma_t(z) - \sigma_t(0))$ (kg m^{-3}) across the Guinean shelf obtained with the nonlinear analysis for the given tidal driving and stratification I. Arrows indicate direction of wave propagation.

illustration of the generation process in the area under study. The position of the maximum of the generating force is within section AB at the 350–450 m isobaths, where the bottom slope is steepest. Positions A and B represent the initial evolutionary stage of the wave perturbations propagating from the place of their generation. At some distance from the source of generation they form wave packets marked by C and D.

Let us consider the process of the formation and evolution of the wave packets more thoroughly, and compare the results of the numerical modeling with the measured data. Figure 6.16 displays the computed density field for the spring tide at the same isobaths and tidal phases as those observed in Figure 6.14. Comparison of these profiles confirms that a solitary internal wave of depression is actually formed between the isobaths at 350 and 450 m depth. The shoreward propagating wave, position A in Figure 6.15, has a wavelength of about 2 km, which closely coincides with the measured value of 2.2 km, Figure 6.14(a). However, it has a somewhat smaller amplitude, namely $\cong 9$ m. Estimations of the parameters of the nonlinearity and dispersion (Section 5.1) yield the values $\varepsilon \cong 0.02$ and $\mu \cong 0.1$ (we estimated the wavelength at the level 0.5 of the amplitude). Such values of ε and μ indicate comparable manifestations of dispersion and nonlinearity effects. If dispersion predominates nonlinearity, then a solitary wave traveling shoreward begins to transform gradually; a packet of shorter waves develops at its rear slope. For example, four to five oscillations are clearly seen at the isobaths 230–370 m (Figure 6.16(b)); their maximum amplitude is 12 m, and the wavelength is about 500 m. Similar values are also disclosed from the measured waves: 12–15 m for the amplitude and 500–600 m for the wavelength. So, the initial stage of the generation and evolution of a solitary internal wave is in good agreement with the recorded data.

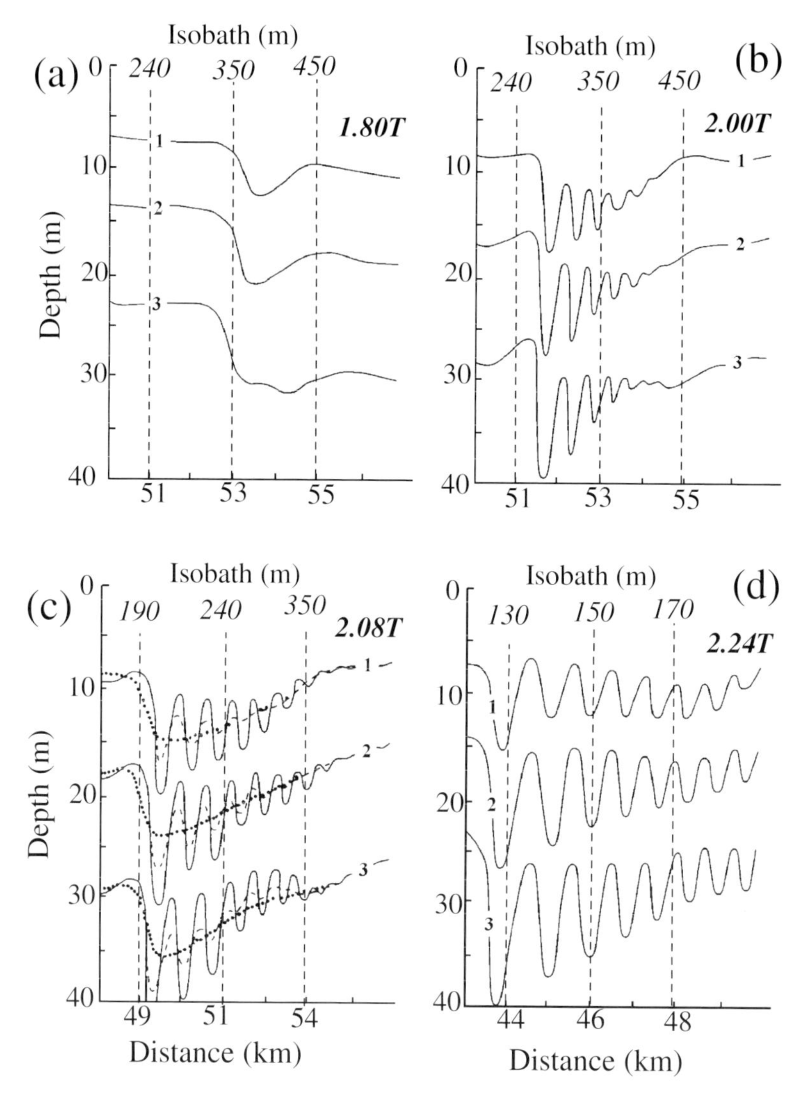

Figure 6.16. Density field $p(z) - p(0)$ (kg m^{-3}) at different stages of the solitary wave evolution. Calculations were performed for the first type of fluid stratification (I in Figure 6.13(b)). T denotes the tidal period, and $1.8T$, $2.0T$, ... mark the time slices after wave generation. For a detailed explanation, see the text.

During the ongoing shoreward propagation of the wave packet over the sloping bottom, the number of oscillations increases, and the wave amplitudes grow. The increase of the amplitudes is caused by the nonlinear energy transfer from an initial solitary wave to oscillations of lesser scales, as well as by a decrease of the width (depth) of the wave guide. Between the 190 and 300 m isobaths, the calculated wave train has all the characteristics similar to the measured isopycnal time series.

Indeed, in the numerical experiment, a wave packet has five to seven identifiable oscillations with wavelengths of about 600 m and a maximum amplitude of 12–15 m, see Figure 6.16(b). Qualitatively, the same wave packet was recorded in the measurements, Figure 6.14(b).

The further evolution of both the numerically calculated and the measured wave packets is somewhat different. The experimentally recorded wave train traveling shoreward proceeded to disintegrate into waves of still smaller scales, with wavelengths of 150–200 m at the 150 to 200 m isobaths (Figure 6.14(d)) and amplitudes of about 15–17 m. By contrast, the numerical counterparts are stretched, due to the dispersion effects with wavelengths $\cong$ 700 m at the 130 to 200 m isobaths. However, the number of waves in the trains coincides with those observed.

This difference in the behavior of the wave packets at the final stage of their evolution may be explained by overestimated values of the coefficients of the turbulent viscosity, A^{H} and K^{H} (see (1.79) and (1.81)) for the shelf area in the numerical model. An example of the influence of the magnitude of the turbulent viscosity coefficients on wave dynamics is shown in Figure 6.16(c). Here, the solid lines correspond to the wave profile calculated with $A^{\mathrm{H}} = K^{\mathrm{H}} = 2.0\,\mathrm{m^2\,s^{-1}}$; the dashed lines were obtained with $A^{\mathrm{H}} = K^{\mathrm{H}} = 20\,\mathrm{m^2\,s^{-1}}$; and the dotted lines with $A^{\mathrm{H}} = K^{\mathrm{H}} = 200\,\mathrm{m^2\,s^{-1}}$. It is seen that a tenfold increase of the viscosity coefficients does not change the structure of the wave packet, but significantly decreases its amplitude. However, at $A^{\mathrm{H}} = K^{\mathrm{H}} = 200\,\mathrm{m^2\,s^{-1}}$, no wave packet is formed at all. All nonlinear wave properties are suppressed by dissipation.

Next, let us analyze the generation of internal waves for the buoyancy frequency profile II depicted in Figure 6.13(b) by the dashed line. With an uprising of the pycnocline, the location of the generating force, i.e. segment AB in Figure 6.15, shifts towards the shelf zone; see also Figure 6.12. As a result, a solitary wave is not formed directly at the 400 to 500 m isobaths, as it was for the type I stratification; it is generated 20 km closer to the coast at the 60 to 70 m isobaths, Figure 6.17(a), and the depth of the pycnocline, H_{p}, is located in between the free surface and the bottom. The generated internal wave travels shoreward from the area where the undisturbed pycnocline is located closer to the bottom than to the surface. As a result, either a propagating negative wave of depression with amplitude ~8 m (Figure 6.17(a)), will have to change its polarity or it will evolve in some way differently, from a transformation into a soliton. Estimations of the parameters ε and μ give values $\varepsilon = 0.15$, $\mu = 0.01$. In this case there is no balance between nonlinearity and dispersion. Consequently, the wave shown in Figure 6.17(a) is not a soliton; it will evolve into a wave packet over a slowly changing bottom. The initial stage of this evolution is shown in Figures 6.17(b), (c), and (d). In contrast with the version considered above, Figure 6.16, when the relationship between ε and μ was opposite and a wave packet appeared behind the abrupt slope of a solitary

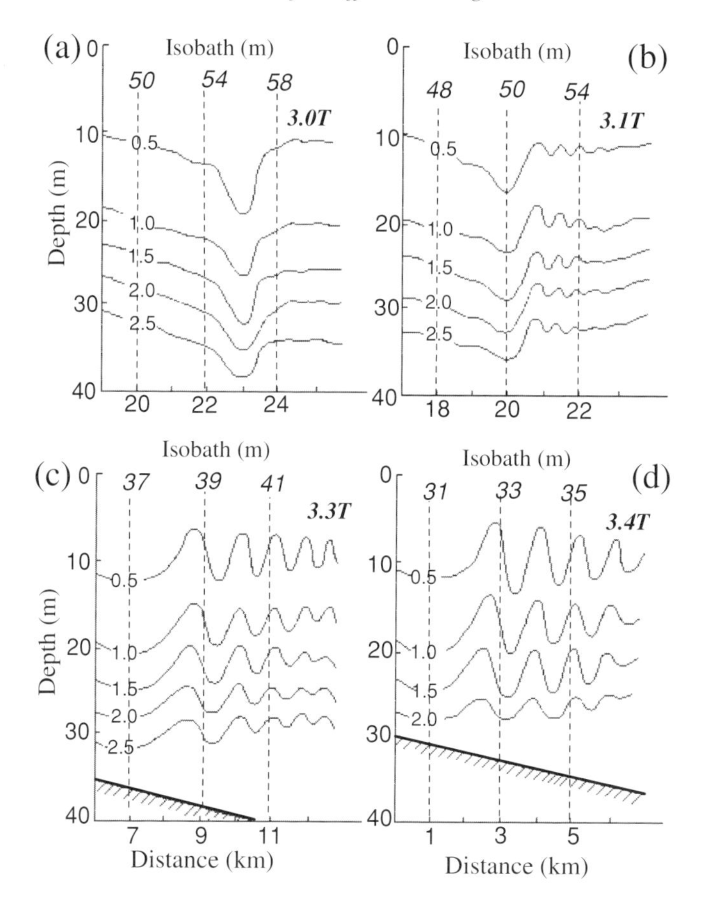

Figure 6.17. Density field $p(z) - p(0)$ (kg m^{-3}) at different stages of the solitary wave evolution. Calculations were performed for the second type of fluid stratification (dashed line (II) in Figure 6.13(b)).

wave (Figure 6.16(b)) the front slope now becomes more gentle. The wave turns into a baroclinic bore plus a dispersive packet of short waves formed behind its slope, Figure 6.17(b). The wave amplitudes of a train grow as it approaches the shore. This is caused by the nonlinear energy transfer from the low-frequency to the high-frequency component, as well as by the narrowing of the wave guide exhibited by the decrease of the basin depth.

6.2.2 Effect of horizontal density gradients

In Chapter 3 we considered, within a linear formulation, the influence of horizontal density gradients on the generation of internal tides. Let us now estimate the effect of a frontal zone on the characteristics of nonlinear baroclinic tides; the basis is

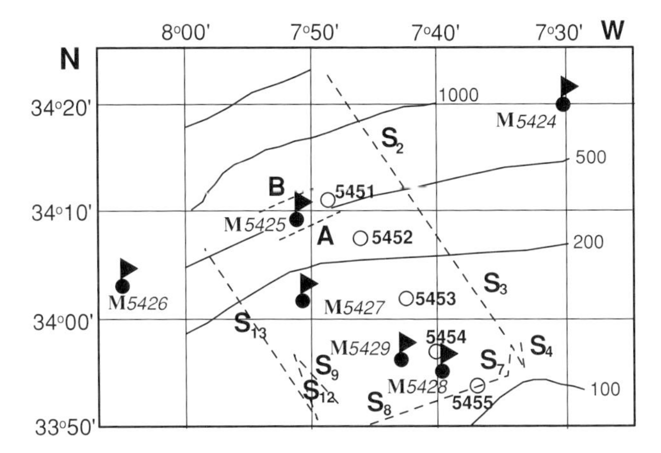

Figure 6.18. Schematic pattern of the test area in the Morocco shelf zone. Open circles indicate moorings; flagged circles are CTD stations; S_m denote tracks, where m indicates the track number.

ref. [258]. This study was motivated by observational data collected in the shelf zone of Morocco, where nonlinear internal tides were measured during cruise 48 of the R/V *Mikhail Lomonosov* in August and September, 1987. A wave experiment was conducted in an area of 40 × 60 square miles extent, 33°55′–34°30′ N and 7°30′–8°10′ W; see Figure 6.18. It included the deployment of six moorings (M5424–M5429) with current meters placed at the isobaths from 100 to 1000 m, four CTD stations (5451–5454), and the towing of a thermistor chain along the seven tracks depicted by the dashed lines in Figure 6.18.

The CTD survey preceded the basic wave experiment, and identified a frontal zone near the shore, which was characterized by a relatively strong horizontal density gradient: the pycnocline depth H_p decreased quite remarkably in the shoreward direction; see Figure 6.19. A strong background current flowing in the southwestward direction was recorded at the moorings. As revealed by the analysis of the mooring data, this current was exposed to horizontal periodic oscillations of tidal origin.

Figure 6.20 displays the temperature time series at moorings M5429 (top) and M5428 (bottom). Comparison of the two series indicates that they correlate rather well. The only basic distinction between them is a time delay of 2 to 2.5 h, which can be explained in terms of the considered topographic generation mechanism by assuming that the shoreward propagating internal tidal waves are generated at the shelf break.

Qualitative analysis of the recorded time series also shows a very interesting regularity: time intervals, marked in Figure 6.20 as T_{ij}, are very accurately equal to 24 ± 0.5 h; those marked as τ_{ij} equal 15 ± 0.5 h, when j is odd, and 9 ± 0.5 h, when

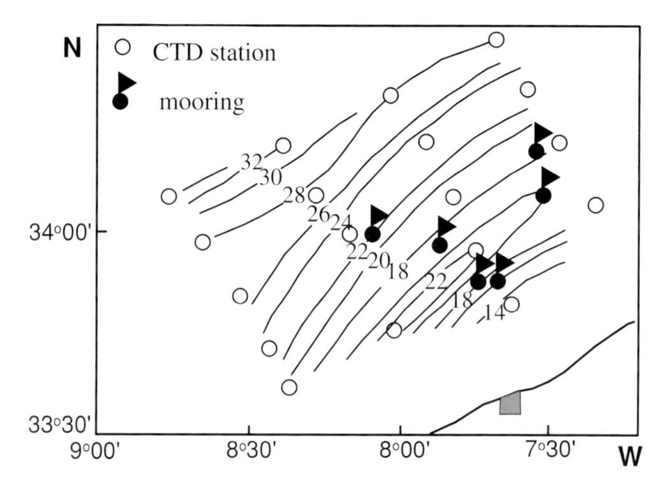

Figure 6.19. Depths (in meters) of the upper boundary of the high-frequency wave guide (frequency = 10 cycles h^{-1}).

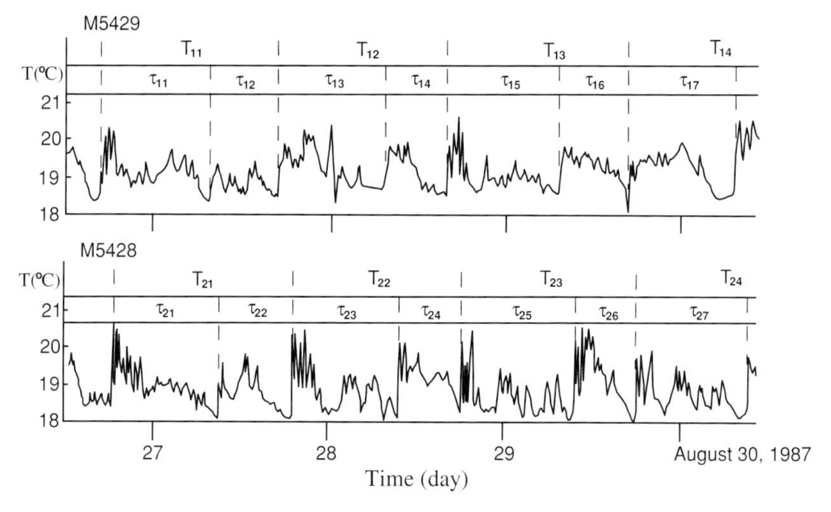

Figure 6.20. Time series of temperature observed at moorings M5429 (top) and M5428 (bottom).

j is even ($i = 1$ indicates mooring M5429 and $i = 2$ refers to mooring M5428; $j = 1$ to 7 denotes the period number). The temporal variation with 9 and 15 h periods in the registration of the wave trains at the moorings allows us to assume that the source of generation of the internal wave changes its location periodically with time, resulting in the different times of wave arrival at the mooring.

Note that the inertial period for the considered area is close to 24 hours. So, we may expect that the position of the frontal zone could oscillate with inertial periodicity, and the difference in the arrivals could be due to the meandering of the along

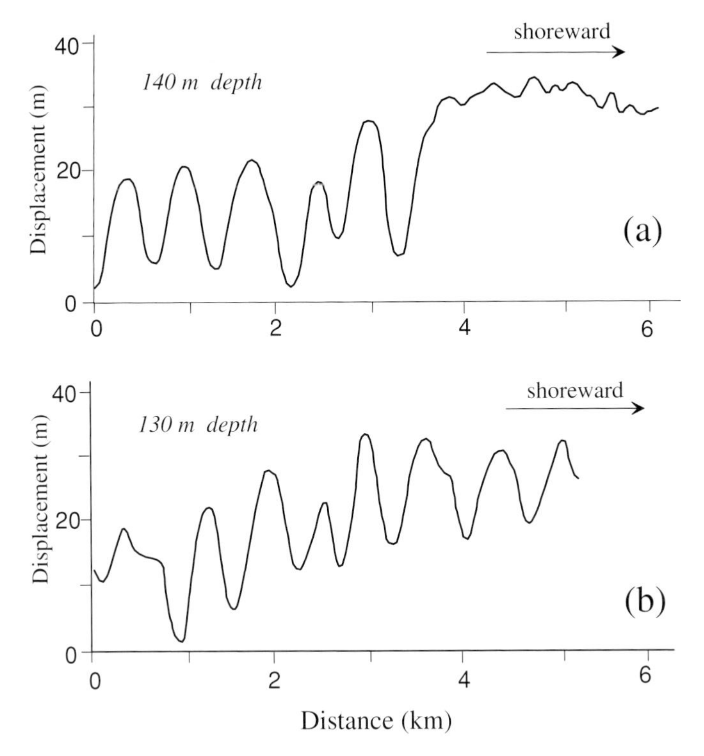

Figure 6.21. Oscillations of the thermocline obtained by a towed thermistor chain at the shelf. Wave profiles at (a) the 140 m and (b) 130 m isobaths are shown.

shore background current. As a consequence, the conditions of the water stratification in the studied area will also change with the diurnal period, and the location of the maximum of the generating force (6.7) will change with the same periodicity. In other words, the place of generation of internal waves moves periodically back and forth from the shore (between the dashed lines A and B in Figure 6.18).

The above reasoning may serve as a defensible explanation of the temporal variability of the arrival of the internal wave packets at the moorings. These waves are of tidal nature, but the place of their generation changes permanently. However, on the shelf they possess all the specific features inherent in the tidally generated internal waves studied above. For instance, Figure 6.21 shows the wave profile recorded by a thermistor chain at the two shelf sections. An internal tidal wave, recorded at the 140 m isobath (Figure 6.21(a)), represents a typical wave profile during the disintegration of a baroclinic tidal bore: the thermocline suddenly sank 25 m to the bottom followed by several oscillations with amplitudes 15–20 m. Wavelengths of these oscillations were of the order of 400–800 m. Figure 6.21(b) shows the wave profile of the thermocline closer to the shore, at the 130 m isobath. The number of waves in the train increased during the evolution, but the wave

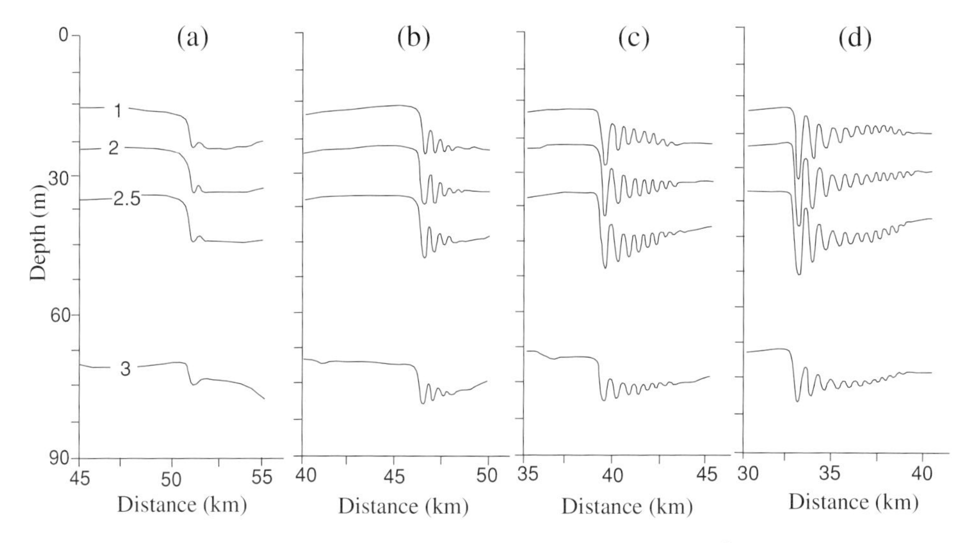

Figure 6.22. Model-predicted density fields $p(z) - p(0)$ (kg m^{-3}). Results are obtained for regions with (a) 200–450 m, (b) 150 m, (c) 140 m, and (d) 130 m depth.

amplitudes decreased due to dissipation, and the characteristic step-like structure disappeared.

Thus, the analysis of Figures 6.20 and 6.21 reveals that the generation and evolution of the internal waves in the presence of a density front are nonstationary in time and nonuniform in space. The presence of horizontal density gradients influences the generation stage of the wave process (in form of the generating force (6.7)), whilst the evolutionary stage remains basically the same as discussed above.

The numerical model was used to check the suppositions inferred from the analysis of the experimental data. The bottom profile and the unperturbed fluid stratification were prescribed in keeping with the actual bathymetry and hydrological survey data compiled in the area under study during the experimental campaign (Figures 6.18 and 6.19).

Figure 6.22(a) shows the initial phase of the generation, which has not been experimentally observed, and the evolution of the wave packet, i.e. baroclinic tidal bore between the 200 and 450 m isobaths, which starts to disintegrate and transform into an internal wave packet. Its vertical scale can be estimated as 14 m. While propagating shoreward, this bore gradually transforms into a packet of short waves (Figures 6.22(a)–(c)) with amplitudes of 2–12 m and wavelengths of 200–600 m. The appropriate values derived experimentally are 20 m and 300–600 m, respectively (Figures 6.21(a) and (b)). As the wave train propagates towards the shore, the number of oscillations in the packet increases, just as in the observed data.

6.3 Baroclinic tides over steep bottom features: "mode" and "beam" approaches

The linear models of baroclinic tides in Chapters 2 and 3 were developed on the basis of two different ideas, which conventionally can be called the "mode" and "beam" approaches. The *mode* approach assumes that the generated wave field consists of a sum of elementary baroclinic modes (1.66), which are the solution of the BVP (1.67). As a result, beyond the underwater obstacle we obtain an infinite number of baroclinic modes, propagating in both directions from it, and the resulting wave field is their superposition. The problem is reduced to finding the amplitudes of the generated modes.

Alternatively, the *beam* approach is based on the application of the method of characteristics which is used for the solution of the hyperbolic wave equation (2.37). The basic result obtained within the framework of the beam approach is that near to critical or supercritical bottom features (in terms of the inclination of the characteristic lines $dz/dx = \pm\alpha$) the wave "beam" emanating directly from the shelf contains the basic part of the baroclinic tidal energy. It was shown in Section 2.5 that in the linear theory these two methods are equivalent and supplement each other.

It should, however, be noted that in the real ocean the amplitudes of high baroclinic modes attenuate rather quickly with the distance from the source of generation. This fact can be simply illustrated by the following reasoning. Consider the linear wave motions in a nonrotating viscous ocean. In the Boussinesq approximation system, (1.29)–(1.30) can be readily reduced to a single equation for the stream function:

$$\Delta\psi_{tt} + N^2(z)\psi_{xx} - (A^{\mathrm{H}} + K^{\mathrm{H}})\Delta\psi_{xxt} + A^{\mathrm{H}}K^{\mathrm{H}}\Delta\psi_{xxxx} = 0. \tag{6.8}$$

(For simplicity, we put the coefficients of vertical viscosity A^{V} and diffusion K^{V} as equal to zero.) We are looking for a solution that is periodic in time, i.e.

$$\psi(x, z, t) = \overset{*}{\psi}(x, z)\exp(\sigma t), \tag{6.9}$$

where σ is complex valued, so $\mathrm{Im}(\sigma)$ is the frequency and $\mathrm{Re}(\sigma)$ is the coefficient of wave attenuation. Substituting (6.9) into (6.8) we obtain

$$\overset{*}{\psi}_{zz} + \left[\frac{N^2(z) - (\sigma)^2)}{\sigma^2}\right]\overset{*}{\psi}_{xx} - \left(\frac{A^{\mathrm{H}} + K^{\mathrm{H}}}{\sigma}\right)\Delta\overset{*}{\psi}_{xx} + \frac{A^{\mathrm{H}}K^{\mathrm{H}}}{\sigma^2}\overset{*}{\psi}_{xxxx} = 0. \tag{6.10}$$

If we write the spatial part of the stream function as

$$\overset{*}{\psi}(x, z) = \overset{0}{\psi}\exp\imath(k_{\mathrm{h}}x + k_{\mathrm{v}}z),$$

the dispersion relation

$$\sigma^2 + \sigma(A^{\mathrm{H}} + K^{\mathrm{H}})k_{\mathrm{h}}^2 + N^2(z)k_{\mathrm{h}}^2/k^2 + A^{\mathrm{H}}K^{\mathrm{H}}k_{\mathrm{h}}^4 = 0 \tag{6.11}$$

is obtained, where k_{h} and k_{v} are the horizontal and vertical components of the wave vector $\mathbf{k}$ ($k = |\mathbf{k}| = (k_{\mathrm{h}}^2 + k_{\mathrm{v}}^2)^{1/2}$). The solution of (6.11) can be expressed as

$$\sigma = -\frac{A^{\mathrm{H}} + K^{\mathrm{H}}}{2}k_{\mathrm{h}}^2 \pm \imath N(z)\frac{k_{\mathrm{h}}}{k}\left[1 - \frac{(A^{\mathrm{H}} - K^{\mathrm{H}})^2 k_{\mathrm{h}}^2 k^2}{4N^2(z)}\right]^{1/2}, \tag{6.12}$$

from which we read off

$$\mathrm{Re}(\sigma) = -\tfrac{1}{2}(A^{\mathrm{H}} + K^{\mathrm{H}})k_{\mathrm{h}}^2.$$

This expression states that the wave attenuation is linearly dependent on the coefficients of viscosity and diffusivity, and it is proportional to k_{h}^2. Therefore, the first baroclinic mode with the smallest wavenumber is usually registered far from bottom roughnesses, whilst the higher modes may have died out. Recalling also the result from Section 2.5 that the wave beam is simply the superposition of the baroclinic modes, it is not surprising that at some distance from an obstacle it becomes less pronounced and gradually disappears. In conclusion, the beam approach is likely to be useful for the interpretation of oceanographic data in close proximity to an underwater obstacle, but the mode approach is more appropriate at some distance from it.

Let us estimate how both methods developed in the linear theory can be applied for data interpretation in the nonlinear case. To this end, we consider the dynamics of baroclinic tides in the shelf break area on the basis of observational data obtained in the region of the New York Bight during JUSREX-92 [260]; see Figure 4.1. In this case, the parameter of nonlinearity, ε_1, see (4.1), can be estimated to be 0.05.

Figure 6.23 shows the evolution of the density field during two tidal periods in the area of the M1 mooring. In most parts of this figure wave formations are discernible which are inherent in the second and higher baroclinic modes with counterphase isopycnal displacements above and below the pycnocline (for instance, the positions marked A and B). The varicose narrowings and widenings of the pycnoclines are marked by circled numbers 1–10. It is quite difficult to find the exact horizontal scales of these wave formations from the data *in situ*. The distances between two adjacent crests and troughs are in the range 9–12 km. According to the linear theory, the wavelength of the second baroclinic mode is $\lambda_2 = 11.3$ km. It should also be noted that these formations are not stationary. We failed to determine the velocity of their propagation directly from Figure 6.23, since we know neither the velocity of the tidal flow at the site of towing, nor the varying background water characteristics.

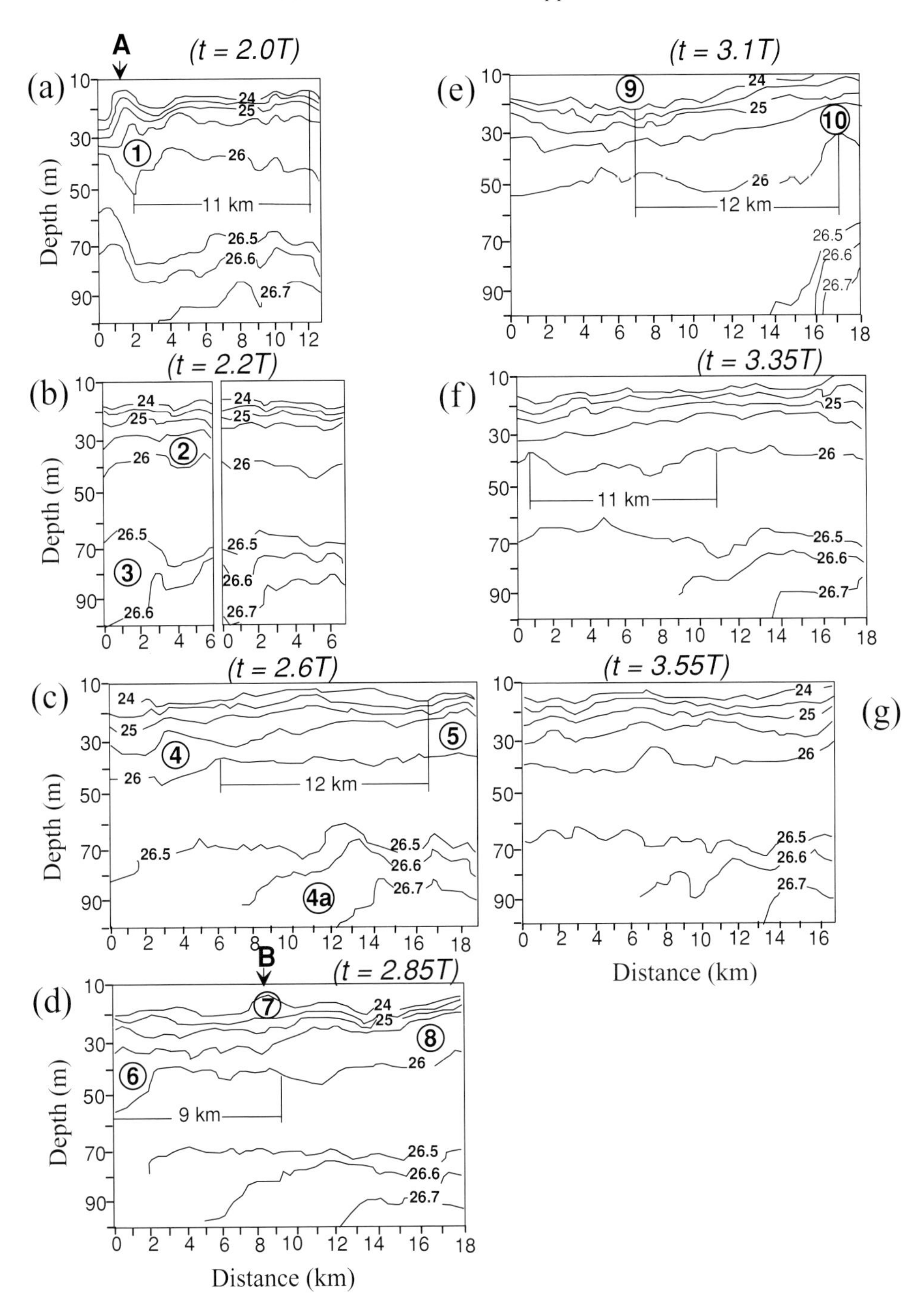

Figure 6.23. Conventional density fields (kg m^{-3}) recorded by ship tracks (line *d* in Figure 4.1) in the New York Bight during JUSREX-92.

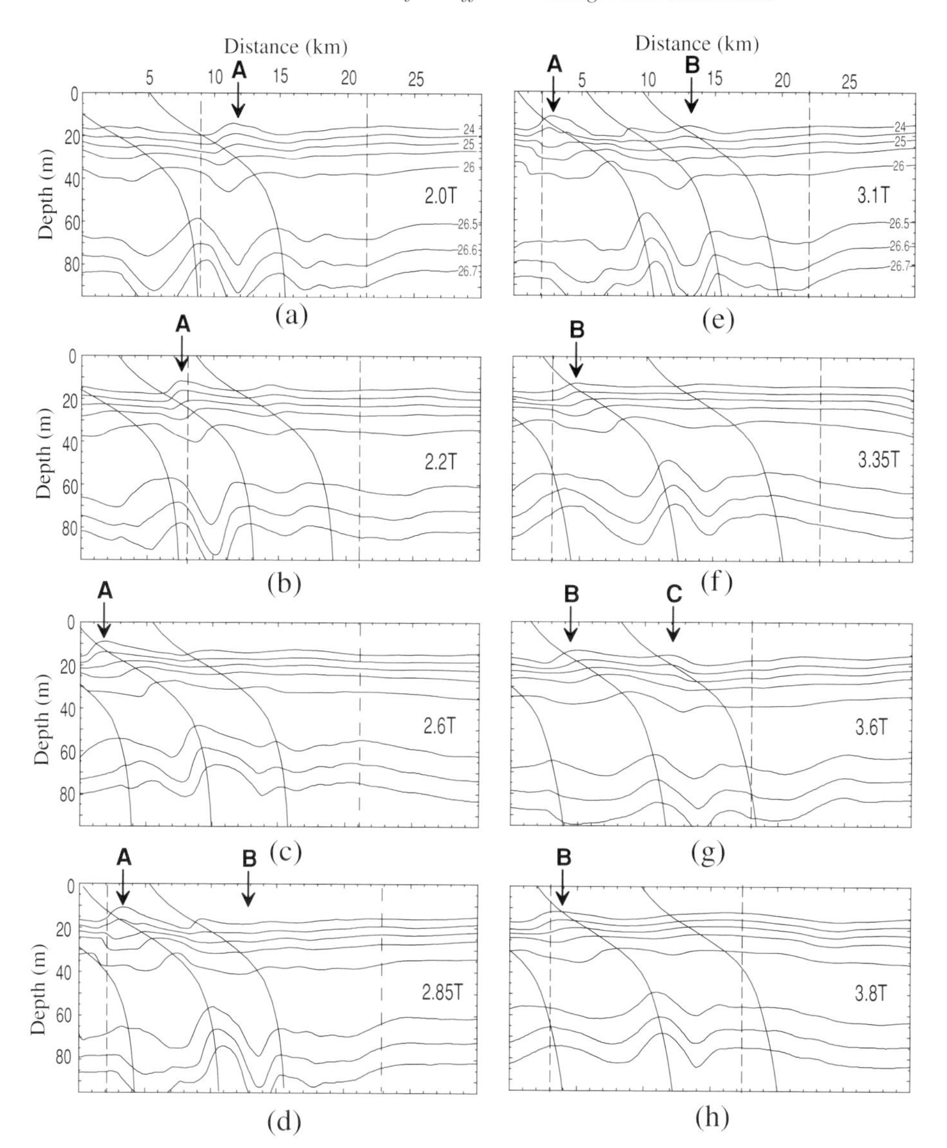

Figure 6.24. Numerical results of the evolution of the conventional density fields in the shelf break area of the New York Bight during two tidal cycles. Isolines are given in kg m^{-3}.

The situation becomes much clearer if the results of the mathematical modeling are taken into account. The density fields calculated by the model are presented in Figure 6.24 (the development of the wave process in the studied area for $t \leq 2T$ was shown in Figure 4.7). They exhibit the same tidal phases as the observed fields in Figure 6.23 (we consider the process after $t = 2T$ when all initial transient processes in the area of generation were terminated). The vertical dashed lines in Figures 6.24(a)–(h) denote the boundaries of the beginning and the end of the towings shown in Figures 6.23(a)–(g).

Let us consider the isopycnal disturbance marked in Figure 6.24(a) by the letter A. Its formation was completed by the time $t = 2T$, and it looks like a second mode. The displacement of the isopycnals from their stationary position is 10 m in the upper layer and 20 m in the lower layer. Figures 6.24(b)–(f) show that wave A moves shoreward to the left, but its velocity is not constant in time. During the first half of the tidal cycle ($t = 2$–$2.5T$), when the tidal flow is directed to the coast, wave A propagates a distance of ~10 km (Figures 6.24(b), (c)). During the ebb phase ($t = 2.5$–$3T$), when the tidal flux moves from the shelf to the right, wave A is trapped by the current, and even propagates backward a little (Figures 6.24(d), (e)). The value of its phase velocity, c_p, averaged over the tidal period is 0.21 m s^{-1}. The linear theory gives a value of $c_p = 0.24$ m s^{-1} for the second baroclinic mode. Good coincidence of these two values and the qualitative conformity of the structure of the experimental and theoretical waves A to the structure of the second baroclinic mode suggest that the mode approach is able to predict this phenomenon adequately.

The wavelength of the second baroclinic mode calculated from the wave fields and obtained numerically is similar to that found from the linear wave theory, $\lambda_2 = 11.3$ km. Actually, during the tidal cycle $t = 2$–$3T$, another wave, similar to A, marked B in Figures 6.24(d) and (e), is formed. The distance between these waves varies with time in the range 10–12 km, which is very close to the theoretical value of the wavelength of the second baroclinic mode, 11.3 km. During the next tidal cycle, $t = 3$–$4T$, the next second baroclinic mode, marked by the letter C in Figure 6.24(g), is generated, and the distance between B and C is the same as that between A and B.

Scrutiny of Figure 6.24 thus indicates periodic excitation of wave formations, propagating to the coast, with characteristics close to those of the second baroclinic mode. This conclusion, namely the generation of a large number of baroclinic modes above steep underwater obstacles, was previously claimed within the framework of the linear theory (see Section 2.4). This result is also valid for the nonlinear generation process.

Let us try to analyze now the results of the numerical modeling on the basis of the beam approach. The curved lines which extend from the upper left to the lower right in the parts of Figure 6.24 are the characteristic lines of the hyperbolic equation (1.42). Their positions are connecting the crests of the isopycnals in the lower layer for $z < -50$ m. It is seen that, when arranged in such a manner, they cross the local wave formations (varicose widening of the pycnocline) in the upper layer as well. It is interesting to note that the wave beam is exposed to the tidal current and moves with it. However, near-surface wave perturbations move simultaneously with those in the bottom layers, so that all of them are related to one and the same characteristic line. In other words, the wave rays exist all the time, but their positions change in the horizontal direction.

The beam character of the baroclinic tides in the area of the shelf break is also seen in Figure 6.25, where the temperature, salinity, and density fields are shown for the New York Bight as a result of the towing *d* in Figure 4.1. The seasonal thermocline outside the shelf break is located between the 10 and 25 m depths. It spreads in the shelf break area, where the isotherms are deepened. For example, the local depth of the 13 °C isotherm was 25 m, and that of the 11 °C isotherm was 40 m. Such a significant lowering of isotherms indicates that, in the area of the shelf break, there is an active thermocline ventilation and that heat is transferred downward from the upper layers. The analysis of the salinity field, Figure 6.25(b), discloses the opposite picture: isohalines upwell significantly over the shelf break. For instance, the local uplift of the 34°/$_{oo}$ isohaline is 30 m. Thus, the salinity field is also subjected to active dynamic processes that take place in the region of the shelf edge and result in significant upwelling of saline waters. The density field, Figure 6.25(c), is deformed to a lesser extent: isopycnal displacements do not exceed 10 m. However, the presence of an intense diapycnic heat and salt exchange between the bottom and surface waters results in a significant local pycnocline widening over the shelf break. The most probable reason for this phenomenon is hydrodynamic instability and breaking of waves above the shelf slope, which increases the values of the vertical turbulent exchange coefficients.

The Richardson number, Ri, is a measure of the stability of the stratification. The distribution of the Ri values in the shelf break area obtained numerically during one tidal cycle is shown in Figure 6.26. It is seen that, in the area of the shelf break, where the large vertical shifts of the speed of the generated internal waves take place, the Richardson number is below 0.25. This zone with small Richardson numbers originates at the edge of the shelf break and extends upward along the shelf. It can be seen that the instability area follows the direction of the characteristic line, which is shown by the thick line. The zone of instability is thus coincident with the zone of the wave beam where baroclinic perturbations are most intensive.

Comparison of Figures 6.25 and 6.26 shows that the zone of intensive diapycnic mixing, which was found experimentally, also originates in the area of the shelf break and extends towards the coast. Its inclination and width are close to those of the wave beam.

6.4 Strong high-mode baroclinic response over steep bottom topography

In the previous section it was illustrated that both "mode" and "ray" approaches, originally developed for infinitesimal waves, can also be applied for the

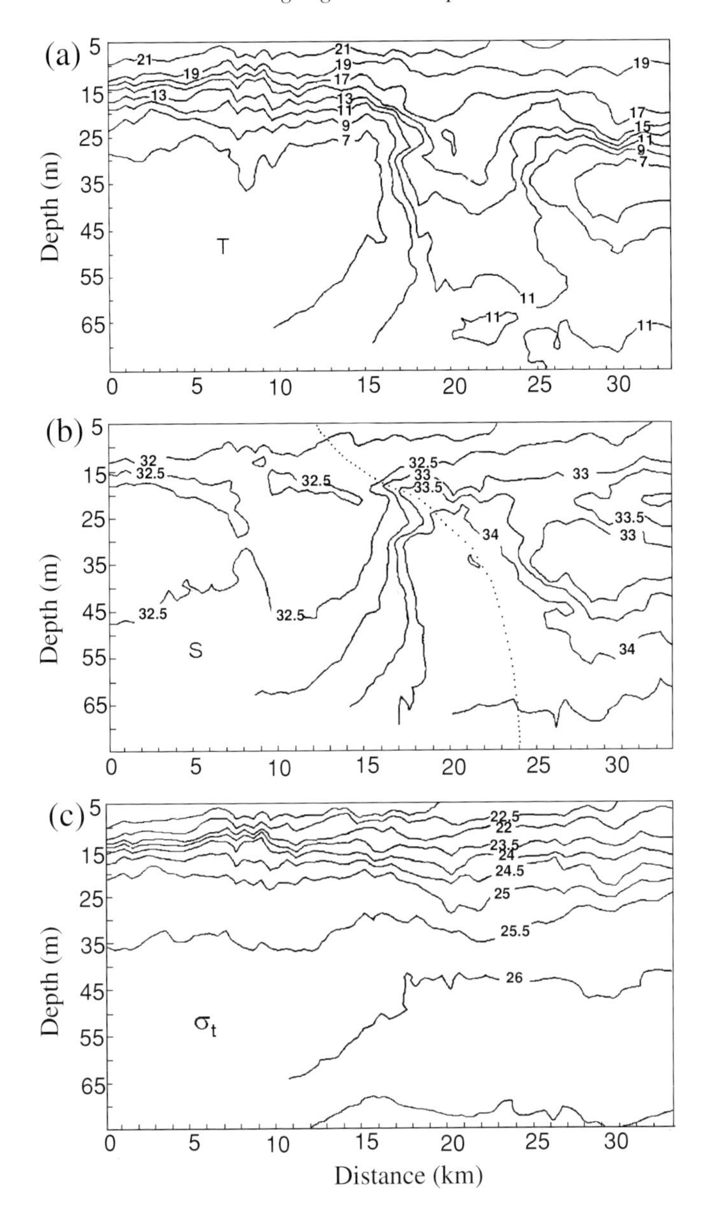

Figure 6.25. (a) Temperature (°C), (b) salinity (°/$_{\circ\circ}$), and (c) conventional density (kg m^{-3}) fields recorded during JUSREX-92 at the New York Bight shelf break area. The dotted line in (b) is the characteristic line of the wave equation.

nonlinear case to help interpretation of the baroclinic tides near steep bottom features. The parameter of nonlinearity, ε_1, for the New York Bight was about 0.05, and the bottom profile was close to critical, albeit still subcritical. In such a case, as demonstrated in Section 6.3, generation of the second baroclinic tidal mode is

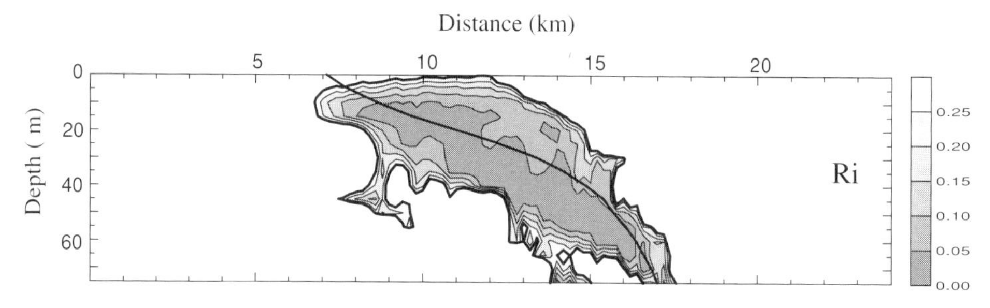

Figure 6.26. Model-predicted distribution of the Richardson number in the shelf break area. The inclined line is the characteristic line of the wave equation, and the shaded area marks regions prone to instability.

quite evident. We can expect a yet stronger high-mode baroclinic response if the nonlinearity is larger and the bottom topography steeper than in the previous case. Let us investigate now the joint effect of the steepness of an underwater obstacle and the intensity of the tidal forcing when both these parameters, γ and ε_1, are quite large, as in ref. [246]. In such circumstances, the essential contribution to the generated wave field will combine the two effects: (i) nonlinearity, resulting in the formation of short-periodic internal wave packets; (ii) processes connected with the excitement of higher order baroclinic modes.

Conditions close to those described above prevail in the region of the Portuguese Shelf slope shown in Figure 6.27. Data from field measurements [109] and the results of remote sensing [49], [57] show that the Iberian Shelf is a place of regular manifestation of short-periodic internal waves generated by the tides. The buoyancy frequency profile is characterized by a well stratified water layer below a depth of 100 m with a value of the buoyancy frequency of $N \sim 0.006\,\mathrm{s}^{-1}$; see Figure 6.28(b). This, along with the fact that the bottom topography is quite steep – we consider the slope area around the area marked by letter X in Figure 6.27 – leads to a supercritical regime of generation of baroclinic tides (in terms of the characteristic lines). Effectively, the inclination of the characteristic lines is roughly inversely proportional to the buoyancy frequency, i.e. $\alpha \sim 1/N$. Thus, the larger the value of N, the smaller the value of α. Strong stratification in the abyss causes gently sloping characteristic lines. Together with the steep bottom topography in the area under study, this results in the situation presented in Figure 6.28(a), in which the bottom configuration and the position of the characteristic line emanating from the shelf break are shown. The conditions of stratification and the parameters of the bottom relief are as follows: in the top part, above 250 m depth, the continental slope is subcritical, and the bottom steepness γ is less than α, but below this level the bottom topography is steeper than the characteristic line, so conditions are supercritical.

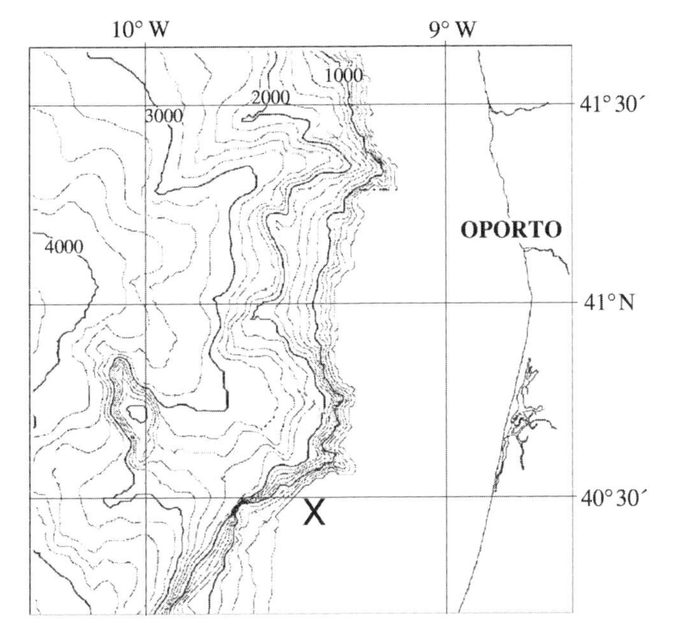

Figure 6.27. Chart of the Portuguese (Iberian) Shelf slope area.

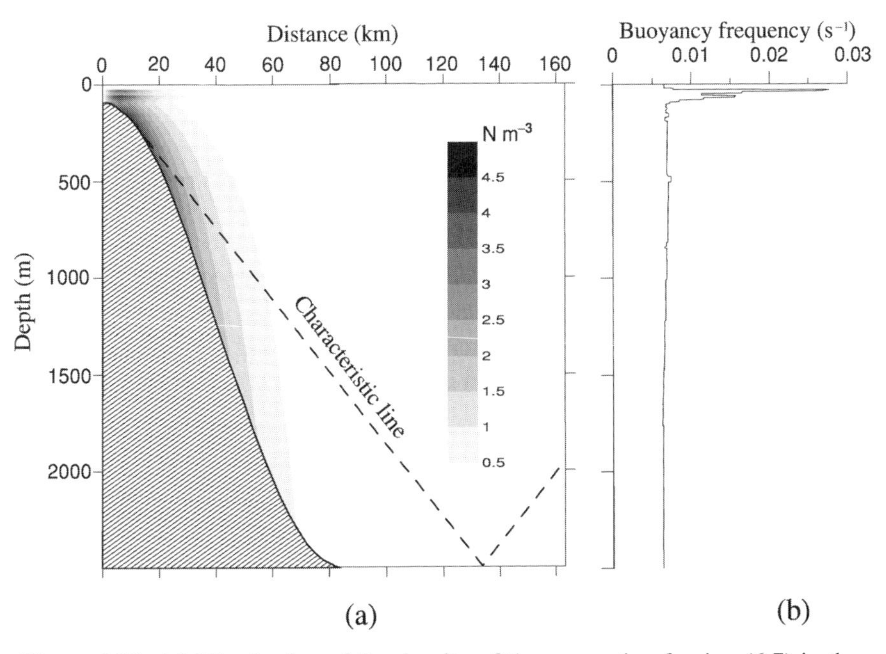

Figure 6.28. (a) Distribution of the density of the generating forcing (6.7) in the area of steep bottom topography. The characteristic line emanating from the shelf break is shown by the dashed line. (b) Vertical distribution of the buoyancy frequency as measured in the Iberian Shelf slope area.

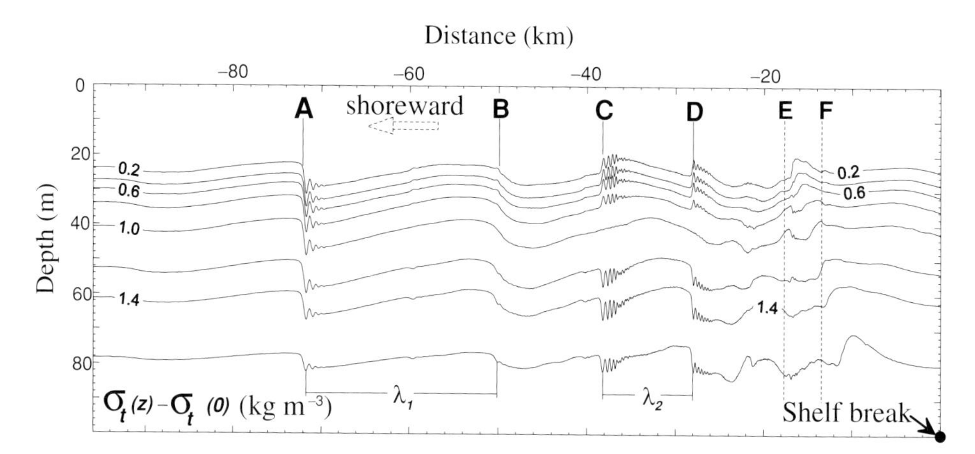

Figure 6.29. Model-predicted field of the referenced conventional density (kg m^{-3}) at $t = 3.25T$.

In the linear case, i.e. for infinitesimal waves, this would be ideal conditions for the generation of high baroclinic modes, which, in turn, create a wave beam like that presented in Figure 2.19. However, strong nonlinearity can introduce new effects. The field of the tidal forcing presented in Figure 6.28, and calculated for a barotropic tidal flux $\Psi_0 = 40\,\text{m}^2\,\text{s}^{-1}$, is typical for the studied area and indicates that such effects can be relatively strong. The parameter ε_1, estimated for such a value of Ψ_0, is about 0.1, or even larger if we consider only the upper part of the bottom topography.

Let us summarize the results of the numerical modeling. The following values of the initial parameters were used in the calculations: $H_1 = 100\,\text{m}$, $H_3 = 2500\,\text{m}$, $2l = 80\,\text{km}$, $\Psi_0 = 40\,\text{m}^2\,\text{s}^{-1}$, $T = 12.4\,\text{h}$, $\varphi = 41^\circ$. Figure 6.29 shows a vertical section of the referenced conventional density in the shelf zone, a snapshot taken 3.25 tidal cycles after commencement of the motion. The direction of the barotropic tidal flow is shown by the dotted arrow. The position of the shelf edge is marked by a dark dot in the lower right corner.

Scrutiny of this figure reveals that the generated wave field at a large distance from the shelf break can be considered as a superposition of different oscillations. Position A in Figure 6.29 represents the process of disintegration of a baroclinic bore into a packet of short-periodic internal waves (see above). Position B shows the next moving shoreward baroclinic bore generated one tidal cycle later. The distance between A and B is equal to 22 km and correlates well with the wavelength λ_1 of the first baroclinic mode. Positions C and D represent two packets of short-periodic internal waves with characteristics of the second baroclinic mode. The distance between them equals 10 km, which coincides with λ_2. Wave packets A and C are

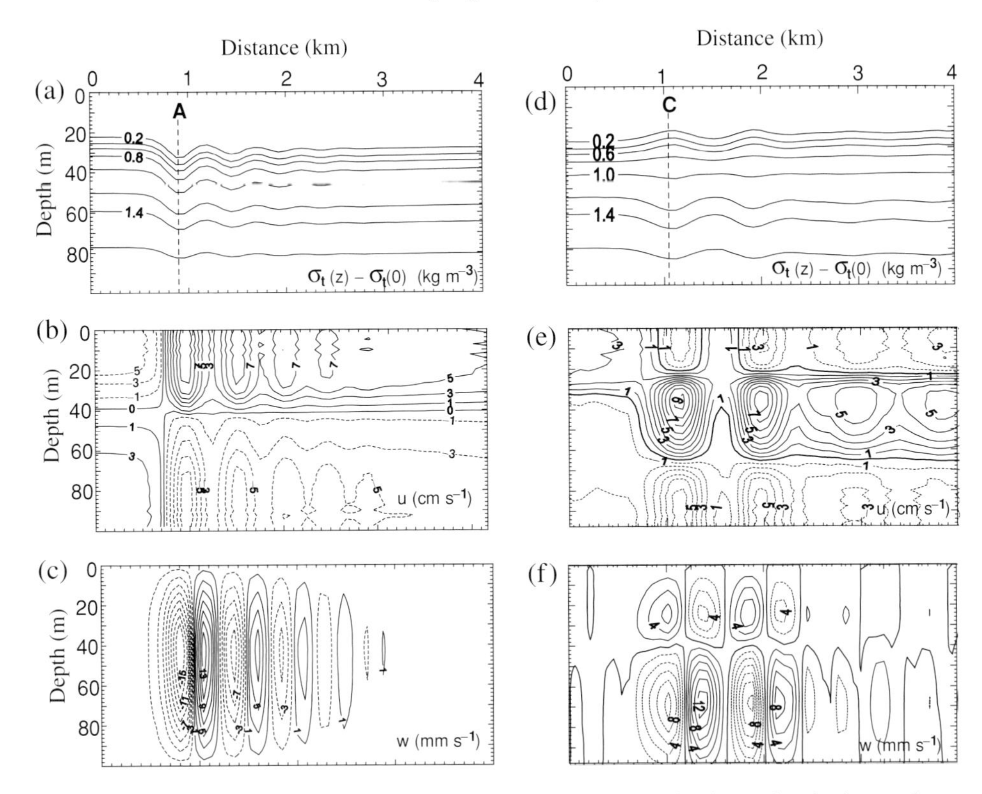

Figure 6.30. Vertical sections of (a), (d) density, (b), (e), horizontal velocity, and (c), (f) vertical velocity of the wave packets marked by letters A and C in Figure 6.29.

arranged by the wave amplitude. The leading waves in both packets possess the largest amplitudes. Due to the larger phase speed, because of nonlinear dispersion, they outstrip other waves, detach from the wave packets, and transform into solitary internal waves.

Let us analyze the vertical structure of these two wave trains. Packet A is shown in Figure 6.30(a) in greater detail. The appropriate horizontal and vertical velocities are shown in Figures 6.30(b) and (c), respectively. It is clearly seen that the isopycnals in packet A have everywhere co-phase vertical displacements, which are peculiar to the first baroclinic mode. The structure of the horizontal and vertical velocity fields confirms this statement: the horizontal velocity changes its sign in the layer of the pycnocline, and the vertical motion is in phase through the total water column.

Now let us analyze the structure of wave packet C, which is shown in Figure 6.30 along with its horizontal and vertical velocity fields. It is seen that the

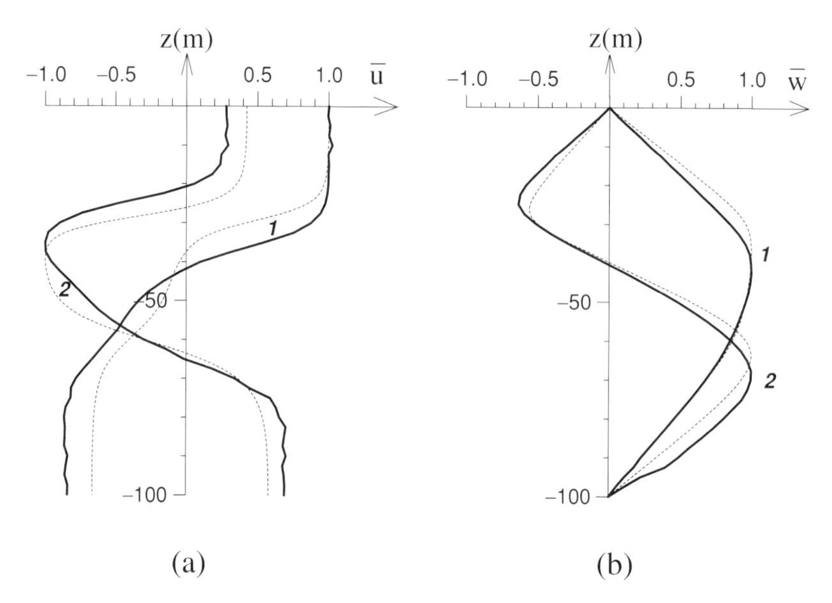

Figure 6.31. Normalized vertical profiles of (a) horizontal and (b) vertical velocities for the first waves from packets A (solid lines 1) and C (solid lines 2) depicted in Figure 6.30. The corresponding profiles for the K–dV solitons are shown by the dotted lines.

isopycnals in the surface and bottom layers, above and below the depth 40 m, are displaced in opposite directions; Figure 6.30(d). Note also that the isopycnal $\sigma_t(z) = \sigma_t(0) + 0.9\,\text{kg m}^{-3}$ is practically undisturbed. The mentioned counterphase displacements of the isopycnals and the associated structure of the horizontal and vertical velocity fields, see Figures 6.30(e) and (f), confirm the supposition that, qualitatively, the internal waves in packet C belong to the second baroclinic mode.

A more convincing confirmation of the conclusion that the leading packets, A and C, of internal waves can be interpreted as SIWs of the first and second baroclinic modes can be obtained from a careful comparative analysis of their horizontal and vertical structures with the same characteristics of the analytical solitons as in Chapter 5. The normalized vertical profiles of the u and w velocities of the two leading waves are shown in Figure 6.31, along with similar profiles for the K–dV solitons. Numbers 1 and 2 correspond to the mode numbers.

It is clearly seen that the first- and second-mode waves obtained numerically exhibit a good qualitative and quantitative agreement with the analytical solution. Some deviations can be explained by the fact that the considered waves propagate on the background of some other wave motions but not in an undisturbed medium.

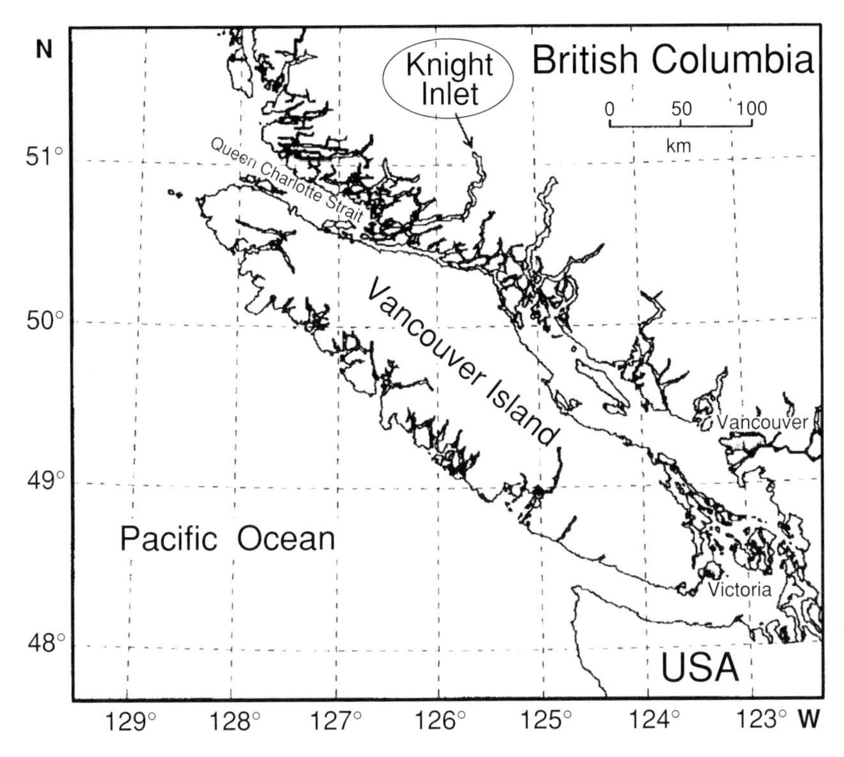

Figure 6.32. Southern coastline of British Columbia, showing the location of Knight Inlet.

6.5 Generation mechanism at large Froude numbers

Throughout this book we have considered the mechanism of topographic generation of baroclinic tides for the cases when the Froude number is below or slightly above unity. Such situations are typically observed in most sites of the World Ocean. However, there are isolated places where the generation conditions are definitely supercritical when the tidal velocity substantially exceeds the phase speed of the generated waves. Such an interesting place is Knight Inlet in British Columbia, Canada (Figure 6.32), a long stratified fjord with a 60 m sill separating two deep basins [59], [60]. The field measurements described, for instance, in ref. [59] were performed with the use of towed and profiling temperature–salinity sensors, acoustic Doppler and echo-sounder imaging, and aerial photographs. The basic aim of the experiment was to show that the generation mechanism over abrupt bottom topography at large Froude numbers differs from that under subcritical conditions. As inferred from the experiment, intensive solitary internal waves are formed above the sill crest during the flood tide. The waves are trapped immediately downstream of the internal hydraulic control at the sill crest. As the tidal current slackens, the large mass of nearly stationary fluid that has accumulated above, and just downstream of, the sill crest escapes, and solitary waves propagate upflow.

Let us estimate the maximum Froude number over the sill crest on the basis of the measured water stratification and data of the tidal motion. According to ref. [221] the amplitude of the M_2 tide in a nearby point of the sill is equal to 1.5 m. So, we can estimate the value of the barotropic tidal flux as $\Psi_0 = 90\,\mathrm{m^2\,s^{-1}}$. Consequently, for tidal currents above the sill we have a maximum value of $1.5\,\mathrm{m\,s^{-1}}$. On the other hand, the BVP (1.67) gives a phase speed c_p of linear internal waves equal to $0.66\,\mathrm{m\,s^{-1}}$ (we used the background fluid stratification from ref. [59]). So, an estimate of the Froude number is about 2.3, well in the strong supercritical regime.

We use the numerical model of Chapter 4 with input parameters for the bottom topography and background stratification typical of Knight Inlet to reproduce a strong flow–topography interaction. The results of the calculations are displayed in Figure 6.33. Analysis of the hydraulic response to a slowly changing tidal current shows that the nonlinear internal bore appears just upstream of the hydraulic jump, which occurs at the lee side of the sill when $t = 0.1T$ (Figure 6.33(a)). During the next $0.05T$, the bore disintegrates into a packet of solitary waves and becomes trapped by the tidal current. At the lee side of the sill, the mixing in the hydraulic jump is clearly seen in Figure 6.33(b). The packet of internal waves is transported to the lee side of the sill by the intensifying tidal current and then arrested at $t = 0.4T$, Figure 6.33(c). It moves upstream again, when $t = 0.45T$, and the tidal flow slackens. An estimation of the value of the phase speed of the packet between $t = 0.45T$ and $t = 0.55T$ yields a value of the nonlinear phase speed equal to $0.85\,\mathrm{m\,s^{-1}}$, corresponding to the real Froude number of 1.76.

Thus, to conclude this part we can state that the internal waves generated in Knight Inlet possess all the features inherent in the lee waves described in Chapter 4. They are generated near the top of the sill, are arrested by tidal flux at the lee side of the topography, and escape when the conditions become subcritical. These waves are relatively short, with horizontal scales close to those of the bottom topography variations, as it should be for lee waves.

It is clear that such short waves and the mechanism of their generation do not feel the effects of the rotation of the Earth. To prove this we repeated the above numerical experiment under identical conditions but omitting the Coriolis parameter. We reproduced superinertial conditions above the critical latitude ($\varphi = 81^\circ$ instead of $\varphi = 51^\circ$), where, according to the linear theory, baroclinic tides cannot exist. For comparison, the results of both calculations are presented in Figure 6.34. The conformity of these two fields is quite evident. This provides support for the statement that the generation mechanism at large Froude numbers can be considered without accounting for the rotation of the Earth.

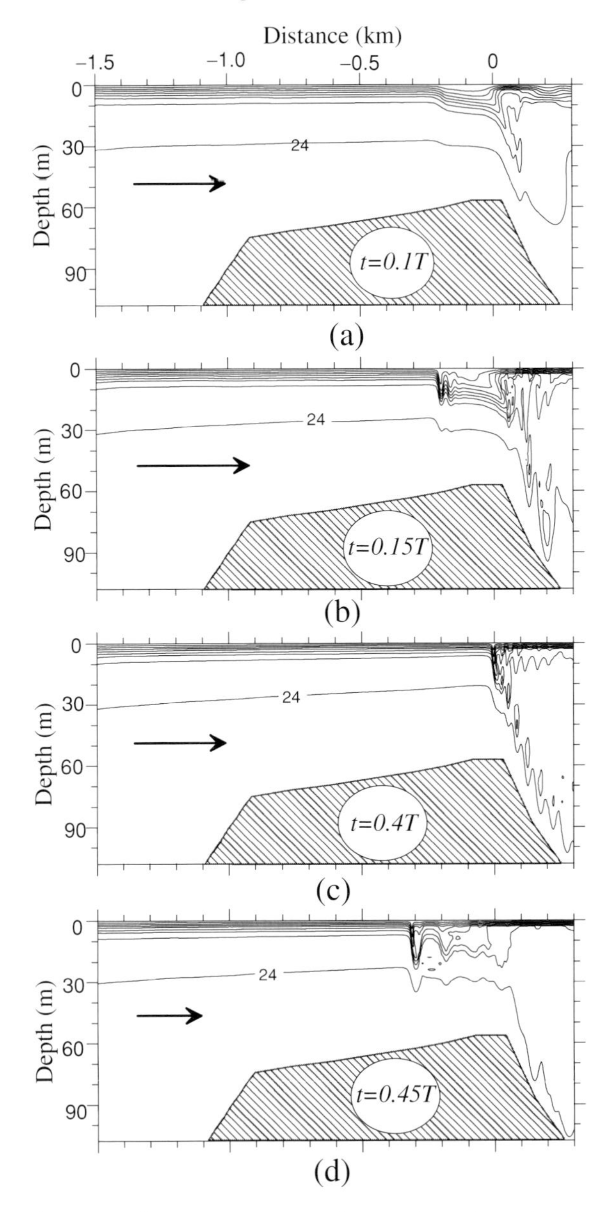

Figure 6.33. Model-predicted evolution of the density field σ_t (contour interval is $2\,\mathrm{kg\,m^{-3}}$) in Knight Inlet. The direction of the barotropic tidal flux is shown by the arrows.

The bottom topography in Knight Inlet is quite steep ($\alpha > \gamma$). Therefore, we should check the following: lee waves have horizontal scales matching the bottom topography where they are generated; therefore, near a gently sloping topography, are lee waves much wider and, very likely, also much weaker? The answer to this

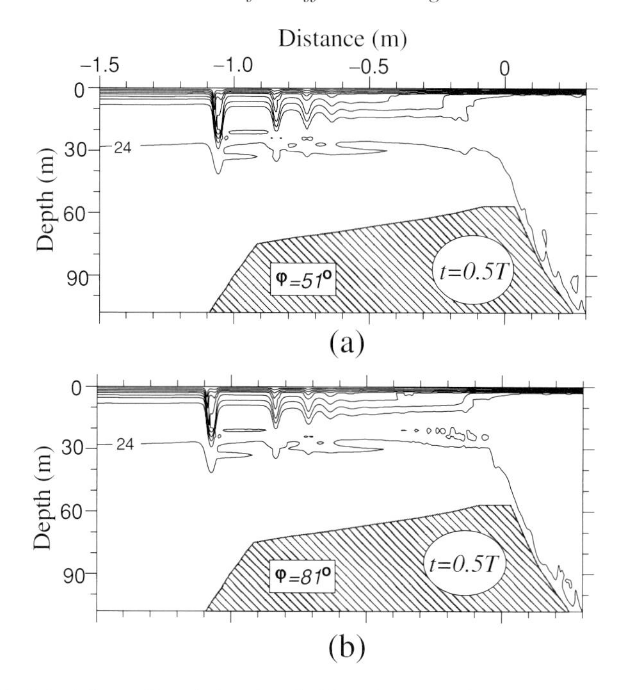

Figure 6.34. Model-predicted density field σ_t (contour interval 2 kg m^{-3}) for $t = 0.5T$ (a) when $\varphi = 51^\circ$ and Coriolis effects are accounted for and (b) when $\varphi = 81^\circ$ and Coriolis effects are dropped.

question can be found in Figure 6.35, where modeling results are presented for a sill which is ten times wider than the real sill in Knight Inlet.

It is evident that the hydraulic jump is formed only at the lee side of the sill at $t = 0.4T$. Its frontal face becomes steeper with time, and a baroclinic bore is fully formed by $t = 0.5T$. It is much weaker than that for the real steep topography, and it does not disintegrate into a packet of internal waves directly at the place of generation (above the top of the bottom topography).

6.6 Summary of generation mechanism

In this short section we shall try to summarize the paramount conclusions obtained regarding the generation mechanisms of baroclinic tides, as they take place in a stratified ocean during flow–topography interactions. The basic parameters which control the qualitative structure and quantitative characteristics of the generated wave are as follows:

- the law of vertical fluid stratification, in particular the positions of the permanent and seasonal pycnoclines relative to the tops of the bottom topographic obstruction, which define the degree of the water stratification in the abyss;

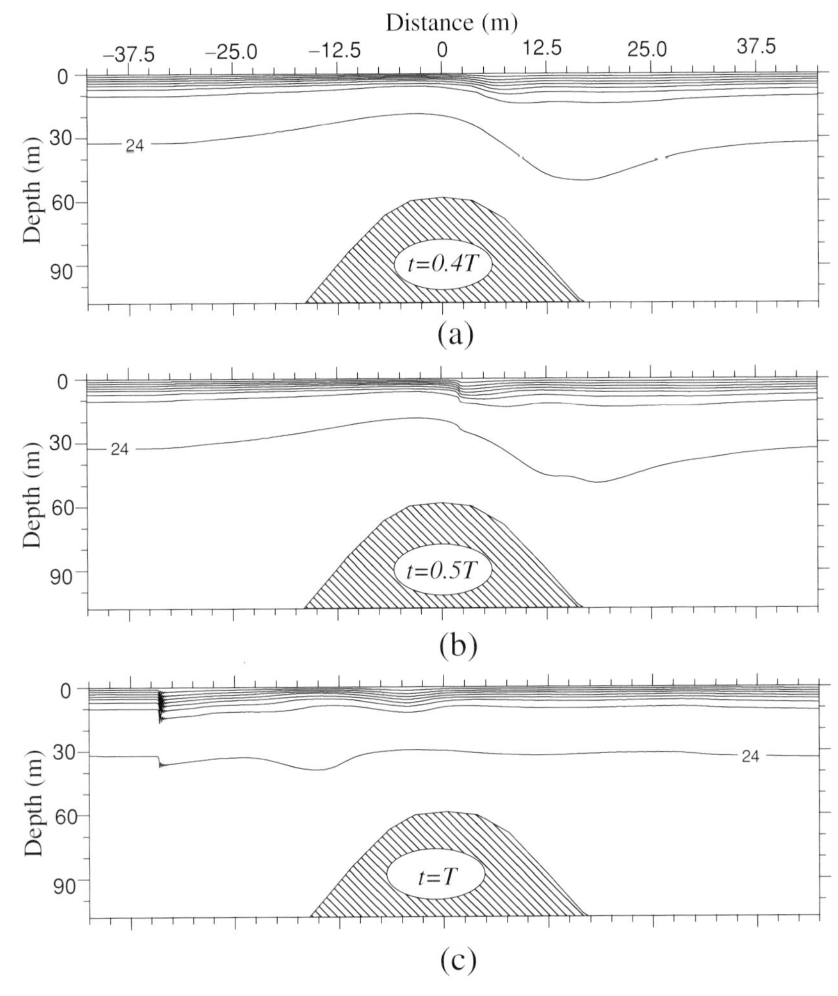

Figure 6.35. Model-predicted evolution of the density field σ_t (contour interval $2\,\mathrm{kg\,m^{-3}}$) over a flat topography (ten times wider than the sill in Knight Inlet).

- the shape and extent of bottom topographic features (profiles, widths, and heights);
- the intensity of the barotropic tidal forcing, i.e. the maximum of the velocity of the barotropic tidal flux.

Depending on the combination of these parameters, a large variety of baroclinic tides are created in the World Ocean. Let us review the basic possible situations. Figure 6.36 contains a short summarizing table with brief helpful descriptions. With respect to the bottom inclination, all generation areas are divided into two large categories: flat and steep bottom topographic variations. The steepness of the bottom profile is understood here as the proportion of the bottom topographic inclination, dH/dx, to the steepness of the characteristic lines of the wave equation, α.

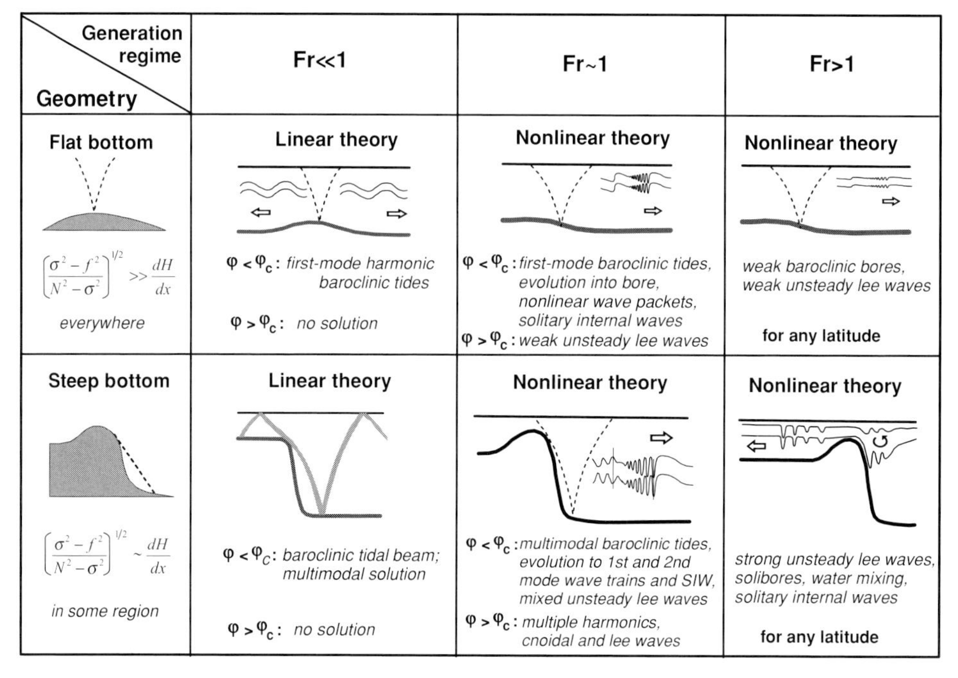

Figure 6.36. Scheme of generation mechanisms of tidally induced internal waves for different oceanic conditions.

In each range of the intensity of the external tidal forcing, three different regimes of internal wave generation can be distinguished; they are classified by the value of the Froude number. If the maximum of the velocity of a barotropic tidal flux is much less than the phase speed of the generated waves, then the linear theory is a very good approximation for the description of baroclinic tides. Consequently, a weak barotropic tide generates basically harmonic first-mode baroclinic tidal waves over flat topography, propagating from the source of generation. Over steep bottom topographies, a barotropic tide generates a large number of baroclinic tidal modes, which in composition create a tidal beam attached to the sharp bottom edges. Note that above the critical latitude no solution exists in the framework of the linear theory.

At intermediate Froude numbers – the phase speed of the generated waves is of the order of the tidal flux – nonlinear effects become evident. Due to the nonlinear steepening, the generated first-mode internal tidal wave, which radiates from the source, transforms into a baroclinic bore; it gradually disintegrates into a packet of short nonlinear internal waves and SIWs. If the bottom steepness is large, high baroclinic modes are also effectively generated. Not only first-mode, but also high-mode wave packets and SIWs are stimulated. At high latitudes, above φ_c, the generation of baroclinic tidal waves is suppressed by the Earth's rotation; therefore, only unsteady lee waves in the form of multiple harmonics and cnoidal waves can be generated.

The most crucial regime of the baroclinic tidal generation occurs at strong supercritical flow–topography interactions, when $\mathrm{Fr} > 1$. In such a case, the generated waves are trapped by the strong tidal flux at the lee side of the bottom topography, and they extract energy from the background current and grow to high-amplitude waves. They are released from the generation area when the tidal flow slackens and propagate upstream as a solibore[1] or as a packet of strongly nonlinear internal waves. Intensive vertical water mixing can take place in such a regime behind the bottom topographic obstruction. Because the generated waves possess horizontal scales which are comparable to those of the bottom topography variations, the generation mechanism over the relatively short bottom features does not depend on the rotation of the Earth.

Of course, the descriptive scheme presented above reflects only the dominant features of the generation mechanism, and cannot describe the whole variety of the specific oceanic conditions; neither are they a universal tool. However, we believe that this description is a very useful tool, which furthers our understanding of most cases of oceanic tidal generation.

[1] A solibore is an undulating group consisting of pulse-like displacements of isopycnals, close to a series of solitary waves.

7
Three-dimensional effects of baroclinic tides

In the previous chapters we have only considered the two-dimensional nature of baroclinic tides. This simplification of the three-dimensional behavior is a valid approximation in shelf zones, at least as a tendency, due to the oblongness of the continental margins. It is, however, often violated when one tries to study the wave-generation process near three-dimensional bottom features like oceanic banks or abrupt changes of the shelf break topography in the along shore direction. Such places constitute a remarkable sink of barotropic tidal energy into baroclinic wave components if only a substantial along slope tidal flux interacts with along shore bottom variations. Under such conditions, the correct prediction of the total scattering of the barotropic tidal energy into internal tides is impossible without using three-dimensional global tidal models, which can predict the dynamics in those shelf areas where such a remarkable along slope forcing of internal tides may exist. Figure 7.1 illustrates this point. The broad gray bands show regions where the tide, derived from the barotropic model of Schwiderski [214], propagates along the coast as a Kelvin-type wave, and where the amplitude of its elevation is larger than 0.4 m. If the depth of the ocean near shore is estimated typically as 2000 m, then the value of the maximum barotropic tidal flux along the slope is of the order of 60 $\mathrm{m^2\,s^{-1}}$. The narrower "sausage-like" bands are the 11 internal tidal hotspots identified by Baines [11] using two-dimensional forcing. Black rings indicate most of those locations on continental shelves and shelf edges where large internal tides or nonlinear internal waves have been reported in the scientific literature. They exceed the number of observations of significant internal tides reported by Ostrovsky and Stepanyants [183].

Recent advances in numerical modeling (see, e.g., refs. [47], [101], and [270]) began to make three-dimensional modeling of internal tides a viable alternative to the slice approach, at least at the local scale. In a global sense, the three-dimensional models are used for estimations of the sink of tidal energy to internal waves and further to mixing processes. Sjöberg and Stigebrandt [219] were the first to show

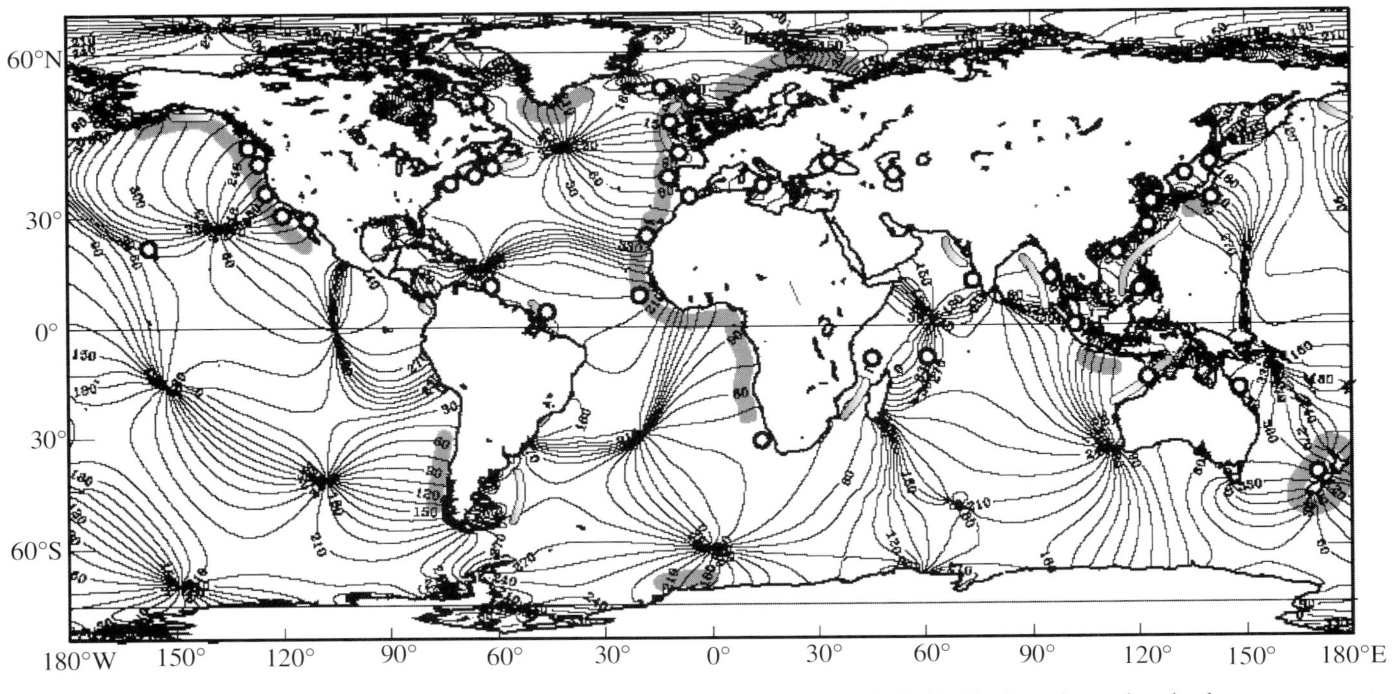

Figure 7.1. Phase lag of the global M_2 barotropic tide elevation from Schwiderski [214]. The broad gray bands along some coasts show where the tide appears to have a Kelvin wave form with an amplitude larger than 0.4 m. Some of these places may have along slope forcing of internal tides. The narrow "sausage-like" shapes indicate regions that were identified by Baines [11] as having across slope forcing. Rings indicate published observations of large internal tides or SIWs on the continental slope.

the global distribution of the conversion rate from the surface to the internal tidal energy based on simplified linear analytical theory, using a gridded database for bathymetry and a "Levitus database" for stratification. The more advanced three-dimensional global baroclinic tidal models are usually developed in the framework of primitive equations taking account of real bottom topography and real stratification (e.g. [152], [177]). However, such models usually apply the hydrostatic pressure approximation and thus are not appropriate for the description of the essentially nonhydrostatic phenomena as, e.g., SIWs and lee waves generated by strong tides.

In the framework of the present book, we are not going to study all three-dimensional effects that may take place in the generation mechanism of baroclinic tides by topographic features. This is quite a difficult task, and a separate book would have to be devoted to this topic. However, in this chapter we will try to show that some three-dimensional effects occurring in flow–topography interactions can be explained on the basis of the already developed two-dimensional models. We will illustrate this statement using examples of wave refraction at the Portuguese shelf [218] and effects of variable cross-sections in the generation mechanism of baroclinic tides in narrow channels, with a specific application to the Trondheim Fjord [257].

7.1 Influence of wave refraction

7.1.1 Observations of SIWs on the Portuguese Shelf

Observations of tidally generated strong SIWs were performed in August, 1994, on the Portuguese Shelf, using thermistor chains, moorings, and remote sensing techniques; see Figure 7.2 [109]. Data obtained by the ship-mounted thermistor chains revealed the appearance of thermocline depressions at different points of the shelf with a vertical range as large as 45 m, Figure 7.3. From ship-borne radar soundings, SIWs could be seen to propagate along 73° (i.e. to the north of a line drawn normal to the shelf edge). Obervations from a month long thermistor chain mooring deployed in August, 1994 (at mooring M in Figure 7.2), showed that packets of large SIWs occurred in every tidal cycle. The measured packets of SIWs were also observed in a SAR image captured by the ERS-1 satellite on August 8, Figure 7.2.

The presence of large SIWs suggests that there must be a region of large internal forcing nearby. The most reasonable supposition is to assume an interaction of the barotropic tidal flux with the shelf break perpendicular to the shelf, as has been discussed many times above. This leads to the obvious conclusion that the source of the internal waves must be located above the slope somewhere off mooring M; see Figure 7.2. However, the estimations performed by a two-layer nonlinear internal

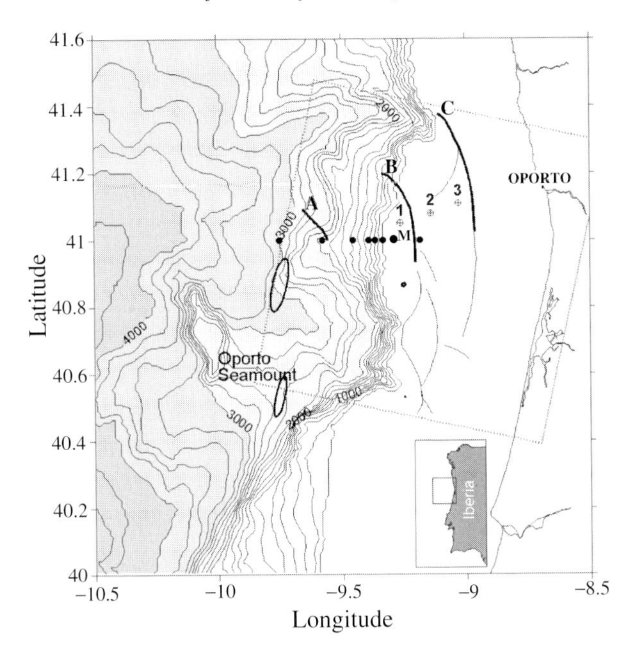

Figure 7.2. Part of the Portuguese Shelf with detailed bathymetry. The inset shows the position of the map in relation to the Iberian Coast. The three dark lines, labeled A, B, and C, indicate the wave packets from the ERS-1 SAR image. Short-term thermistor chain stations observed SIWs at 1, 2, and 3. A long-term thermistor chain mooring was deployed at M. The black dots indicate the positions of CTD probes. Tidal fluxes are shown by ellipses, with $80\,\mathrm{m^2\,s^{-1}}$ volume flux as the semi-major-axis of the largest one.

tide-generation-slice model, based on a numerical representation of K–dV dynamics [73], predicted an essentially linear internal tidal response with an amplitude of 1.5 m. To achieve a realistic response as observed in the measurements, the velocity of the barotropic tidal flux had to be increased artificially by a factor of six [109]. These investigations lead us to believe that the observed SIWs could not have been generated at the nearby shelf edge. Thus, the following question remains. Why are the internal waves observed on the Portuguese Shelf at 41° N many times larger than expected from the two-dimensional theory? A reasonable explanation can be found if we analyze the distribution of the tidal forcing in the studied area and assume that the location of the SIWs' true source may have been between 50 and 60 km from the mooring M (Figure 7.2) rather than at the shelf break, which was only 9 km away. In this section we demonstrate that along slope interactions can be a major source of internal tidal energy and may, indeed, explain why the signal observed at 41° N was so large.

To find the answer to our question, the horizontal distribution of the tidal forcing in (6.7) was integrated through the water column and calculated according to the

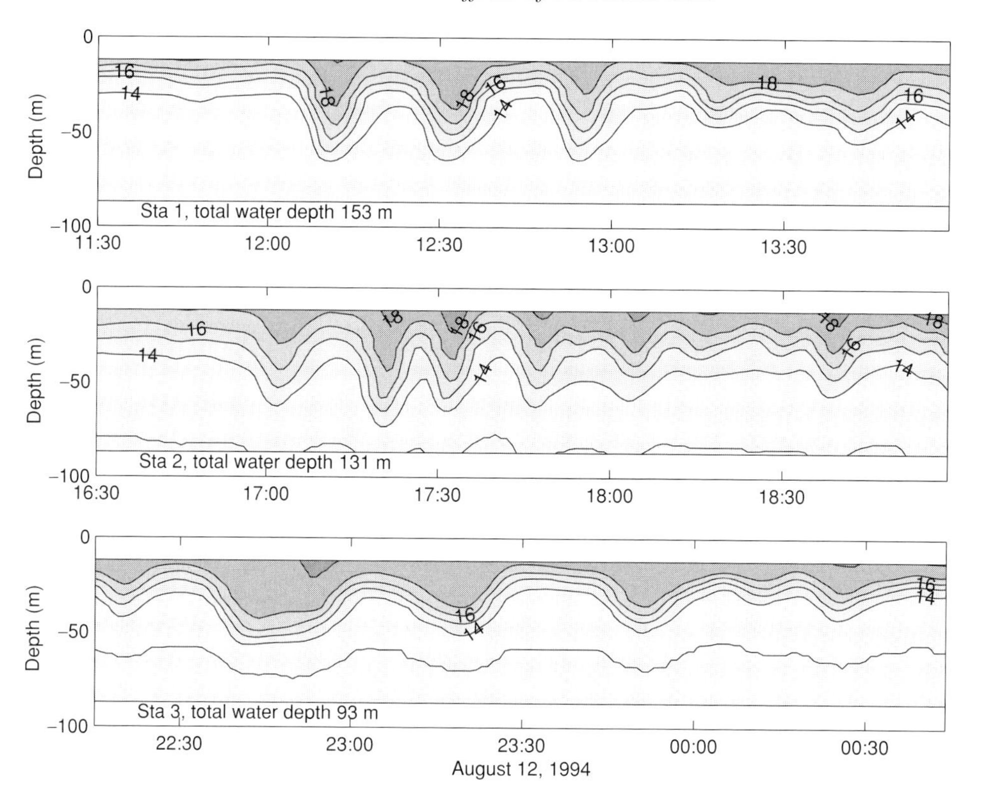

Figure 7.3. Successive observations of a packet of SIWs propagating from stations 1 to 3 (see Figure 7.2) on August 12, 1994. Wave packets of similar size were seen in every tidal cycle. Temperatures are in degrees Celsius.

formula

$$\mathscr{F}(x, y) = (\bar{\rho}_0/\sigma) \int_{H_{(x,y)}}^{0} N^2(z)z[(\Psi_{0x}\, dH^{-1}/dx)^2 + (\Psi_{0y}\, dH^{-1}/dy)^2]^{1/2}\, dz. \tag{7.1}$$

Here, Ψ_{0x} and Ψ_{0y} are the zonal and meridional components of the barotropic tidal flux. They were computed on a 1 km grid with the bathymetry and domain given in Figure 7.2, and forced along the boundary by elevations derived from ref. [214]. Equation (7.1) gives an upper bound for the magnitude of $\mathscr{F}$ rather than its true value, because for simplicity it assumes that Ψ_{0x} and Ψ_{0y} are in phase.

Figure 7.4 shows that the waves marked A and B crossed a part of the shelf where the forcing is relatively small, and that forcing over the Oporto Seamount is also weak. Stronger forcing can be found along the slope to the south of latitude 40.8° N, with the strongest portion being along the part that extends westward to the south of

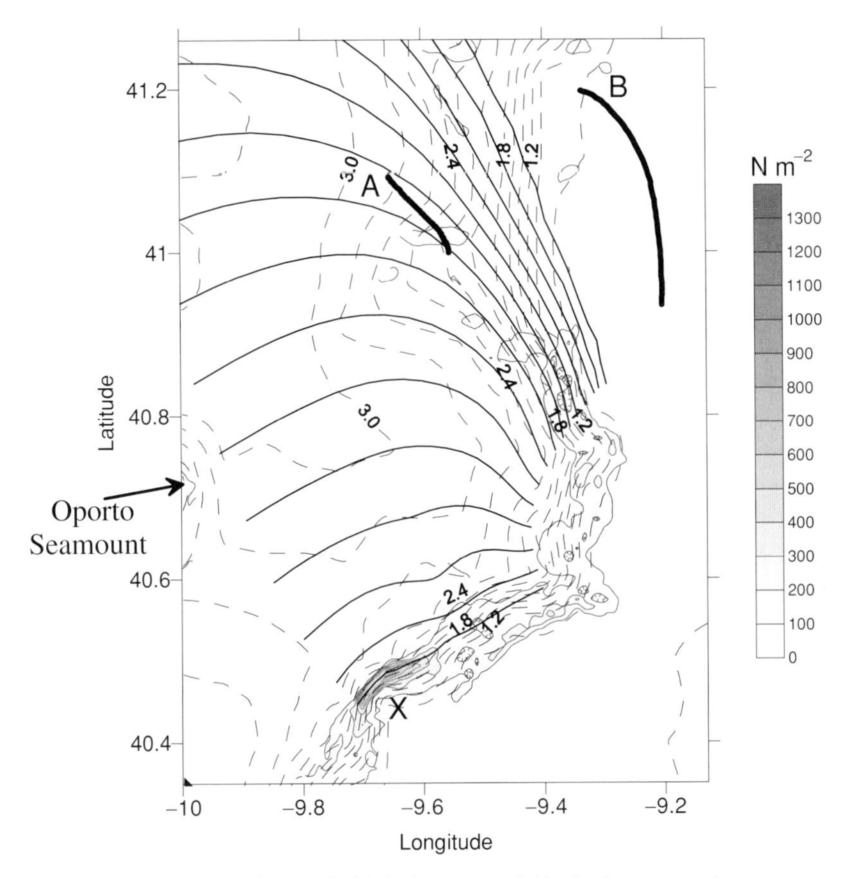

Figure 7.4. M_2 mode 1 internal tidal phase speed (dashed contours in meters per second) overlain with the positions, at successive hourly intervals, of an internal tidal wave crest emanating from the main generation site south of 41° N. The shaded region over the slope shows the M_2 internal tidal forcing $\mathscr{F}$ (N m^{-2}). X is close to the region of largest forcing. Wave packets A and B are indicated.

40.6° N. In the region near the point marked in Figure 7.4, $\mathscr{F}$ exceeds 1000 N m^{-2}. However, this latter site is a significant distance from the observations, and before investigating it further it is necessary to check whether those nearer sources (Oporto Seamount) could have generated the waves.

7.1.2 Generation of waves at the Oporto Seamount

From a cursory glance at the relationship between bathymetry and the internal wave crests, one might be tempted to think that the waves were generated by the Oporto Seamount (Figure 7.2), even though they only rise about one-third of the ocean depth above the sea floor. Note also that this possible source of generation is essentially three-dimensional. So, to take into account all three-dimensional effects (place of generation, radial attenuation), we should develop and apply a three-dimensional tidal model. For rough estimations of wave amplitudes, we can use the linear equations for a two-layer fluid with the hydrostatic pressure approximation in the following form. In the upper layer,

$$\left.\begin{aligned} \frac{\partial u_+}{\partial t} - f v_+ &= -\frac{1}{\rho_+}\frac{\partial P_+}{\partial x}, \\ \frac{\partial v_+}{\partial t} + f u_+ &= -\frac{1}{\rho_+}\frac{\partial P_+}{\partial y}, \\ \frac{\partial(\xi_+ - \xi_-)}{\partial t} + h_+\left(\frac{\partial u_+}{\partial x} + \frac{\partial v_+}{\partial y}\right) &= 0, \\ P_+ &= P_\mathrm{a} + g\rho_+(\xi_+ - z). \end{aligned}\right\} \tag{7.2}$$

And in the lower layer,

$$\left.\begin{aligned} \frac{\partial u_-}{\partial t} - f v_- &= -\frac{1}{\rho_-}\frac{\partial P_-}{\partial x}, \\ \frac{\partial v_-}{\partial t} + f u_- &= -\frac{1}{\rho_-}\frac{\partial P_-}{\partial y}, \\ \frac{\partial \xi_-}{\partial t} + \frac{\partial[(H - h_+)u_-]}{\partial x} + \frac{\partial[(H - h_+)v_-]}{\partial y} &= 0, \\ P_- = P_\mathrm{a} + g\rho_+(\xi_+ - \xi_- + h_+) &+ g\rho_-(\xi_- - z). \end{aligned}\right\} \tag{7.3}$$

Subscripts "+" and "−" correspond to the upper and lower layers, respectively. Introducing the new variables $\phi_+ = \xi_+$, $\phi_- = \epsilon\xi_- + \chi\phi_+$ ($\chi = \rho_+/\rho_-$, $\epsilon = 1 - \chi$), and using the tidal periodicity of the motion (1.41), equations (7.2)–(7.3) can be reduced to the following form:

$$\left.\begin{aligned} h_+\left(\frac{\partial^2 \overset{*}{\phi}_+}{\partial x^2} + \frac{\partial^2 \overset{*}{\phi}_+}{\partial y^2}\right) + \frac{\sigma^2 - f^2}{g\epsilon}(\overset{*}{\phi}_+ - \overset{*}{\phi}_-) = 0, \\ \frac{\partial}{\partial x}\left(H\frac{\partial \overset{*}{\phi}_-}{\partial x}\right) + \frac{\partial}{\partial y}\left(H\frac{\partial \overset{*}{\phi}_-}{\partial y}\right) + \frac{\imath f}{\sigma}\left(\frac{\partial H}{\partial x}\frac{\partial \overset{*}{\phi}_-}{\partial y} - \frac{\partial H}{\partial y}\frac{\partial \overset{*}{\phi}_-}{\partial x}\right) \\ + \frac{\sigma^2 - f^2}{g\epsilon}(\overset{*}{\phi}_- - \chi\overset{*}{\phi}_+) = 0. \end{aligned}\right\} \tag{7.4}$$

Normally, this homogeneous system of equations is used to describe freely propagating surface and interfacial tidal waves in a two-layer fluid. External forcing is introduced in system (7.4) if one considers the problem of scattering of a barotropic tidal wave on local variations of the bottom topography. In this case, the solution of system (7.4) can be represented as the superposition of an incident barotropic tidal wave propagating in the horizontal plane with a direction inclined with respect to the x-axis by the angle Θ. With the scattered (η_+) and interface (η_-) surfaces, these waves can be represented as

$$\left.\begin{aligned} \overset{*}{\phi}_+ &= \eta_+ + A_0 \exp[\imath k(x\cos\Theta + y\sin\Theta)], \\ \overset{*}{\phi}_- &= \eta_- + A_0[1 - k^2 g h_+ \epsilon/(\sigma^2 - f^2)]\exp[\imath k(x\cos\Theta + y\sin\Theta)]. \end{aligned}\right\} \quad (7.5)$$

Here, A_0 and $A_0[1 - k^2 g h_+ \epsilon/(\sigma^2 - f^2)]$ are the amplitudes of the barotropic tidal wave at the free surface and interface, respectively, and k is its wavenumber. With good accuracy, $k = [(\sigma^2 - f^2)/gH]^{1/2}$ (Section 3.1), corrections to this formula due to stratification are negligibly small for barotropic tides.

After substitution of (7.5) into (7.4), the wave disturbances η_+ and η_- can be found numerically. The inhomogeneous elliptic system of equations was solved with a finite discretization technique, and the resulting matrix equation was solved with the Gauss–Zeidel iteration method. A square computational area (500×500 km^2) and a rectangular grid with a space step of 1 km were used. An attenuation of the wave amplitude with distance from the seamount equivalent to a radial divergence of the wave energy was assumed, and zero boundary conditions for η_+ and η_- at the lateral fluid boundaries of the computational area were employed.

The following parameters were used for the calculations: $h_+ = 30$ m, $H = 3200$ m, $\chi = 0.998$, and $A_0 = 0.6$ m. The Oporto Seamount rises more than 1 km above the sea bed, its top is located 2 km below the surface, and its shape is approximated using the topography shown in Figure 7.2. The internal tide radiated in all directions from the seamount along the interface with an initial amplitude of 1 m, which, by the time it had traveled sufficiently far to reach the shelf edge, had fallen to 0.25 m. The equivalent values for the internal tidal energy flux dropped from about 10 W m^{-1} at the seamount to 0.6 W m^{-1} at the shelf edge. This value is much smaller than the observed flux.

This exercise demonstrated that the internal wave energy flux observed on the shelf is unlikely to have originated from the Oporto Seamount.

7.1.3 Far-field generation from a shelf edge

The above investigation of the role played by a seamount highlights the point that energy flowing from a simple localized oceanic source can be severely diminished by radial spreading. What is needed is a large source associated with a wave guide. If, as Dale *et al.* suggest [50], internal tides on an eastern boundary in the northern hemisphere tend to propagate polewards, then there is reason to suspect that the source will be found to the south of 41° N.

Part of the energy that is generated along the slope region near site X in Figure 7.4 will initially propagate into the ocean, and part of it will travel onto the shelf. The energy propagating onto the shelf is almost certainly locally dissipated. That which propagates into the ocean, however, is likely to travel a long distance, and may also undergo refraction in the horizontal plane due to local variations in phase speed. Rays were drawn to investigate the refraction of the internal tide propagating into the ocean on the assumption that a significant part of its energy was contained in mode 1.

The idea of the ray theory is based upon the assumption of local behavior of a wave,

$$\psi(\mathbf{x}, t) = A(\mathbf{x}, t)\exp[\imath S(\mathbf{x}, t)], \tag{7.6}$$

in which the amplitude $A(\mathbf{x}, t)$ changes through space and time more slowly than the phase function $S(\mathbf{x}, t)$. For small intervals of space $\Delta\mathbf{x}$ and time Δt, the function $S(\mathbf{x}, t)$ can be expanded in Taylor series and truncated at the linear terms; i.e.

$$S(\mathbf{x}, t) = S_0 + \nabla S \cdot \Delta\mathbf{x} + \frac{\partial S}{\partial t}\Delta t + o(\Delta\mathbf{x}, \Delta t).$$

The local values of the wave vector and frequency are now defined by

$$\mathbf{k} = \nabla S, \qquad \sigma = -\frac{\partial S}{\partial t}. \tag{7.7}$$

The velocity of propagation of constant phase, $S(\mathbf{x}, t) = \text{const.}$, may be found from the condition $dS(\mathbf{x}, t) = 0$ or

$$\frac{\partial S}{\partial t}dt + \nabla S \cdot d\mathbf{x} = 0. \tag{7.8}$$

Thus, the phase speed

$$\mathbf{c_p} = d\mathbf{x}/dt \tag{7.9}$$

of the surface $S = \text{const.}$ has the following connection with the frequency:

$$\sigma = \mathbf{k} \cdot \mathbf{c_p}.$$

Equation (7.9) was taken as a basis for the construction of a phase pattern of baroclinic tides near the Portuguese Shelf. In order to build the rays, the wave front was assumed to originate from a line that coincided with the position of the maximum slope along the spur near site X (see Figure 7.4). The internal tide was assumed to advance in a direction normal to this front by a distance $\Delta\mathbf{x} = \Delta t\mathbf{c_p}$,

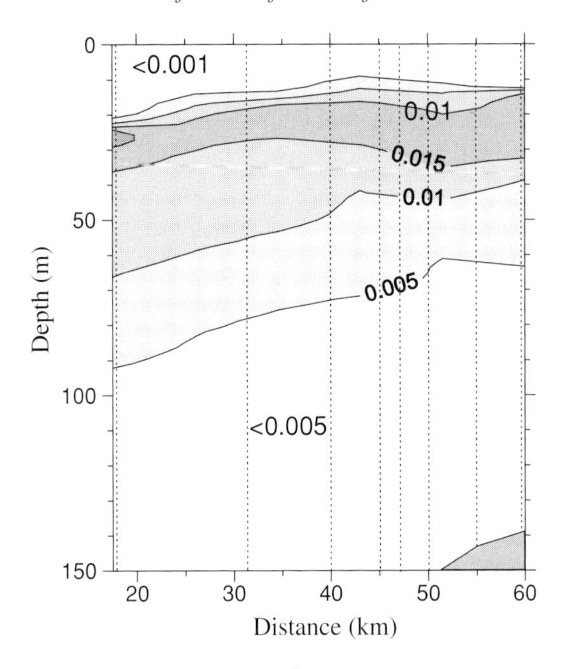

Figure 7.5. Buoyancy frequency field N (s^{-1}) obtained by CTD sampling along 41° N.

where $\mathbf{c_p}$ is measured in meters per second and t is 3600 s. At each time step, Δt, a new wave front was drawn and the process repeated.

The value of the phase speed, c_p, was found from the BVP (1.67) on the basis of *in situ* data for $N(x, y, z)$ and bottom topography $H(x, y)$. The buoyancy frequency $N(x, z)$ in the upper 150 m along 41° N decreased towards the shelf (Figure 7.5), and the pycnocline was wider and deeper in the ocean than it was at the shelf edge. The existence of this frontal zone is explained by seasonal upwelling. The larger water depth beyond the shelf edge, combined with the stronger stratification, led to a four-fold decrease in c_p for the first-mode internal tide between the open ocean and the shelf break. Note that, although the CTD measurements were restricted to the cross-section along 41° N, in Figure 7.4 it was assumed that the tendency for a shoreward decrease in stratification and uplift of the pycnocline took place everywhere.

The position of the advancing wave front for the first 21 hours is shown in Figure 7.4. Since the phase speed c_p decreases substantially towards the shore, it is not surprising that the internal tide, which initially radiates seaward from the slope, is subsequently refracted back towards the shelf edge at about 41° N.

It should be noted that ray theory is formally valid if c_p changes slowly compared with the wavelength. In our case, the wavelength of a mode 1 internal tide is comparable to the scale of the bottom changes. However, as studied above, freely propagating intensive internal tidal waves usually disintegrate into a sequence of short-wavelength SIWs, for which the ray theory can be applied. The reason for this effect is nonlinear steepening and dispersive disintegration of steep baroclinic bores through a mechanism described in Chapter 4 and shown in Figures 4.7 and 4.8.

Figure 4.9 shows that internal tides propagating seaward from the Iberian Shelf break could indeed disintegrate into packets of SIWs. The wave train contains six or seven waves, with the leading ones having a peak-to-trough amplitude of about 40–50 m, i.e. close to the observed amplitude on the Iberian Shelf (see Figure 7.3). Thus, the SIWs observed at A to C and stations 1 to 3 (Figure 7.2) were internal tides that had initially radiated seaward from the shelf slope near X, and had subsequently been refracted and transformed in the manner shown in Figure 7.4.

It is noticeable that the surface expression of the wave at B is considerably longer than that at A. This is probably because the amplitude of the internal tide that radiated seaward from X was not constant along its front. The nonlinear disintegration of a baroclinic bore into a series of solitary waves takes longer when its amplitude is smaller.

7.2 Baroclinic tides in narrow channels and straits

Two-dimensional models can also be applied to study other three-dimensional effects of baroclinic tides. As an example, we consider the tidal dynamics in narrow channels and straits of variable cross-section [257]. The narrowness here is understood in terms of the proportion between the baroclinic Rossby radius of deformation and the width of the channel. The latter must be considerably smaller than the former to avoid the effect related to the rotation of the Earth. If this is the case, we can exclude the Coriolis force from the governing system and build the tidal model on the basis of the laterally integrated Reynolds equations (1.29). Such an approach is very suitable for the investigation of the water exchange processes between two adjacent basins connected by narrow straits, and has been used for the Bosphorus [111] and for the Strait of the Dardanelles [224], which connect the Black Sea and the Mediterranean Basin. This idea can also be exploited for the investigation of the water exchange exerted by the tides in fjords. To this end, the Trondheim Fjord in Norway was selected (Figure 7.6(a)) [257]; its innermost basin (Beitstad Fjord) is connected by a narrow strait (Skarnsund) to the remaining water body; see Figure 7.6(b). From the point of view of water renewal within the innermost basin, it is very important to understand which processes affect the water exchange and mixing in the strait. We focus attention on the role of the tidally generated internal waves, because the tidal activity is rather pronounced there [239].

Trondheim Fjord is a typical Norwegian fjord located on the coast of central Norway, between 63°40′ N, 09°45′ E and 64°45′ N, 11°30′ E (Figure 7.6(a)). The

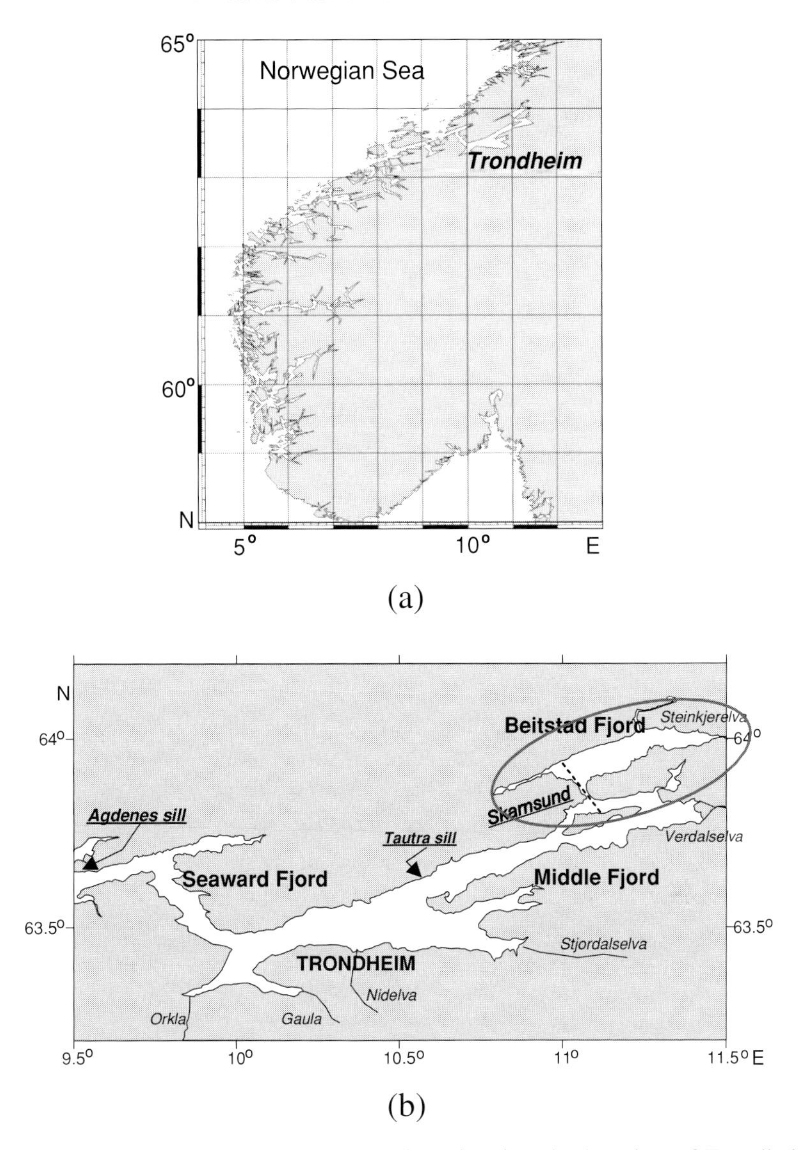

Figure 7.6. (a) Chart of the Norwegian Sea showing the location of Trondheim Fjord. (b) Map of the Trondheim Fjord. The principal landmarks cited in the text are located on this diagram. The area under investigation is marked by the ellipse.

fjord is neither the longest, the widest, nor the deepest fjord in Norway, but with its 120 km length, 1420 km^2 surface area, and almost 600 m depth it is fairly large (Figure 7.6(b)). The currents in the fjord are driven by winds, tides, and fresh water inflow from several major rivers; they are modified by topography, stratification, and the rotation of the Earth. The fjord is divided by sills and narrows essentially into three basins: Seaward Fjord basin, Middle Fjord basin and Beitstad Fjord. The Agdenes sill with a depth of 195 m connects the fjord to the Norwegian Sea.

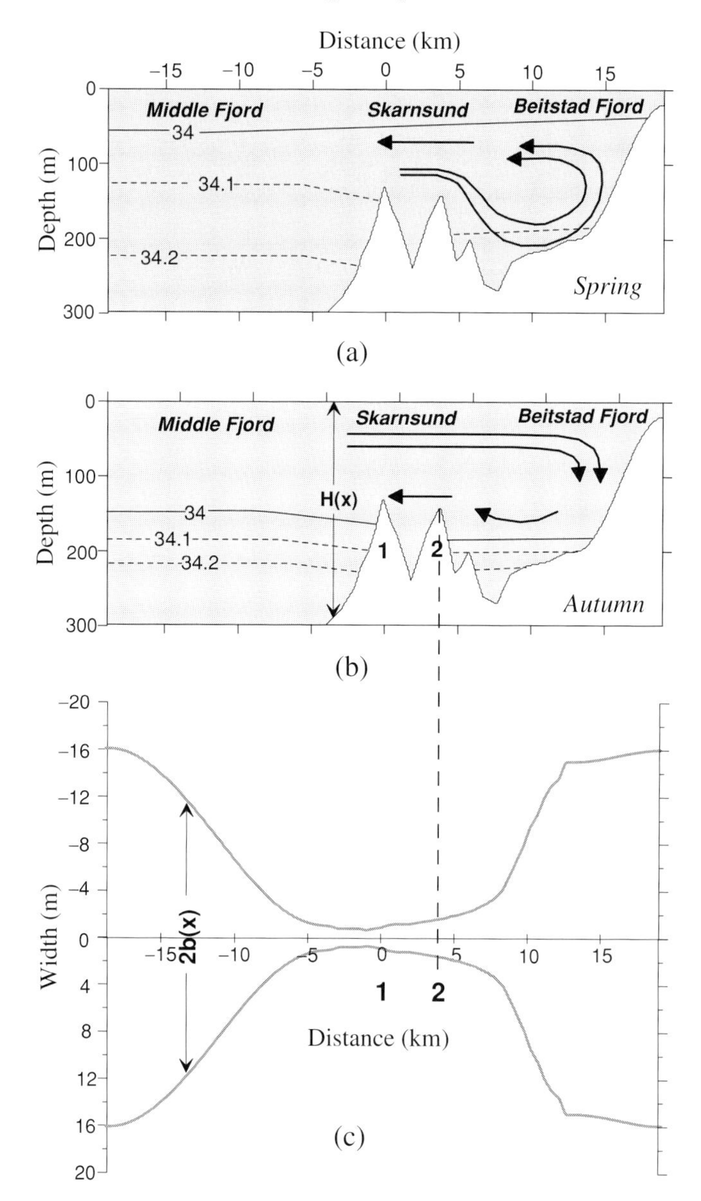

Figure 7.7. Schematic diagram of the fjord, showing the salinity (°/∘∘) distribution for the Middle Fjord and the Beitstad Fjord and the circulation patterns for (a) spring and (b) autumn [108]. (c) One-dimensional fjord representation.

The Seaward basin and the Middle basin are connected by the Tautra sill with a depth of 100 m, and the Middle basin joins the Beitstad Fjord by the very narrow Skarnsund strait (minimum width ∼1000 m) containing sills, of which the highest is peaking at a depth of 130 m. Thus, Beitstad Fjord is the farthest part of Trondheim Fjord from the ocean; consequently, deep water renewal in this part of the Trondheim system is important. From the measured mean yearly salinity distributions obtained during 1963–75 [108] it is seen that deep water in all parts of the fjord become less saline with increased distance from the ocean. Moreover, in every basin the salinity distribution with depth changes from season to season.

The detailed measurements of the mean currents carried out at the Skarnsund sills show more complicated mean velocity profiles representing a three- to four-layer flow structure [108] with maximum values of about 0.2 m s^{-1}, as shown in Figures 7.7(a) and (b). The observed maxima of instantaneous, not averaged, velocities were 0.78 m s^{-1}, 10 m below the brackish layer on the east side of the Skarnsund sill, and 1.04 m s^{-1} at 80 m depth on the west side.

The complex structure of the mean flow in the strait highlights the importance of the baroclinic effects at depth and their influence on the mixing processes in the intermediate and deeper water masses between the two water bodies.

7.2.1 Modification of the model for straits

Beitstad Fjord, Skarnsund, and Middle Fjord are approximated by a single straight channel with variable width and depth and rectangular cross-section. Actually, the channel axis is curved and a curvilinear orthogonal coordinate system is used for which the x-axis agrees with the channel axis, the y-axis is normal to it but horizontal, and the binormal direction is vertical. All effects of curvilinearity of this coordinate system in the balance laws of mass and momentum are assumed to be negligibly small so that operationally the (x, y, z)-coordinates act as if they were Cartesian (Figure 7.7(c)). The sill positions are marked 1 and 2. The origin, 0, of the along-channel axis $0x$ is placed at the center of the strait, and it coincides with the position of the highest sill labeled 1. The z-coordinate is directed vertically upwards, with $z = 0$ at the free surface. The area has irregular bottom topography, $z = -H(x)$, and the variable breadth of the channel is denoted by $2b(x)$; see Figure 7.7(c).

To simplify the three-dimensional system (1.29), we integrate all equations across the channel according to

$$\grave{F}(x, z, t) = \frac{1}{2b(x)} \int\limits_{-b(x)}^{b(x)} F(x, y, z, t)\, dy.$$

Thus, the laterally averaged Reynolds equations describing the generation process of the internal waves in the fjord have the following form [257]:

$$\left.\begin{aligned} \grave{u}_t + \grave{u}\grave{u}_x + \grave{w}\grave{u}_z &= -\grave{P}_x/\rho_0 + [(2bA^{\mathrm{H}}\grave{u}_x)_x + (2bA^{\mathrm{V}}\grave{u}_z)_z + \tau^u]/2b, \\ \grave{w}_t + \grave{u}\grave{w}_x + \grave{w}\grave{w}_z + g\grave{\rho}/\rho_0 &= -\grave{P}_z/\rho_0 \\ &\quad + \left[(2bA^{\mathrm{H}}\grave{w}_x)_x + (2bA^{\mathrm{V}}\grave{w}_z)_z + \tau^w\right]/2b, \\ (2b\grave{u})_x + (2b\grave{w})_z &= 0, \\ \grave{\mathsf{S}}_t + \grave{u}\grave{\mathsf{S}}_x + \grave{w}\grave{\mathsf{S}}_z &= [(2bK^{\mathrm{H}}\grave{\mathsf{S}}_x)_x + (2bK^{\mathrm{V}}\grave{\mathsf{S}}_z)_z]/2b, \\ \grave{\mathsf{T}}_t + \grave{u}\grave{\mathsf{T}}_x + \grave{w}\grave{\mathsf{T}}_z &= [(2bK^{\mathrm{H}}\grave{\mathsf{T}}_x)_x + (2bK^{\mathrm{V}}\grave{\mathsf{T}}_z)_z]/2b. \end{aligned}\right\} \quad (7.10)$$

Here $\grave{u}$ and $\grave{w}$ are the width-averaged components of the velocity in the x- and z-directions; $\grave{P}$, $\grave{\rho}$, $\grave{\mathsf{T}}$, and $\grave{\mathsf{S}}$ are the width-averaged pressure, density, temperature, and salinity, respectively; $\tau^u = A^{\mathrm{H}}u_y(b) - A^{\mathrm{H}}u_y(-b)$ and $\tau^w = A^{\mathrm{H}}w_y(b) - A^{\mathrm{H}}w_y(-b)$ are the sidewall stresses. The density ρ is calculated from the equation of state [238]. The vertical turbulent kinematic viscosity A^{V} and diffusivity K^{V} were determined by the Richardson number dependent parameterizations (5.62).

We now introduce the stream function $\grave{\psi}$ and vorticity $\grave{\omega}$ according to

$$\left.\begin{aligned} \grave{u} = \grave{\psi}_z/2b, \quad \grave{w} = -\grave{\psi}_x/2b, \\ \grave{\omega} = \grave{\psi}_{xx}/2b + \grave{\psi}_{zz}/2b - b_x\grave{\psi}_x/2b^2. \end{aligned}\right\} \quad (7.11)$$

With this system, (7.10) takes the form

$$\left.\begin{aligned} \grave{\omega}_t + J(\grave{\omega}, \grave{\psi})/2b - b_x\grave{\omega}\grave{\psi}_z/2b^2 &= g\grave{\rho}_x/\rho_0 \\ + \left[(A^{\mathrm{H}}\grave{\psi}_{xz})_{xz} + (A^{\mathrm{H}}\grave{\psi}_{xx})_{xx} + (A^{\mathrm{V}}\grave{\psi}_{xz})_{xz} + (A^{\mathrm{V}}\grave{\psi}_{zz})_{zz}\right]&/2b \\ - \left\{b_x[(A^{\mathrm{H}}\grave{\psi}_{xx})_x + (A^{\mathrm{V}}\grave{\psi}_{xz})_z + \tau^w]/2b + \tau^u_z - \tau^w_x\right\}&/2b, \\ \grave{\mathsf{S}}_t + J(\grave{\mathsf{S}}, \grave{\psi})/2b &= [(2bK^{\mathrm{H}}\grave{\mathsf{S}}_x)_x + (2bK^{\mathrm{V}}\grave{\mathsf{S}}_z)_z]/2b, \\ \grave{\mathsf{T}}_t + J(\grave{\mathsf{T}}, \grave{\psi})/2b &= [(2bK^{\mathrm{H}}\grave{\mathsf{T}}_x)_x + (2bK^{\mathrm{V}}\grave{\mathsf{T}}_z)_z]/2b. \end{aligned}\right\} \quad (7.12)$$

Along with the boundary conditions (1.84), (1.85b), (1.35), and (1.38), it is solved numerically with the method presented in Chapter 4.

As the fluid stratifications in the cold and warm seasons are quite different, so the parameters of the generated internal waves will be different too. To answer the question about the seasonal variability of the generated waves, two extreme stratifications are considered. The first corresponds to the season with large fresh water runoff (spring, summer) when a very distinct surface layer develops in the fjord. This is why the buoyancy frequency $N(z)$, obtained from salinity and temperature values presented in ref. [108], has a very pronounced pycnocline (Figure 7.8(a)). The other profile is the autumn stratification when $N(z)$ is almost constant through the depth in comparison with that of the spring season (Figure 7.8(b)).

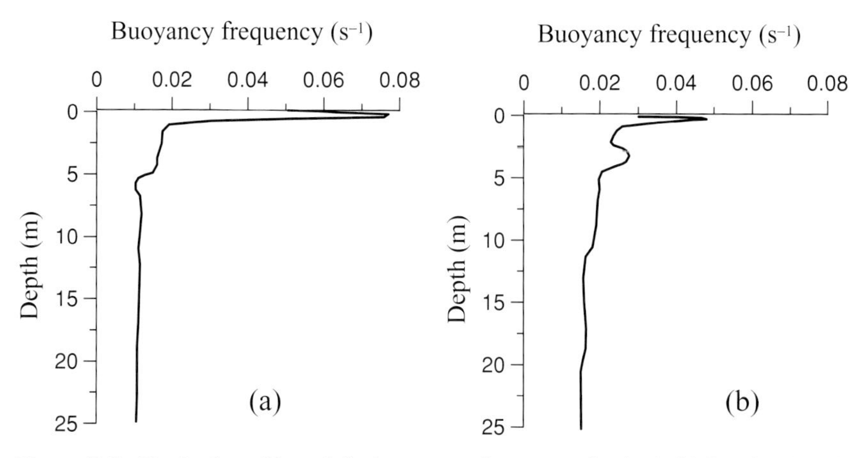

Figure 7.8. Vertical profiles of the buoyancy frequency in the Middle Fjord (a) in spring, (b) in autumn.

According to the observations [108], the tides in the Trondheim Fjord are of semidiurnal nature; therefore, such a tide was used in the modeling (hereafter, $T = 12.4$ h is used as the time scale). The *in situ* and theoretical estimates of the barotropic currents in the Skarnsund, derived from the tidal prism method, give a value of 0.3 m s^{-1} over the first sill. From this we found the amplitude of the stream function Ψ_0. In the basic case runs, we used typical values of the horizontal viscosity and diffusivity, 10 m^2 s^{-1}.

7.2.2 Dynamics of internal waves in the Skarnsund Strait

Let us now consider the generation of internal waves in the middle of the Skarnsund; see Figures 7.6(b) and 7.7. Figure 7.9 shows the evolution of the conventional density σ_t together with the vertical velocity w around the tops of the sills during the first tidal cycle. The stratification of the fluid in this run was taken as typical for the autumn season (Figure 7.8(b)).

The generation process of intensive short waves is clearly visible in Figure 7.9. During the first half of the tidal period, the barotropic current flows rightward with a maximum speed of 0.3 m s^{-1} over the top of the left sill at $t = 0.25T$. It produces upward fluxes on the upstream side of the sills and downward fluxes on the downstream side (Figure 7.9(a)). These vertical flows elevate the isopycnals on the left side, and depress them on the right side of the sills. During the whole first half of the tidal period, these small-scale disturbances are located at nearly the same place over the first sill. As can be seen from Figures 7.9(b)–(d), their positions relative to the sill top are almost stable during the flood phase (compare positions A_1–A_3). Taking into account the existence of the background tidal flow, one can

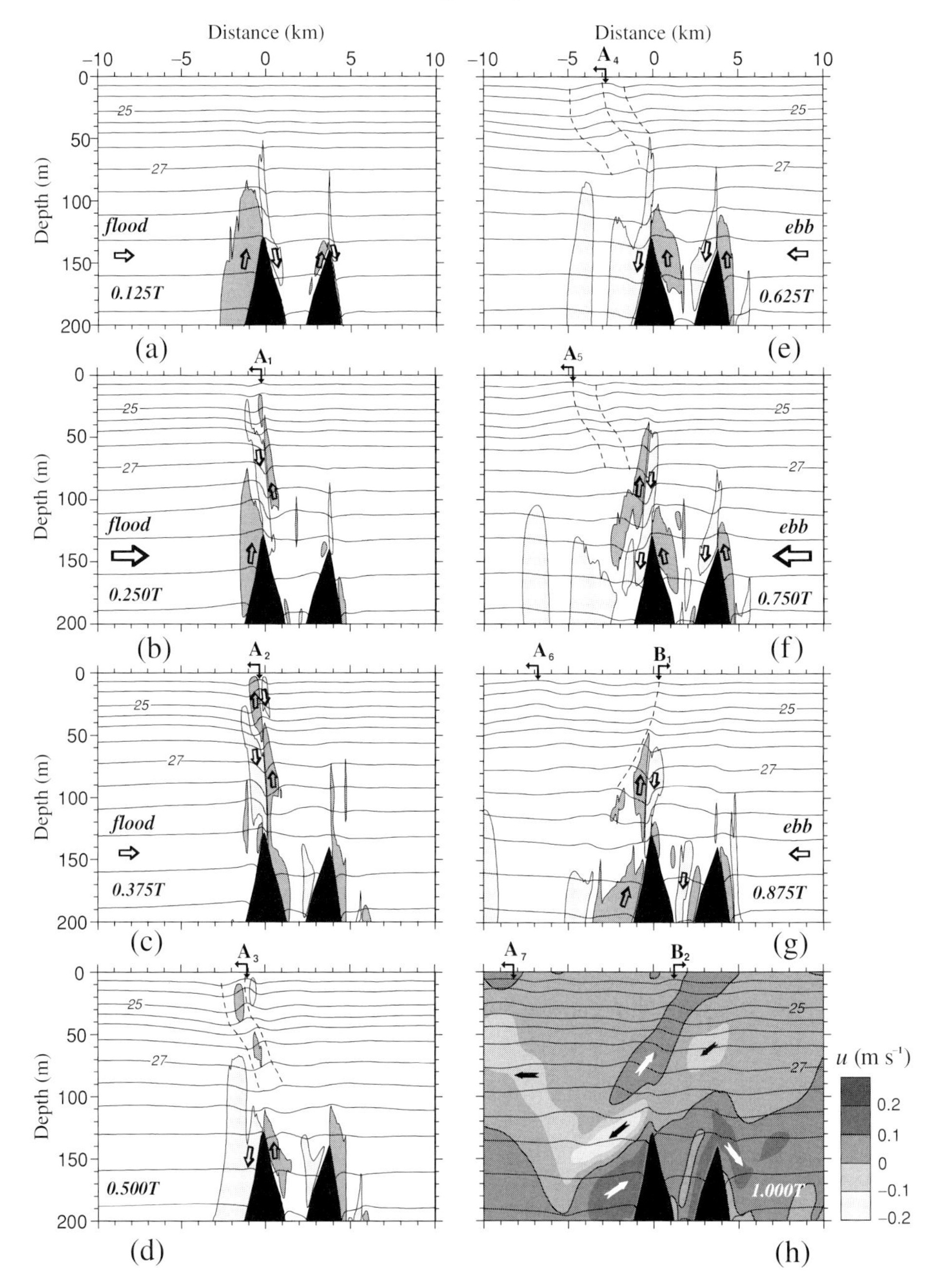

Figure 7.9. Evolution of the conventional density σ_t (isolines shown as solid lines) and vertical velocity w in the Skarnsund strait during the first tidal cycle: after (a) $0.125T$, (b) $0.25T$, (c) $0.375T$, (d) $0.5T$, (e) $0.625T$, (f) $0.75T$, (g) $0.875T$. Vertical velocities greater than $0.003\ \mathrm{m\ s^{-1}}$ are shaded in dark gray and those less than $-0.003\ \mathrm{m\ s^{-1}}$ are shaded in light gray. The density contour interval is $0.5\ \mathrm{kg\ m^{-3}}$. (h) Field of the conventional density σ_t and horizontal velocity u at the end of the first tidal cycle. The thick line represents the zero isopleth.

conclude that these disturbances in fact represent short internal waves propagating upstream. Thus, they are trapped by a topography, and, as was mentioned in Chapter 4, can extract energy from the mean flow. As a result, their amplitudes increase. By the end of the first half of the tidal period, the wave amplitude has reached a value of 8 m in the upper layer; see Figure 7.9(c). The sudden depression of the isopycnals above the first sill is 14 m.

During the second part of the tidal period, when the tidal flow changes direction, the generated waves are released and propagate leftward as free waves from the source of generation; positions A_4–A_7 illustrate the propagation of the wave crest. The amplitude of the propagating crest decreases due to the widening of the channel; see Figure 7.7(c). Estimation of the phase speed, obtained from Figure 7.9, without influence of the tidal flow, gives a magnitude of 0.33 m s^{-1}. So, this value almost coincides with the maximum velocity of the tidal flux (the maximum of the barotropic tidal velocity equals 0.30 m s^{-1} over the first ridge at $t = 0.25T$). Thus, the generated internal waves really are trapped waves.

It is worth mentioning here that the first three baroclinic tidal modes (monochromatic waves having the tidal frequency) have phase speeds equal to 1.40, 0.74, and 0.49 m s^{-1}, respectively. So they cannot be trapped by such a weak tidal flow. Moreover, their wavelengths are 60.5, 31.9, and 21.5 km, respectively. So they are much longer than the short-scale lee waves in Figure 7.9.

The next interesting feature, which further supports the fact that the generated internal waves are lee waves, is the inclination of the lines of the co-phase displacement of the isopycnals. This may be seen from the dashed lines in Figures 7.9(d)–(g), which connect depressions and elevations of isopycnals. Such inclined zones of upwelled and downwelled water are present in all parts of Figure 7.9. To illustrate this fact more clearly, the field of horizontal, instead of vertical, velocity is presented in Figure 7.9(h). Evidently, the inclined co-phase lines of the density disturbances coincide with the isopleths of the horizontal velocity field.

So far we have discussed the generation of internal waves over the first sill. A similar mechanism also occurs over the second sill. However, the internal waves are much weaker here because of the smaller value of the barotropic tidal velocity at the location of the second sill (the maximum barotropic velocity equals 0.11 m s^{-1} instead of 0.30 m s^{-1} at the first sill).

One conclusion from Figure 7.9 is that *the intense small-scale disturbances generated over the sills have the characteristics of unsteady lee waves rather than internal tides.*

To check this idea further, and to obtain confirmation of the conclusion that the generated waves are basically unsteady lee waves, and that the internal tides make a very small input to the total wave field, an additional numerical run was performed.

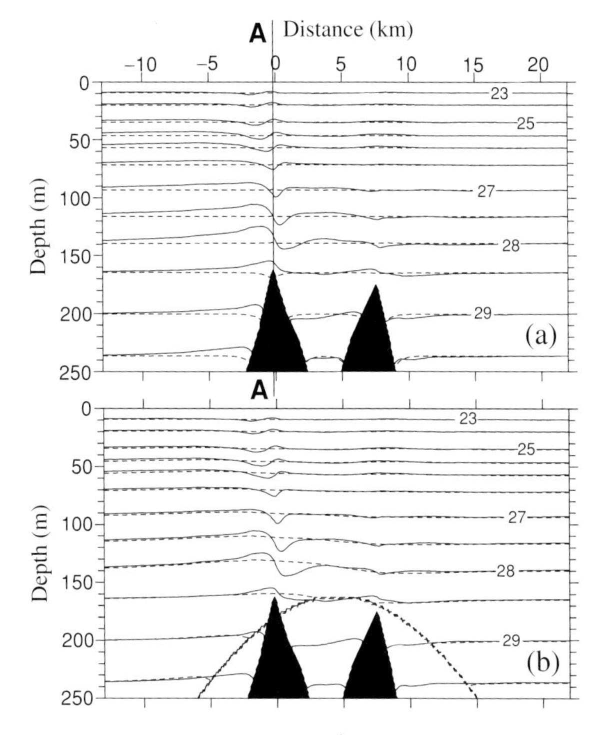

Figure 7.10. (a) Density sections σ_t (kg m^{-3}) at $t = 0.25T$, obtained for the real bottom topography with nonlinear (solid lines) and without nonlinear (dashed lines) terms in (7.12). (b) Comparison of the density fields obtained with the nonlinear system (7.12) at $t = 0.25T$ for the real "steep" (solid lines) and idealized "smoothed" (dashed lines) bottom topography.

The evolution of the wave field was computed in the absence of the nonlinear terms in the system of equations (7.12), i.e. with zero Jacobian.

Figure 7.10(a) displays this comparison for the density fields obtained in the linear (dashed lines) and nonlinear (solid lines) cases at $t = 0.25T$. This pattern makes clear that, in the absence of the nonlinear terms, the wave disturbances, which represent only the baroclinic tides, are much weaker. For example, the maximum elevations of the isopycnal $\sigma_t = 28$ kg m^{-3} from the undisturbed positions are only 1 m, in comparison with 16 m in the nonlinear case. Thus, the advective effects play a decisive role in the formation of the wave field in the Skarnsund where the generated waves have the character of unsteady lee waves. The internal tides, which are usually investigated in such problems, do not play any essential role in the total internal wave energy budget.

The next numerical run was carried out to check whether the characteristics of the unsteady lee waves – wavelength, amplitude, spatial structure – would depend on the curvature of the bottom topography. In Section 6.6 it was shown that particularly strong unsteady lee waves are generated where essential spatial variability of the forcing exists. Such places are connected with abrupt changes of the water depth.

Figure 7.10(b) illustrates this fact, with two density fields overlaid in one graph. The runs were performed for the bottom topography in the Skarnsund (solid lines) and for an idealized "smooth" sill with the same values of height and width. It is clearly seen from Figure 7.10(b) that the waves generated over the idealized "smooth" sill are long and weak.

7.2.3 Residual currents produced by nonlinear waves

It is known that propagating nonlinear waves (in contrast with linear waves) can lead to a residual mass transport. For instance, the first-mode solitary internal wave of depression drags the water in the near-surface layer in the direction of its propagation and below the pycnocline in the opposite direction. Because of the permanent character of the tidal motions in fjords, such a mechanism which produces residual currents by unsteady lee waves can be very significant in the total water budget of different fjord basins. Such investigations are useful in correctly understanding the formation and changing of the thermohaline structure of the water masses and water renewal in the inner parts of fjords, which, in turn, have a significant effect on the biological processes in such systems. All other mechanisms (wind driven circulation, intense spring-water runoff, mixing, buoyancy fluxes, and others) are also effective, but basically have casual character, and are therefore beyond the scope of our considerations.

Figure 7.11 illustrates the predicted influence of the unsteady lee waves on the water exchange between the Middle Fjord and the Beitstad Fjord through the Skarnsund strait. Figures 7.11(a) and (b) show the distributions of the residual currents produced by the nonlinear waves generated in the Skarnsund strait over the sills in autumn and spring, respectively. These fields were built by an averaging of the total wave fields (consisting of barotropic and baroclinic tides and unsteady lee waves) during a tidal period $9T \leq t \leq 10T$. To remove the manifestation of internal waves, the same averaging was performed with the salinity fields for the autumn (Figure 7.11(c)) and spring (Figure 7.11(d)), respectively.

It is clearly seen from Figures 7.11(a) and (b) that the residual currents generated by the internal waves are nonzero (in a linear model they should be zero). The structure of the residual currents in the strait is rather complicated. The velocity pattern in every vertical section has a three- to four-layered structure. This is in agreement with the experimental measurements of mean currents carried out over the Skarnsund sill [108]. The main finding from Figure 7.11 is *the existence of the residual water flux from the Beitstad Fjord into the Middle Fjord in the upper near-surface layer. Concurrently, over the tops of the sills, there is a residual flux*

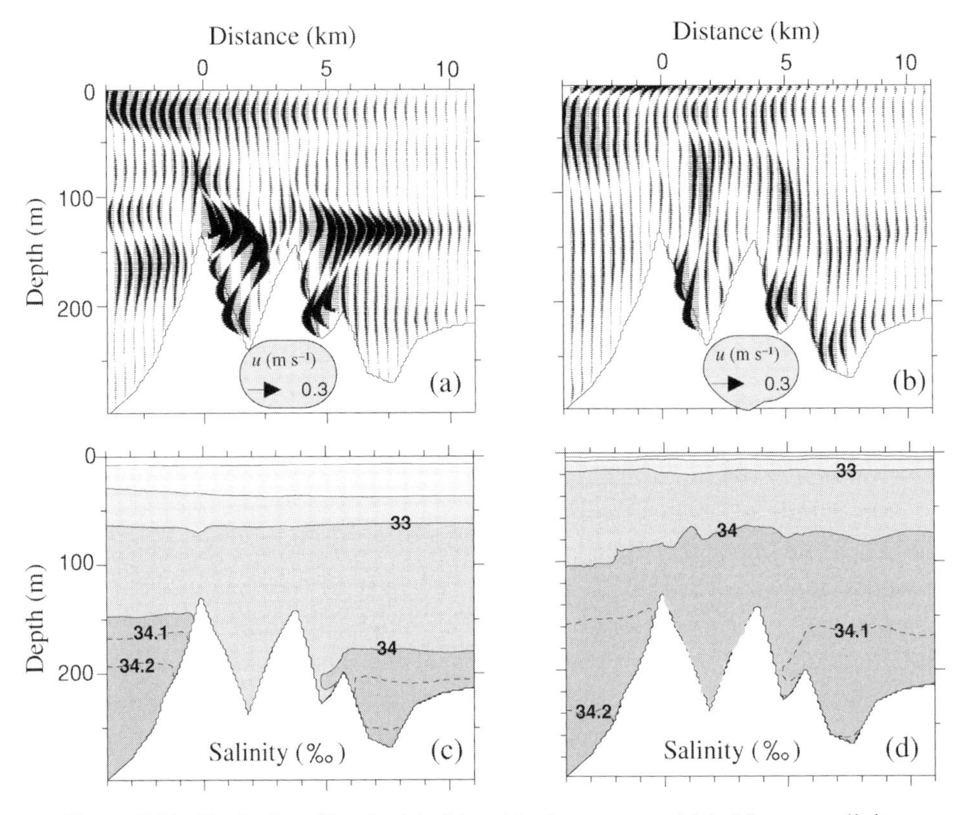

Figure 7.11. Vertical profiles for (a), (b) residual currents and (c), (d) mean salinity field in Skarnsund after ten tidal periods: (a), (c) autumn; (b), (d) spring.

in the opposite direction. This result is important for the understanding of the water renewal in the semienclosed Beitstad Fjord.

The vertical structure of the residual currents with manifestation of third and fourth baroclinic modes reflects the peculiarities of unsteady lee waves. The co-phase lines of these waves are the inclined dashed lines in Figure 7.9; this explains why one can find both elevations and depressions in every vertical section of iso-pycnals; see, for instance, section A–A in Figure 7.10(a).

One should also note that the maximum of the residual currents reaches a value of $0.27\ \mathrm{m\,s^{-1}}$ in autumn (Figure 7.11(a)). Thus, these currents, which are permanently active as the tidal activity does not terminate, can affect the thermohaline structure of the waters in the inner basin. In comparison with the "inverse estuarine" circulation, Figure 7.7(b), this mechanism of circulation is likely to be more appropriate for autumn, as was suggested in ref. [108].

The influence of the water circulation in the Beitstad Fjord due to internal waves on the resulting salinity field is shown in Figure 7.11(c). The penetration of relatively fresh water in the intermediate layers from the Middle Fjord into the Beitstad Fjord constitutes the refreshing mechanism of the near bottom water masses. For instance, the isohaline 34 $^{\circ}/_{oo}$ sank by approximately 30 m due to the penetration of less salty water into the Beitstad Fjord. In addition, mixing processes due to the short internal waves in the region between the two sills led to the formation of almost homogeneous water there.

The process of the water renewal in the Beitstad Fjord in spring is not as pronounced as it is in autumn. In spring the generated lee waves are less intense and the water circulation is less developed. The presence of the upper, strongly stratified layer, caused by a river run-off, leads to a shift of the extrema of the eigenfunctions towards the free surface and further from the tops, a circumstance which damps the generation process of the lee waves. As a result, the residual currents in Skarnsund do not exceed 0.18 m s^{-1} (Figure 7.11(b)).

Of course, the illustrated mechanism of water renewal is not the only possible one. The other mechanisms mentioned above are equally active. For an overall investigation of water dynamics, one should consider all the mechanisms acting together. Nevertheless, the processes considered here play a key role in the water exchange in the area under study.

7.2.4 Experiments on the dynamics of a passive admixture

This section is devoted to the investigation of the dynamics of a passive admixture as a tracer. Such scenarios can show more explicitly the existence of water exchange through the channel caused by nonlinear internal waves. One can trace the penetration of different water masses into each other using the results.

The numerical experiments were carried out during 20 tidal cycles for two typical fluid stratifications, in spring and autumn. An additional equation describing the dynamics of a passive admixture of a substance C was added to the governing system (7.12); the equation is the same as for the temperature or salinity. The boundary conditions and coefficients for horizontal and vertical turbulent diffusivity were chosen to be similar to those for the temperature and salinity.

At $t = 0$ the following initial conditions for C were assigned: to the left of the vertical dashed line in Figure 7.12, in the Middle Fjord, the concentration C was taken to be constant and equal to unity. To the right of this line, in the Skarnsund strait and in the Beitstad Fjord, the concentration was set to zero.

Figure 7.12 displays the evolution of the concentration field C for the autumn season. The correlation between Figure 7.12 and Figure 7.11(a) is obvious. The residual wave currents, excited above the tops of the sills and directed from the Skarnsund strait into Beitstad Fjord, drag the passive admixed tracer with them. With every tidal cycle the substance penetrates further and propagates into the deep

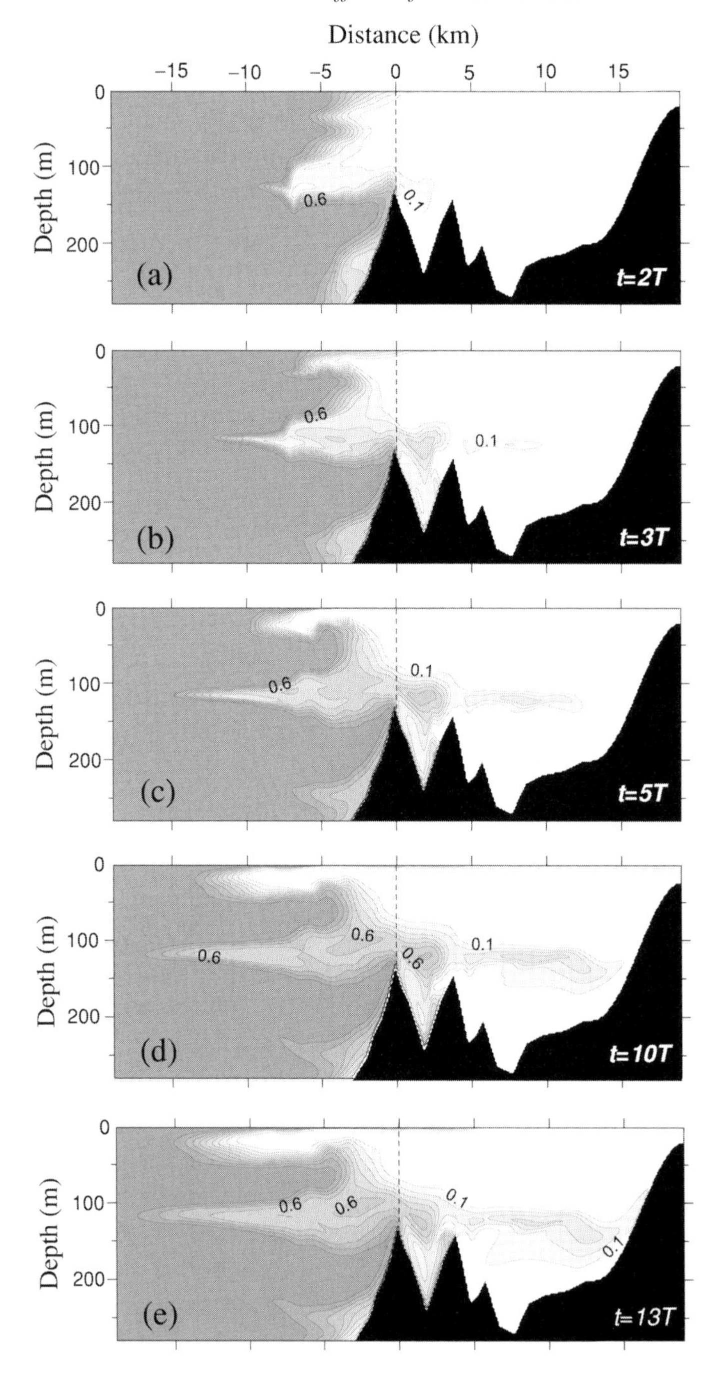

Figure 7.12. Evolution of a passive admixture field in the Skarnsund strait and Beitstad Fjord during 13 tidal cycles in autumn. At the beginning of the numerical run, the concentration C was equal to unity in the Middle Fjord (left from the vertical dashed line) and zero to the right of this line. Darker shading corresponds to greater values of the concentration C. The contour interval for C is 0.1.

part of the strait to the opposite boundary of the innermost basin. This process can be observed in Figure 7.12, where the field of concentration for different tidal cycles is presented. The maximum value of the residual current in Skarnsund is $0.27\,\mathrm{m\,s^{-1}}$, and the tongue of the Middle Fjord waters reaches the opposite side of the basin during the 13th tidal cycle. At the same time, according to Figure 7.11(a), the upper, less saline waters are carried away from the fjord by the near-surface residual currents. Using Figure 7.12, it is possible to estimate the distance of their propagation. By the end of the 13th tidal cycle, the frontal boundary of the water with concentration $C = 0.1$ moved a distance approximately 10 km from its initial position.

Figure 7.13 displays results for a similar numerical experiment conducted in spring. The main difference between the water transport for spring and autumn stratifications is that the tongue of the Middle Fjord waters is wider, it propagates more slowly, and reaches the right lateral boundary of the Beitstad Fjord after 17 tidal cycles.

These results suggest the following.

The described water exchange mechanisms, caused by tidally generated lee waves, are different in different seasons; peculiarities of the water stratification may introduce variations in the lee wave pattern and the water transport across the sill.

For instance, the presence of a less saline near-surface layer in spring (compare Figures 7.8(a) and (b)) is responsible for the formation of thinner layers of the residual near-surface current and counter-current. This feature can be explained by the behavior of the eigenfunctions of the standard boundary value problem with a given buoyancy frequency profile. For a thin, strongly stratified upper layer, they have their extrema closer to the surface, where the eigenfunctions change more dramatically.

A further question which so far remains unanswered is: Why does the deep water penetrate from the Skarnsund strait into the Beitstad Fjord, which is obviously very important for the ecosystem of the Beitstad Fjord, and why does subsurface water propagate into the Middle Fjord, but not vice versa? The answer to this question is almost obvious: it is because of the asymmetry of the bottom topography and coastline of the Skarnsund strait, and consequently the unequal conditions for the generation of lee waves at different places. In reality, the bottom profile of the strait is such that the two largest sills are set to one side from the narrowest place of the strait; see Figure 7.7(c). That is why the conditions for the generation of lee waves on different slopes of the sills are different, even when taking into account that they have almost the same parameters including height, shape, and steepness. The analysis of evolution of the density field has also shown that the largest waves are generated on the left side of the first sill; the strait is narrower and the barotropic

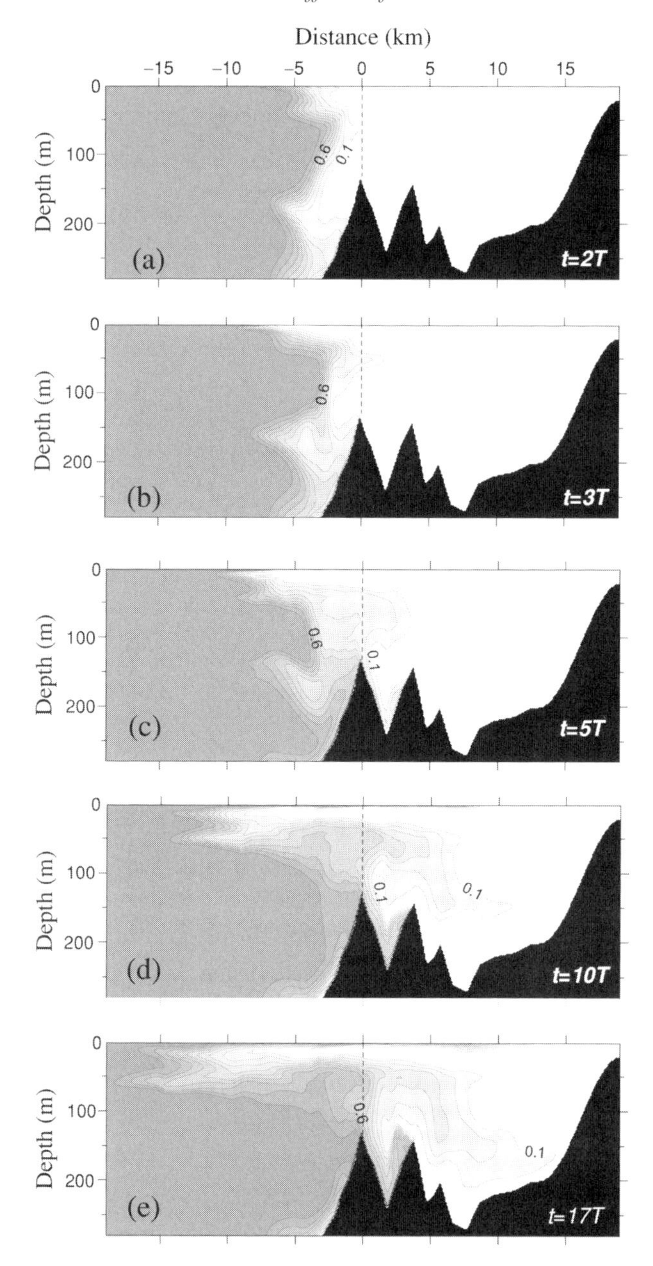

Figure 7.13. As for Figure 7.12, but in spring.

tidal flow is strongest here. Such results were partly presented in Figure 7.9; they showed that the more intense waves propagate leftward.

Nevertheless, even under conditions of absolute symmetry the tidally excited nonlinear lee waves are able to affect the thermohaline water structure. To show this,

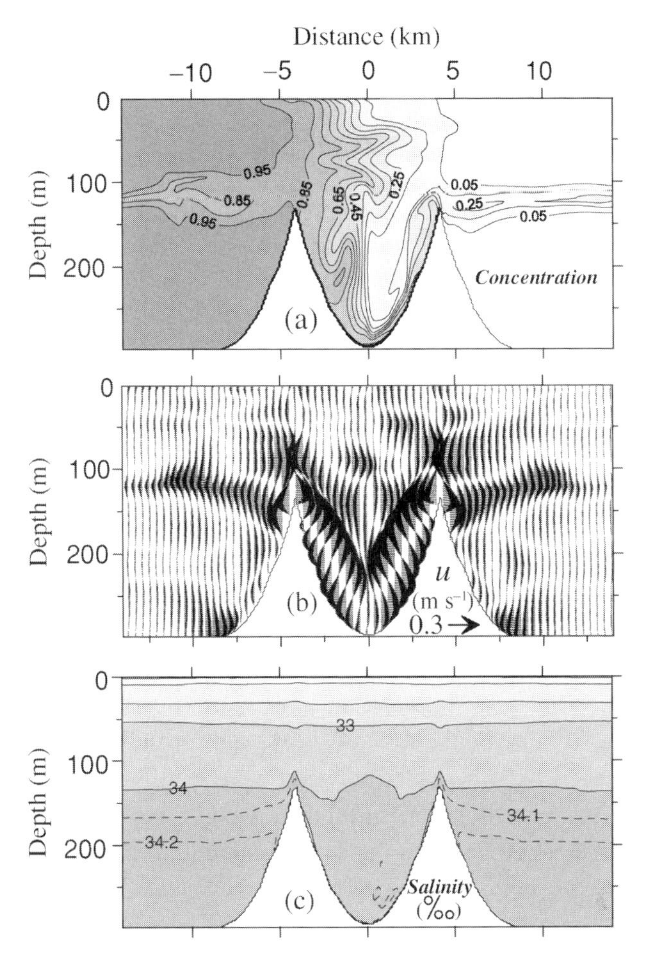

Figure 7.14. Patterns (a) of concentration, C, of passive admixture, (b) residual currents, and (c) mean salinity obtained in the symmetric channel with symmetric bottom topography.

we carried out an additional numerical experiment. The boundaries of a channel were changed to resemble ideal symmetry, with the minimum cross-section at $x = 0$. The heights and the shapes of the two sills were identical, and they were placed at the same distance from the center of the strait. The basic autumn stratification was used, and the depth of the channel and its minimum width were close to those of the Skarnsund strait. The concentration of passive admixture of a tracer C was taken to be constant, equal to 1 for $x < 0$ and 0 for $x > 0$. The basic goal of this numerical experiment was to consider the hypothetical situation with complete symmetry in the model, and to check whether the circulation presented in Figure 7.11 (and resulting in the water renewal in the Beitstad Fjord) is the result of the asymmetric model configuration or not.

Of course, no absolute symmetry can exist in such models because of the use of initial conditions with the barotropic flux directed for the flood ($0 < t < 0.5T$) from left to right. Nevertheless, the influence of the initial conditions diminishes with time when the initial disturbances leave the center of the computation area. At a definite stage of evolution, the model solution is almost independent of the initial conditions. So, it makes sense to consider all fields at the moment of time when all transition processes will cease.

A situation approaching this is presented in Figure 7.14. It shows the result of the evolution of the field of concentration C (Figure 7.14(a)), the residual currents (Figure 7.14(b)), and the field of salinity (Figure 7.14(c)) after six tidal cycles. Deviations from complete symmetry can be explained in accordance with the above-mentioned remarks.

Figure 7.14 shows that a mass transport by nonlinear internal waves also exists in the symmetrical model. However, in this case the pattern of residual currents produced by unsteady lee waves is almost symmetrical. These currents caused the water circulation presented in Figure 7.14(b). The main peculiarity of this circulation is that the excited short-scale internal lee waves transport the water in the near-surface layer from the two reservoirs to the strait, where it mixes with surrounding water masses, sinks down, and returns to the adjacent basins with counter-currents. Two vertical circulating cells are formed in the strait. As can be seen from Figure 7.14(c), the water between the two sills is overmixed due to the wave activity. Thus, one should mention that, even in the symmetric model, water from the left basin penetrates into the right basin due to the described circulation and mixing processes between the sills.

References

[1] Alpers, W. (1985) Theory of radar imaging of internal waves. *Nature*, **314**, 245–247.

[2] Alpers, W., & Salusti, E. (1983) Scylla and Charybdis observed from space. *Journal of Geophysical Research*, **88**, 1800–1808.

[3] Althaus, A.M., Kunze, E., & Sanford, T.B. (2003) Internal tide radiation from Mendocino Escarpment. *Journal of Physical Oceanography*, **33**, 1510–1527.

[4] Apel, J.R., Byrne, H.M., Prony, J.R., & Charnell, R.L. (1975) Observations of oceanic internal and surface waves from the Earth resources technology satellite. *Journal of Geophysical Research*, **80**, 865–881.

[5] Apel, J.R., Holbrook, J.R., Liu, A.K., & Tsai, J.J. (1985) The Sulu Sea internal soliton experiment. *Journal of Physical Oceanography*, **15**, 1625–1651.

[6] Armi, L., & Farmer, D.M. (1988) The flow of Atlantic water through the strait of Gibraltar. *Progress in Oceanography*, **21** (1), 1–105.

[7] Baines, P.G. (1971) The reflection of internal/inertial waves from bumpy surfaces. *Journal of Fluid Mechanics*, **46**, 273–291.

[8] Baines, P.G. (1971) The reflection of internal/inertial waves from bumpy surfaces. Part 2: Split reflection and diffraction. *Journal of Fluid Mechanics*, **49**, 113–131.

[9] Baines, P.G. (1973) The generation of internal tides by flat-bump topography. *Deep-Sea Research*, **20** (2), 179–206.

[10] Baines, P.G. (1974) The generation of internal tides over steep continental slopes. *Philosophical Transactions of the Royal Society of London Series A*, **277**, 27–58.

[11] Baines, P.G. (1982) On internal tide generation models. *Deep-Sea Research*, **29** (3A), 307–339.

[12] Baines, P.G. (1983) Tidal motion in submarine canyon: a laboratory experiment. *Journal of Physical Oceanography*, **13** (2), 310–328.

[13] Baines, P.G. (1995) *Topographic Effects in Stratified Flows*. Cambridge: Cambridge University Press.

[14] Balmforth, N.J., Ierley, G.R., & Young, W.R. (2002) Tidal conversion by subcritical topography. *Journal of Physical Oceanography*, **32**, 2900–2914.

[15] Barbee, W.B., Dworski, J.G., Irish, J.H., Larsen, L.H., & Rattray, M. (1975) Measurements of internal waves of tidal frequency near continental boundary. *Journal of Geophysical Research*, **80** (15), 1965–1974.

[16] Beardsley, R.C. (1970) An experimental study of internal waves in a closed cone. *Studies in Applied Mathematics*, **49**, 187–196.

[17] Bell, T.H. (1975) Topographically generated internal waves in the open ocean. *Journal of Geophysical Research*, **80** (3), 320–327.

[18] Bell, T.H. (1975) Lee waves in stratified flows with simple harmonic time dependence. *Journal of Fluid Mechanics*, **67** (4), 705–722.
[19] Benjamin, T.B. (1966) Internal waves of finite amplitude and permanent form. *Journal of Fluid Mechanics*, **25**, 241–270.
[20] Benney, D.J. (1966) Long nonlinear wave in fluid flows. *Journal of Mathematical Physics*, **25**, 241–270.
[21] Benney, D.J., & Ko, D.R.S. (1978) The propagation of long large amplitude internal waves. *Studies in Applied Mathematics*, **59** (3), 187–199.
[22] Boguki, D., & Garrett, C. (1993) A simple model for the shear-induced decay of internal solitary waves. *Journal of Physical Oceanography*, **23**, 1767–1776.
[23] Brandt, P., Rubino, A., Alpers, W., & Backhaus, J.O. (1997) Internal waves in the Strait of Messina studied by a numerical model and synthetic aperture radar images from the ERS 1/2 satellites. *Journal of Physical Oceanography*, **27**, 648–663.
[24] Brandt, P., Rubino, A., & Fischer, J. (2002) Large-amplitude internal solitary waves in the North Equatorial Counter Current. *Journal of Physical Oceanography*, **32**, 1567–1573.
[25] Briscoe, M. (1984) Tides, solitons and nutrients. *Nature*, **312**, 1–5.
[26] Brown, D.J., & Christie, D.R. (1998) Fully nonlinear solitary waves in continuously stratified incompressible Boussinesq fluid. *Physics of Fluids*, **10**, 2569–2586.
[27] Brunt, D. (1952) *Physical and Dynamical Meteorology*. Cambridge: Cambridge University Press.
[28] Cacchione, D., & Wunsch, C. (1974) Experimental study of internal waves over a slope. *Journal of Fluid Mechanics*, **66**, 223–239.
[29] Cartwright, D.E. (1993) Theory of ocean tides with application to altimetry. In *Satellite Altimetry in Geodesy and Oceanography*, eds. R. Rummel & F. Sansó. Berlin: Springer, pp. 99–141.
[30] Chapman, D.C., Giese, G.S., Collins, M.G., Encarnacion, R., & Jacinto, G. (1991) Evidence of internal swash associated with Sulu Sea solitary waves? *Continental Shelf Research*, **11**, 591–599.
[31] Chen, C.T., & Millero, P.S. (1976) The specific volume of sea water at high pressures. *Deep-Sea Research*, **23** (7), 595–612.
[32] Chen, D., & Beardsley, R.C. (1995) A numerical study of stratified tidal rectification over finite-amplitude bank. Part I: Symmetric bank. *Journal of Physical Oceanography*, **25**, 2090–2110.
[33] Chen, D., & Beardsley, R.C. (1998) Tidal mixing and cross-frontal particle exchange over a finite amplitude asymmetric bank: A model study with application to Georges Bank. *Journal of Marine Research*, **56**, 1163–1201.
[34] Chen, D., Ou, H.W., & Dong, C. (2003) A model study of internal tides in coastal frontal zone. *Journal of Physical Oceanography*, **33**, 170–187.
[35] Chereskin, T.K. (1983) Generation of internal waves in Massachusetts Bay. *Journal of Geophysical Research*, **88** (4), 2649–2661.
[36] Cherkesov, L.V. (1973) *Surface and Internal Waves*. Kiev: Naukova dumka (in Russian).
[37] Cherkesov, L.V. (1976) *Hydrodynamics of Surface and Internal Waves*. Kiev: Naukova dumka (in Russian).
[38] Chiswell, S.M. (2000) Tidal energetics over the Chatham Rise, New Zealand. *Journal of Physical Oceanography*, **30**, 2452–2460.
[39] Choi, W., & Camassa, R. (1999) Fully nonlinear internal waves in a two-fluid system. *Journal of Fluid Mechanics*, **396**, 1–36.

[40] Chuang, W.S., & Wang, D.P. (1981) Effects of density front on the generation and propagation of internal tides. *Journal of Physical Oceanography*, **7** (10), 1357–1374.
[41] Copson, E.T. (1935) *An Introduction to the Theory of Functions of a Complex Variable*. Oxford: Oxford University Press.
[42] Courant, R., & Hilbert, D. (1953) *Methods of Mathematical Physics*. New York: Interscience Publishers.
[43] Cox, C.S., & Sandström, H. (1962) Coupling of internal and surface waves in water of variable depth. *Journal of the Oceanographical Society of Japan*, **20**, 499–513.
[44] Craig, P.D. (1987) Solution for internal tidal generation over coastal topography. *Journal of Marine Research*, **45** (1), 83–105.
[45] Craig, P.D. (1988) A numerical model study of internal tides on the Australian Northwest Shelf. *Journal of Marine Research*, **46** (1), 59–76.
[46] Craik, A.D.D. (1985). *Wave Interactions and Fluid Flows*. Cambridge: Cambridge University Press.
[47] Cummins, P.F., & Oey, L.-Y. (1997) Simulation of barotropic and baroclinic tides off northern British Columbia. *Journal of Physical Oceanography*, **27**, 762–781.
[48] Cummins, P.F., Cherniawsky, J.Y., & Foreman, G.G. (2001) North Pacific internal tides from Aleutian Ridge: altimeter observations and modeling. *Journal of Marine Research*, **59**, 167–191.
[49] Da Silva, J.C.B., Ermakov, S.A., Robinson, I.S., Jeans, D.R.G., & Kijashko, S.V. (1998) Role of surface films in ERS SAR signatures of internal waves on the shelf. 1. Short-period internal waves. *Journal of Geophysical Research*, **103** (C4), 8009–8031.
[50] Dale, A.C., Huthnance, J.M., & Sherwin, T.J. (2001) Coastal-trapped internal waves and tides at near-inertial frequencies. *Journal of Physical Oceanography*, **8**, 2958–2970.
[51] De Witt, L.M., Levine, M.D., & Paulson, C.A. (1982) Internal waves in JASIN: spectra, variability and internal tide. *Journal of Physical Oceanography*, **12**, 1245–1259.
[52] Diebels, S., Schuster, B., & Hutter, K. (1994) Nonlinear internal waves over variable topography. *Geophysical and Astrophysical Fluid Dynamics*, **76**, 165–192.
[53] Djordjevic, V.D., & Redercopp, L.G. (1978) The fission and disintegration of internal solitary waves moving over two-dimensional topography. *Journal of Physical Oceanography*, **8**, 1016–1033.
[54] Dushaw, B.D., Cornuelle, B.D., Worcester, P.F., Howe, B.M., & Luther, D.S. (1995) Barotropic and baroclinic tides in the Central North Pacific Ocean determined from long-range reciprocal acoustic transmissions. *Journal of Physical Oceanography*, **25**, 631–647.
[55] Egbert, G.D., & Ray, R.D. (2001) Estimates of M_2 tidal energy dissipation from TOPEX/Poseidon altimeter data. *Journal of Geophysical Research*, **106** (C10), 22 475–22 502.
[56] Ekman, V.W. (1904) On dead water (Norwegian North Polar Expedition 1893–1896). *Scientific Results*, **5** (15), 125.
[57] Ermakov, S.A., Da Silva, J.C.B., & Robinson, I.S. (1998) Role of surface films in ERS SAR signatures of internal waves on the shelf. 1. Internal tidal waves. *Journal of Geophysical Research*, **103** (C4), 8031–8043.
[58] Farmer, D.M. (1978) Observation of long nonlinear internal waves in a lake. *Journal of Physical Oceanography*, **8**, 63–73.

[59] Farmer, M.D., & Army, L. (1999) The generation and trapping of solitary waves over topography. *Science*, **283** (5399), 188–190.
[60] Farmer, M.D., & Dugan, S.J. (1980) Tidal interaction of stratified flow with a sill in Knight Inlet. *Deep Sea Research*, **27A**, 239–254.
[61] Farmer, M.D., & Freeland, H.J. (1983) The physical oceanography of fjords. *Progress in Oceanography*, **12**, 147–220.
[62] Fedorov, K.N. (1983) *Physical Nature and Structure of Oceanic Fronts*. Leningrad: Gidrometeoizdat (in Russian).
[63] Fedorov, K.N., & Ginsburg, A.I. (1988) *Near Surface Layer of the Ocean*. Leningrad: Gidrometeoizdat (in Russian).
[64] Feng, M., Merrifield, M.A., Pinkel, R., *et al*. (1998) Semidiurnal tides observed in the western equatorial Pacific during the Tropical Ocean-Global Atmosphere Coupled Ocean-Atmosphere Response Experiment. *Journal of Geophysical Research*, **103** (C5), 10 253–10 272.
[65] Franklin, B. (1774) Of the stilling of waves by means of oil. *Philosophical Transactions* **64**, 445–460.
[66] Fu, L.L., & Holt, B. (1982) Internal waves in the Gulf of California: Observations from a space-borne radar. *Journal of Geophysical Research*, **89**, 2053–2060.
[67] Garrett, C. (2003) Internal tides and ocean mixing. *Science*, **301**, 1858–1859.
[68] Garrett, C. (2003) Mixing with latitude. *Nature*, **422**, 477–478.
[69] Garrett, C., & Munk, W. (1972) Space-time scales of internal waves in the open ocean. *Geophysical Fluid Dynamics*, **3** (3), 225–254.
[70] Gasparovic, R.F., Apel, J.R., & Kasichke, E.S. (1988) An overview of SAR internal wave signature experiment. *Journal of Geophysical Research*, **88**, 12 301–12 316.
[71] Gasparovic, R.F., Apel, J.R., Thompson, D.R., & Toscho, J.S. (1986) A comparison of SIR-B synthetic aperture radar data with ocean internal wave measurement. *Science*, **232**, 455–463.
[72] Gear, J.A., & Grimshaw, R. (1983) A second-order theory for solitary waves in shallow fluids. *Physics of Fluids*, **26**, 14–29.
[73] Gerkema, T. (1996) A unified model for the generation and fission of internal tides in a rotating ocean. *Journal of Marine Research*, **54**, 421–450.
[74] Gerkema, T. (2001) Internal and interfacial tides: Beam scattering and local generation of solitary waves. *Journal of Marine Research*, **59**, 227–255.
[75] Gerkema, T., & Zimmerman, J.T.F. (1995) Generation of nonlinear internal tides and solitary waves. *Journal of Physical Oceanography*, **25** (6), 1081–1094.
[76] Gill, A.E. (1982) *Atmosphere-Ocean Dynamics*. New York: Academic Press.
[77] Gordon, R.L. (1978) Internal waves' climate near the coast of Northwest Africa during JOINT-1. *Deep-Sea Research*, **25**, 625–643.
[78] Goryachkin, Yu.N., Grodsky, S.A., Ivanov, V.A., Kudryavtsev, V.N., & Lisichenok, A.D. (1991) Long time observation at the internal wave packet evolution. *Izvestiya, Atmospheric and Oceanic Physics*, **27** (3), 326–334.
[79] Goryachkin, Yu.N., Ivanov, V.A., & Pelinovsky, E.N. (1992) Transformation of internal tidal waves on the shelf of Guinea. *Soviet Journal of Physical Oceanography*, **3** (4), 309–317.
[80] Greenberg, M.D. (1971) *Application of Green's Function in Science and Engineering*. Englewood Cliffs, New Jersey: Prentice Hall.
[81] Gregg, M.C., & Briscoe, M.G. (1979) Internal waves, fine structure, microstructure and mixing in the ocean. *Reviews of Geophysics and Space Physics*, **17** (7), 1524–1547.

[82] Grimshaw, R. (1985) Evolution equations for weakly nonlinear, long internal waves in a rotating fluid. *Studies in Applied Mathematics*, **73**, 1–33.

[83] Grimshaw, R., Pelinovsky, E., & Talipova, T. (1997) The modified Korteweg-de Vries equation in the theory of large-amplitude internal waves. *Nonlinear Processes in Geophysics*, **4**, 237–250.

[84] Grimshaw, R., Pelinovsky, E., & Talipova, T. (1999) Solitary wave transformation in a medium with sign-variable quadratic nonlinearity and cubic nonlinearity. *Physica D*, **4**, 237–250.

[85] Grue, J., Jensen, A., Rusas, P.O., & Sveen, K.J. (2000) Breaking and broadening of internal solitary waves. *Journal of Fluid Mechanics*, **413**, 181–217.

[86] Gustafsson, K.E. (2001) Computations of the energy flux to mixing processes via baroclinic wave drag on barotropic tides. *Deep-Sea Research*, **48**, 2283–2295.

[87] Halpern, D. (1971) Observation of short-period internal waves in Massachusetts Bay. *Journal of Marine Research*, **29**, 116–132.

[88] Haury, L.R., Briscoe, M.G., & Orr, M.H. (1979) Tidally generated internal wave packets in Massachusetts Bay. *Nature*, **278**, 10–14.

[89] Hayes, S.P., & Halpern, D. (1976) Observation of internal waves and coastal upwelling off the Oregon coast. *Journal of Marine Research*, **34**, 245–267.

[90] Heathershow, A.D., New, A.L., & Edwards, P.D. (1987) Internal tides and sediment transport at the shelf break in the Celtic Sea. *Continental Shelf Research*, **7**, 485–517.

[91] Helfrich, K.R. (1992) Internal solitary wave breaking and run-up on a uniform slope. *Journal of Fluid Mechanics*, **243**, 133–154.

[92] Helfrich, K.R., & Melville, W.K. (1986) On long nonlinear internal waves over slope-shelf topography. *Journal of Fluid Mechanics*, **167**, 285–308.

[93] Helfrich, K.R., Melville, W.K., & Miles, J.M. (1984) On interfacial solitary waves over slowly varying topography. *Journal of Fluid Mechanics*, **149**, 305–317.

[94] Hendry, R.M. (1977) Observations of the semidiurnal internal tide in the Western North Atlantic Ocean. *Philosophical Transactions of the Royal Society of London A*, **286**, 1–24.

[95] Hibiya, T. (1988) The generation of internal waves by tidal flow over Stellwagen Bank. *Journal of Geophysical Research*, **93**, 533–542.

[96] Hibiya, T., Niwa, Y., & Fujiwara, K. (1998) Numerical experiments of nonlinear energy transfer within the oceanic internal wave spectrum. *Journal of Geophysical Research*, **103**, 18 715–18 722.

[97] Holloway, P.E. (1983) Internal tides on the Australian North-West Shelf: a preliminary investigation. *Journal of Physical Oceanography*, **13** (8), 1357–1370.

[98] Holloway, P.E. (1987) Internal hydraulic jumps and solitons at a shelf break region on the Australian North-West shelf. *Journal of Geophysical Research*, **92**, 5405–5416.

[99] Holloway, P.E. (1996) A numerical model of internal tides with applications to the Australian North-West shelf. *Journal of Physical Oceanography*, **26** (1), 21–37.

[100] Holloway, P.E., & Barnes, B. (1998) A numerical investigation into the boundary layer flow and vertical structure of internal waves on a continental shelf. *Continental Shelf Research*, **15**, 31–65.

[101] Holloway, P.E., & Merrifield, M.A. (1999) Internal tide generation by seamounts, ridges, and islands. *Journal of Geophysical Research*, **104** (C11), 25 937-25 951.

[102] Holloway, P.E., Pelinovsky, E.P., Talipova, T., & Barnes, B. (1997) A nonlinear model of internal tide transformation on the Australian North-West Shelf. *Journal of Physical Oceanography*, **27**, 871–896.

[103] Holloway, P.E., Pelinovsky, E.N., & Talipova, T. (1999) A generalized Korteweg-de Vries model of internal tide transformation in the coastal zone. *Journal of Geophysical Research*, **104** (C8), 18 333–18 350.

[104] Huthnance, J.M. (1989) Internal tides and waves near the continental shelf edge. *Geophysical and Astrophysical Fluid Dynamics*, **48**, 81–106.

[105] Hüttemann, H., & Hutter, K. (2001) Baroclinic solitary water waves in a two-layer fluid system with diffusive interface. *Experiments in Fluids*, **30**, 317–326.

[106] Hutter, K. (1991) Large scale water movements in lakes. *Aquatic Sciences*, **53**, 100–135.

[107] Ivanov, V.A., Konyaev, K.V., & Serebryany A.N. (1981) Groups of intense internal waves in the shelf zone of the sea. *Izvestiya, Atmospheric and Oceanic Physics*, **17** (12), 1302–1309.

[108] Jacobson, P. (1983) Physical oceanography of the Trondheimsfjord. *Geophysical and Astrophysical Fluid Dynamics*, **26**, 871–896.

[109] Jeans, D.R.G., & Sherwin, T.J. (2001) The variability of strongly non-linear solitary internal waves observed during an upwelling season on the Portuguese Shelf. *Continental Shelf Research*, **21**, 1855–1878.

[110] Jeffreys, H. (1920) Tidal friction in shallow sea. *Philosophical Transactions of the Royal Society of London A*, **221**, 239–264.

[111] Johns, B., & Oĝuz, T. (1990) The modeling of the flow of water through the Bosphorus. *Dynamics of Atmosphere and Oceans*, **14**, 229–258.

[112] Joseph, K.I. (1977) Solitary waves in a finite depth fluid. *Journal of Physics A–Mathematical and General*, **10**, 1225–1227.

[113] Kagan, B.A., & Sündermann, J. (1996) Dissipation of tidal energy, paleotides, and the evolution of the Earth-Moon system. *Advances in Geophysics*, **38**, 179–266.

[114] Kakutani, T., & Yamasaki, N. (1978) Solitary waves on a two-layer fluid. *Journal of the Oceanographical Society of Japan*, **45**, 674–679.

[115] Kantha, L.H., & Clayson, C.A. (2000) *Numerical Models of Oceans and Oceanic Processes*, International Geophysics Series, vol. 66. New York: Academic Press.

[116] Kantha, L.H., & Tierney, C.C. (1997) Global baroclinic tides. *Progress in Oceanography*, **40**, 163–178.

[117] Kao, T.W., Pan, F.-Sh., & Renouard, D. (1985) Internal solitons on the pycnocline: generation, propagation, shoaling and breaking over a slope. *Journal of Fluid Mechanics*, **159**, 19–53.

[118] Khatiwala, S. (2003) Generation of internal tides in an ocean of finite depth: analytical and numerical calculations. *Deep-Sea Research I*, **50**, 3–21.

[119] Konyaev, K.V., & Sabinin, K.D. (1992) *Waves Inside the Ocean*. St. Petersburg: Gidrometeoizdat (in Russian).

[120] Konyaev, K.V., Sabinin, K.D., & Serebryany, A.N. (1995) Large-amplitude internal waves at the Mascarene Ridge in the Indian Ocean. *Deep-Sea Research*, **42** (11/12), 2075–2091.

[121] Kozubskaya, G.I., Konyaev, G.V., Pludeman, A., & Sabinin, K.D. (1999) Internal waves at the slope of the Bear Island from the data of the Barents Sea Polar Front Experiment (BSPF-92). *Oceanology*, **39** (2), 147–154.

[122] Koop, C.G., & Butler, G. (1981) An investigation of internal solitary waves in a two-fluid system. *Journal of Fluid Mechanics*, **112**, 225–251.

[123] Krauss, W. (1966) *Interne Wellen*. Berlin: Gebrüder Bornträger (in German).

[124] Kubota, T.D., Ko, R.S., & Dobbs, L. (1978) Propagation of weakly nonlinear internal waves in a stratified fluid of finite depth. *Journal of Hydronautics*, **12**, 157–165.

[125] Kunze, E. (1985) Near-inertial wave propagation in geostrophic shear. *Journal of Physical Oceanography*, **15**, 544–565.
[126] Kunze, E., & Toole, J.M. (1997) Tidally driven vorticity, diurnal shear, and turbulence atop Fieberling Seamount. *Journal of Physical Oceanography*, **27**, 2663–2693.
[127] Kunze, E., Rosenfeld, L.K., Carter, G.S., & Gregg, M.C. (2002) Internal waves in Monterey Submarine Canyon. *Journal of Physical Oceanography*, **32**, 1890–1913.
[128] Kuznetsov, A.S., Paramonov, A.N., & Stepanyants Yu.A. (1984) Investigation of solitary internal waves in the tropic zone of West Atlantic. *Izvestiya, Atmospheric and Oceanic Physics*, **20** (10), 975–984 (in Russian).
[129] Lamb, K.G. (1994) Numerical experiments of internal wave generation by strong tidal flow across a finite amplitude bank edge. *Journal of Geophysical Research*, **99**, 843–864.
[130] Lamb, K.G., & Yan, L. (1996) The evolution of internal wave undular bores: comparisons of fully nonlinear numerical model with weakly nonlinear theories. *Journal of Physical Oceanography*, **26**, 2712–2734.
[131] Larsen, L.H. (1969) Internal waves incident upon a knife edge barrier. *Deep-Sea Research*, **16**, 411–419.
[132] Le Blond, P., & Mysak, L. (1978) *Waves in the Ocean*. Amsterdam: Elsevier Scientific Publishing Company.
[133] Leaman, K.D. (1980) Some observations of baroclinic diurnal tides over a near-critical bottom slope. *Journal of Physical Oceanography*, **10**, 1540–1551.
[134] Lee, C.-Y, & Beardsley, R.C. (1974) The generation of long nonlinear internal waves in a weakly stratified shear flow. *Journal of Geophysical Research*, **79**, 453–462.
[135] Leonov, A.I. (1981) The effect of the earth's rotation on the propagation of weak nonlinear surface and internal long oceanic waves. *Annals of the New York Academy of Science*, **373**, 150–159.
[136] Leonov, A.I., & Miropol'sky, Yu.Z. (1975) Toward a theory of stationary nonlinear internal gravity waves. *Izvestiya, Atmospheric and Oceanic Physics*, **11** (5), 298–304.
[137] Levitus, S., & Boyer, T.P. (1994) *World Ocean Atlas 1994, Vol. 3: Temperature NOAA Atlas NESDIS 4*. Washington, D.C.: U.S. Government Printing Office.
[138] Levitus, S., Burgett, R., & Boyer, T.P. (1994) *World Ocean Atlas 1994, Vol. 4: Salinity NOAA Atlas NESDIS 3*. Washington, D.C.: U.S. Government Printing Office.
[139] Lighthill, J. (1978) *Waves in Fluids*. Cambridge: Cambridge University Press.
[140] Liu, A.K. (1988) Analysis of nonlinear internal waves in the New York Bight. *Journal of Geophysical Research*, **93**, 12 317–12 329.
[141] Liu, A.K., & Benney, D.J. (1981) The evolution of nonlinear wave trains in stratified shear flows. *Studies in Applied Mathematics*, **64**, 247–269.
[142] Liu, A.K., Holbrook, J.R., & Apel, J.R. (1985) Nonlinear internal wave evolution in the Sulu Sea. *Journal of Physical Oceanography*, **15**, 1613–1624.
[143] Liu, A.K., Chang, S., Hsu, M.-K., & Liang, N.K. (1998) Evolution of nonlinear internal waves in the East and South China Seas. *Journal of Geophysical Research*, **103** (C4), 7995–8008.
[144] Long, R.R. (1965) On the Boussinesq approximation and its role in the theory of internal waves. *Tellus*, **17**, 46–52.
[145] Longuet-Higgins, M.S. (1965) On the reflection of wave characteristics from rough surface. *Journal of Fluid Mechanics*, **37**, 231–250.

[146] Lynch, J.F., Jin, G., Pawlowicz, R., *et al.* (1996) Acoustic travel-time perturbations due to shallow water internal waves and internal tides in the Barents Sea Polar Front: Theory and experiment. *The Journal of the Acoustical Society of America*, **99** (2), 803–821.

[147] Magaard, L. (1962) Zur Berechnung interner Wellen in Meeresräumen mit nichtebenen Böden bei einer speziellen Dichteverteilung. *Kieler Meeresforschung*, **18**, 161–183.

[148] Matsuura, T., & Hibiya, T. (1990) An experimental and numerical study of the internal wave generation by tide-topography interaction. *Journal of Physical Oceanography*, **20** (4), 506–521.

[149] Maxworthy, T., D'Hieres, G.C., & Didelle, H. (1984) The generation and propagation of internal gravity waves in a rotating fluid. *Journal of Geophysical Research*, **89** (4), 6383–6396.

[150] Maze, R. (1987) Generation and propagation of nonlinear internal waves induced by the tide over a continental slope. *Continental Shelf Research*, **7**, 1079–1105.

[151] Mellor, G.L., & Yamada, T. (1982) Development of turbulence closure model for geophysical fluid problems. *Reviews of Geophysics*, **20**, 851–875.

[152] Merrifield, M.A., & Holloway, P.E. (2002) Model estimates of M_2 internal tide energetics at the Hawaiian Ridge. *Journal of Geophysical Research*, **107** (C8), 10.1029/2001JC000996.

[153] Michallet, H., & Barthelemy, E. (1998) Experimental study of interfacial solitary waves. *Journal of Fluid Mechanics*, **366**, 159–177.

[154] Michallet, H., & Ivey, G.M. (1999) Experiments on mixing due to internal solitary waves breaking on uniform slopes. *Journal of Geophysical Research*, **104**, 13 467–13 477.

[155] Miles, J.W. (1979) On internal solitary waves II. *Tellus*, **31**, 456–462.

[156] Millero, F.J., & Poisson, A. (1981) International one-atmosphere equation of state of seawater. *Deep-Sea Research*, **27A**, 255–264.

[157] Miropol'sky, Yu.Z. (2001) *Dynamics of Internal Gravity Waves in the Ocean*. Dordrecht: Kluwer Academic Publishers.

[158] Miyata, M. (1988) Long internal waves of large amplitude. In *Nonlinear Water Waves*, eds. K. Horikawa & H. Maruo. Berlin: Springer, pp. 399–406.

[159] Monin, A., Kamenkovich, V., & Kort, V. (1977) *Variability of the Oceans*. New York: J. Wiley & Sons.

[160] Mooers, C.N.K. (1975) Several effects of a baroclinic current on the cross-stream propagation of inertial-internal waves. *Geophysical Fluid Dynamics*, **6**, 245–275.

[161] Morozov, E.G., & Vlasenko, V.I. (1996) Extreme tidal internal waves near the Mascarene Ridge. *Journal of Marine Systems*, **9**, 203–210.

[162] Morozov, E.G., Vlasenko V.I., Demidova, T.A., & Ledenev, V.V. (1999) Tidal internal waves propagation over large distances in the Indian Ocean. *Oceanology*, **39**, 51–56.

[163] Munk, W. (1981) Internal waves. In *Evolution of Physical Oceanography: Scientific Survey in Honor of Henry Stommel*, eds. B. A. Warren and C. Wunsch. Cambridge, MA: MIT Press, pp. 264–291.

[164] Munk, W. (1966) Abyssal recipes. *Deep-Sea Research*, **13**, 707–730.

[165] Munk, W. (1997) Once again: once again – tidal friction. *Progress in Oceanography*, **40**, 7–35.

[166] Munk, W., & MacDonald, G. (1968) *The Rotation of the Earth*. Cambridge: Cambridge University Press.
[167] Munk, W.H., & Wunsch, C. (1998) Abyssal recipe II: Energetics of tidal and wind mixing. *Deep-Sea Research I*, **45**, 1977–2010.
[168] Müller, P., & Briscoe, M. (2000) Dynamical mixing and internal waves. *Oceanography*, **13**, 98–103.
[169] Müller, P., & Liu, X. (2000) Scattering of internal waves at finite topography in two dimensions. Part I: Theory and case studies. *Journal of Physical Oceanography*, **30**, 532–549.
[170] Müller, P., & Liu, X. (2000) Scattering of internal waves at finite topography in two dimensions. Part II: Spectral calculations and boundary mixing. *Journal of Physical Oceanography*, **30**, 550–563.
[171] Müller, P., & Xu, N. (1992) Scattering of oceanic internal gravity waves off random bottom topography. *Journal of Physical Oceanography*, **22**, 474–488.
[172] Nakamura, T., & Awaji, T. (2000) The growth mechanism for topographic internal waves generated by an oscillatory flow. *Journal of Physical Oceanography*, **31**, 2511–2524.
[173] Nakamura, T., Awaji, T., Hatayama, T., *et al.* (2000) The generation of large-amplitude unsteady lee waves by subinertial tidal flow: a possible vertical mixing mechanism in the Kuril Straits. *Journal of Physical Oceanography*, **30**, 1601–1621.
[174] Nansen, F. (1902) The oceanography of the north polar basin (Norwegian Northern Polar Expedition, 1893–1896). *Scientific Results*, **39**, 22–48.
[175] Nayfeh, A. (1973) *Perturbation Methods*. New York: Wiley.
[176] New, A.L., & Pingree, R.D. (1990) Large-amplitude internal soliton packets in the central bay of Biscay. *Deep-Sea Research*, **37** (3), 513–524.
[177] Niwa, Y., & Hibiya, T. (2001) Numerical study of the spatial distribution of the M_2 internal tide in the Pacific Ocean. *Journal of Geophysical Research*, **106** (C10), 22 441–22 449.
[178] Ollinger, D. (1999) Modellierung von Temperatur, Turbulenz und Algenwachstum mit einem gekoppelten physikalisch-biologischen Modell. Ph.D. Thesis, University of Heidelberg, (Published on the internet.) ISBN 3-933342-38-4.
[179] Ono, H. (1975) Algebraic solitary waves in stratified fluid. *Journal of the Physical Society of Japan*, **39**, 1082–1091.
[180] Osborn, A.R., & Burch, T.L. (1980) Internal solitons in the Andaman sea. *Science*, **208**, 451–460.
[181] Ostrovsky, L.A. (1978) Nonlinear internal waves in a rotating ocean. *Oceanology*, **18**, 119–125.
[182] Ostrovsky, L.A., & Grue, J. (2003) Evolutional equation for strongly nonlinear internal waves. *Physics of Fluids*, **15** (10), 2934–2948.
[183] Ostrovsky, L.A., & Stepanyants, Yu.A. (1989) Do internal solitons exist in the ocean? *Reviews of Geophysics*, **27** (3), 293–310.
[184] Ou, H.W., & Maas, L.R.M. (1988) Tides near a shelf-slope front. *Continental Shelf Research*, **8**, 729–736.
[185] Pacanowski, R.C., & Philander, S.G.H. (1981) Parameterisation of vertical mixing in numerical models of tropical oceans. *Journal of Physical Oceanography*, **11**, 1443–1451.
[186] Parsons, A.R., Bourke, R.H., Muench, R.D., *et al.* (1996) The Barents Sea Polar Front in summer. *Journal of Geophysical Research*, **101** (C6), 14 201–14 221.

[187] Pedlosky, J. (1987) *Geophysical Fluid Dynamics*. Berlin: Springer.
[188] Perry, R.B., & Schimke, G.R. (1965) Large amplitude internal waves observed off the northwest coast of Sumatra. *Journal of Geophysical Research*, **70**, 2319–2324.
[189] Petrovskii, I.G. (1967) *Partial Differential Equations*. London: Iliffe Books.
[190] Phillips, O. (1977) *The Dynamics of the Upper Ocean*, 2nd edn. Cambridge: Cambridge University Press.
[191] Pingree, R.D., & Mardell, G.T. (1985) Solitary internal waves in the Celtic Sea. *Progress in Oceanography*, **14**, 431–441.
[192] Pingree, R.D., & New, A.L. (1989) Downward propagation of internal tidal energy into Bay of Biscay. *Deep-Sea Research*, **36**, 735–758.
[193] Pingree, R.D., & New, A.L. (1991) Abyssal penetration and bottom reflection of internal tidal energy into Bay of Biscay. *Journal of Physical Oceanography*, **21** (1), 28–39.
[194] Pingree, R.D., Griffiths, D.K., & Mardell, G.T. (1984) The structure of internal tides at the Celtic sea shelf break. *Journal of the Marine Biological Association of the United Kingdom*, **64**, 99–113.
[195] Pingree, R.D., Mardell, G.T., & New, A.L. (1989) Propagation of internal tides from the upper slopes of the Bay of Biscay. *Nature*, **321**, 154–158.
[196] Pisarev, S.V. (1996) Low-frequency internal waves near the shelf edge of the Arctic Basin. *Oceanology*, **36** (6), 771–778.
[197] Prinsenberg, S.J., & Rattray, M., Jr. (1975) Effect of continental slope and variable Brunt-Väisälä frequency on the generation of internal tides. *Deep-Sea Research*, **22**, 251–265.
[198] Prinsenberg, S.J., Wilmott, W.L., & Rattray, M., Jr. (1974) Generation and dissipation of coastal internal tides. *Deep-Sea Research*, **21** (4), 263–282.
[199] Rattray, M., Jr. (1960) On the coastal generation of internal tides. *Tellus*, **12**, 54–62.
[200] Rattray, M., Jr., Dworski, J.G., & Kovala, P.E. (1969) Generation of long internal waves at the continental slope. *Deep-Sea Research*, **16** (1), 179–197.
[201] Rayleigh, Lord (1916) On convection currents in a horizontal layer of fluid when the higher temperature is on the under side. *Philosophical Magazine*, **32** (6), 529–546.
[202] Robinson, R.M. (1970) The effects of a corner on a propagating internal gravity wave. *Journal of Fluid Mechanics*, **42** (2), 257–267.
[203] Rubenstein, D. (1988) Scattering of internal waves by rough bathymetry. *Journal of Physical Oceanography*, **18**, 5–18.
[204] Ray, R.D., & Mitchum, G.T. (1996) Surface manifestation of internal tides generated near Hawaii. *Geophysical Research Letters*, **23**, 2101–2104.
[205] Ray, R.D., & Mitchum, G.T. (1997) Surface manifestation of internal tides in the deep ocean: observations from altimetry and island gauges. *Progress in Oceanography*, **40**, 135–162.
[206] Sabinin, K.D. (1992) Internal wave trains above the Mascarene ridge. *Izvestiya Atmospheric and Oceanic Physics*, **28** (6), 625–633.
[207] Sabinin, K.D., Nazarov, A.A., & Serebryany, A.N. (1990) Short period internal waves and currents in the ocean. *Izvestiya, Atmospheric and Oceanic Physics*, **28** (6), 847–854.
[208] Sabinin, K.D., Nazarov, A.A., & Filonov, A.Ye. (1992) Internal wave trains above the Mascarene Ridge. *Izvestiya, Atmospheric and Oceanic Physics*, **9** (1), 32–36.
[209] Sandström, J.W. (1908) Dynamische Versuche mit Meerwasser. *Annals in Hydrodynamic Marine Meteorology*, p. 6.
[210] Sandström, H. (1969) Effect of topography on propagation of waves in stratified fluids. *Deep-Sea Research*, **16**, 405–410.

[211] Sandström, H. (1976) On topographic generation and coupling of internal waves. *Geophysical Fluid Dynamics*, **7** (3/4), 231–270.
[212] Sandström H., & Elliot, J.A. (1984) Internal tide and solitons on the Scotian shelf: a nutrient pump at work. *Journal of Geophysical Research*, **89**, 6415–6426.
[213] Schlichting, H. (1978) *Boundary Layer Theory*, 7th edn. New York, McGraw-Hill.
[214] Schwiderski, E.W. (1979) *Global Ocean Tides, Part II: The Semidiurnal Principal Lunar Tide (M2), Atlas of Tidal Charts and Maps.* Dahlgren, Virginia: Naval Surface Weapons Center Report NSWC TR 79–414.
[215] Segur, H., & Hammack, J.L. (1982) Soliton models of long internal waves. *Journal of Fluid Mechanics*, **118**, 285–304.
[216] Serebryany, A.N., & Shapiro, G.I. (2000) Overturning of soliton-like internal waves: observations on the Pechora Sea Shelf. In *Fifth International Symposium on Stratified Flows*, Vol. II eds. G.A. Lawrence, R. Pieters, & N. Yonemitsu. Vancouver: University of British Columbia, pp. 1029–1034.
[217] Sherwin, T.J. & Taylor, N.K. (1990) Numerical investigations of linear internal tide generation in the Rockall trough. *Deep-Sea Research*, **37**, 1595–1618.
[218] Sherwin, T.J., Vlasenko, V.I., Stashchuk, N.M., Jeans, D.R.G., & Jones, B. (2002) Along-slope tidal generation as an explanation for some unusual large internal tides. *Deep-Sea Research I*, **49**, 1787–1789.
[219] Sjöberg, B., & Stigebrandt, A. (1992) Computations of the geographical distribution of the energy flux to mixing processes via internal tides and associated vertical circulation in the ocean. *Deep-Sea Research*, **39** (2), 269–291.
[220] Smith, S.G.L., & Yong, W.R. (2002) Conversion of the barotropic tide. *Journal of Physical Oceanography*, **32**, 1554–1566.
[221] Stacey, M.W., Pond, S., & Nowak, Z.P. (1995) A numerical model of circulation in Knight Inlet, British Columbia. *Journal of Physical Oceanography*, **25**, 1037–1062.
[222] Stashchuk, N.M. (1990) Internal tide simulation in the Countercurrent frontal zone of the north-western part of the Tropical Atlantic. *Oceanology*, **30** (2), 204–210.
[223] Stashchuk, N.M., & Cherkesov, L.V. (1991) Generation of internal waves resulting from the interaction of a barotropic tide with a horizontally-inhomogeneous density field and bottom topography. *Physical Oceanography*, **2** (2), 79–88.
[224] Stashchuk, N., & Hutter, K. (2001) Modelling of water exchange through the Strait of the Dardanelles. *Continental Shelf Research*, **21**, 1361–1382.
[225] St. Laurent, L., & Garrett, C. (2002) The role of internal tides in mixing in deep ocean. *Journal of Physical Oceanography*, **32**, 2882–2899.
[226] St. Laurent, L., Stringer, S., Garrett, C., & Perrault-Joncas, D. (2003) The generation of internal tides by abrupt topography. *Deep-Sea Research I*, **50**, 987–1003.
[227] Stommel, H. (1961) Thermohaline convection with two stable regimes of flow. *Tellus*, **13** (2), 224–230.
[228] Summer, H.J., & Emery, K.O. (1976) Internal waves of tidal period off southern California. *Geophysical Fluid Dynamics*, **7** (3/4), 231–270.
[229] Terez, D., & Knio, O. (1998) Numerical simulations of large-amplitude internal solitary waves. *Journal of Fluid Mechanics*, **362**, 53–82.
[230] Thorpe, S.A. (1977) Turbulence and mixing in a Scottish Loch. *Philosophical Transactions of the Royal Society of London A*, **286**, 125–181.
[231] Thorpe, S.A., Hall, A.J., & Croft, I. (1972) The internal surge in Loch Ness. *Nature*, **237**, 96–98.
[232] Tolstoy, I. (1973) *Wave Propagation*. New York: McGraw-Hill.
[233] Tomczak, M., & Godfrey, J.S. (2003) *Regional Oceanography: An Introduction*, 2nd edn. Oxford: Pergamon.

[234] Torgrimson, G.M., & Hickey B.M. (1979) Barotropic and baroclinic tides over the continental slope and shelf off Oregon. *Journal of Physical Oceanography*, **9** (5), 945–961.

[235] Turkkington, B., Eydeland, A., & Wang, S. (1982) Large-amplitude internal waves of permanent form. *Studies in Applied Mathematics*, **85**, 93.

[236] Turner, J.S. (1973) *Buoyancy Effects in Fluids*. Cambridge: Cambridge University Press.

[237] Uda, M. (1938) Researches on "siome" or current rip in the seas and oceans. *Geophysical Magazine*, **11** (4), 306–372.

[238] UNESCO (1981) *Tenth Report of the Joint Panel on Oceanographic Tables and Standards*. UNESCO Technical Report in Marine Science no. 36.

[239] Utnes, T., & Brors, B. (1993) Numerical modelling of 3-D circulation in restricted waters. *Applied Mathematical Modelling*, **17**, 522–535.

[240] Väisälä, V. (1925) Über die Wirkung der Windschwankungen auf die Pilotbeobachtungen. *Societas Scienti Finnica, Communications Physicae Mathematicae*, **2** (19), 1–46.

[241] Vlasenko, V.I. (1987) Generation of internal waves in a stratified ocean of variable depth. *Izvestiya, Atmospheric and Oceanic Physics*, **23** (3), 300–308.

[242] Vlasenko, V.I. (1992) Modelling of nonlinear baroclinic tide generation in the north-western part of the Tropical Atlantic. *Izvestiya, Atmospheric and Oceanic Physics*, **28** (3), 234–239.

[243] Vlasenko, V.I. (1992) Non-linear model for the generation of baroclinic tides over extensive inhomogeneities of the seabed relief. *Physical Oceanography*, **3** (6), 417–424.

[244] Vlasenko, V.I. (1993) Modelling of baroclinic tides in the shelf zone of Guinea. *Izvestiya, Atmospheric and Oceanic Physics*, **29** (5), 673–680.

[245] Vlasenko, V.I. (1994) Multimodal soliton of internal waves. *Izvestiya, Atmospheric and Oceanic Physics*, **30** (2), 161–169.

[246] Vlasenko, V.I. (2000) Peculiarity of baroclinic tides in steep bottom areas. *Morskoi Gidrofizicheskii Zhurnal*, **11** (2), 26–37 (in Russian).

[247] Vlasenko, V.I., & Cherkesov, L.V. (1986). Propagation of internal waves in continuously stratified ocean of variable depth. *Morskoi Gidrofizicheskii Zhurnal*, **3**, 24–30 (in Russian).

[248] Vlasenko, V.I., & Cherkesov, L.V. (1986) Application of perturbation method for investigation of interaction of internal waves with underwater ridge. *Morskoi Gidrofizicheskii Zhurnal*, **5**, 3–8 (in Russian).

[249] Vlasenko, V.I., & Cherkesov, L.V. (1987) Generation of baroclinic tides above continental slope. *Morskoi Gidrofizicheskii Zhurnal*, **2**, 3–9 (in Russian).

[250] Vlasenko, V.I., & Cherkesov, L.V. (1990) The generation of baroclinic tides over steep non-uniform bottom topography. *Physical Oceanography*, **1** (3), 161–170.

[251] Vlasenko, V.I., & Hutter, K. (2001) Generation of second mode solitary waves by the interaction of a first mode soliton with a sill. *Nonlinear Processes in Geophysics*, **8**, 223–239.

[252] Vlasenko, V.I., & Hutter, K. (2002) Numerical experiments on the breaking of solitary internal waves over slope-shelf topography. *Journal of Physical Oceanography*, **32**, 1779–1793.

[253] Vlasenko, V.I., & Hutter, K. (2002) Transformation and disintegration of strongly nonlinear internal waves in stratified lakes. *Annales Geophysicae*, **20**, 2087–2103.

[254] Vlasenko, V.I., & Morozov, E.G. (1993) Generation of semidiurnal internal waves near submarine ridge. *Oceanology*, **33** (3), 282–290.

[255] Vlasenko, V.I., & Stashchuk, N.M. (1991) On the use of a two-layer fluid stratification model in studies of topographically-generated baroclinic tides. *Physical Oceanography*, **2** (4), 263–268.
[256] Vlasenko, V.I., Brandt, P., & Rubino, A. (2000) On the structure of large-amplitude internal solitary waves. *Journal of Physical Oceanography*, **30**, 2172–2185.
[257] Vlasenko, V., Stashchuk, N., & Hutter, K. (2002) Water exchange in fjords induced by tidally generated internal lee waves. *Dynamics of Atmosphere and Ocean*, **35**, 63–83.
[258] Vlasenko, V.I., Ivanov, V.A., Krasin, I.G., & Lisichenok, A.D. (1996) Study of intensive internal waves in the shelf zone of Morocco. *Physical Oceanography*, **7** (4), 281–298.
[259] Vlasenko, V., Stashchuk, N., Hutter, K., & Sabinin, K.D. (2003) Nonlinear internal waves forced by tides near the critical latitude. *Deep-Sea Research I*, **50**, 317–338.
[260] Vlasenko, V.I., Golenko, N.N., Paka, V.T., Sabinin, K.D., & Chapman R. (1997) A study into dynamics of baroclinic tides in the region of the shelf edge. *Oceanology*, **37** (5), 599–609.
[261] Vlasenko, V.I., Golenko, N.N., Paka, V.T., Sabinin, K.D., & Chapman, R. (1997) Dynamics of baroclinic tides in the US shelf. *Izvestiya, Atmospheric and Oceanic Physics*, **33** (5), 651–663.
[262] Wallace, B.C., & Wilkinson, D.L. (1988) Run-up of internal waves on a gentle slope in a two-layered system. *Journal of Fluid Mechanics*, **191**, 419–442.
[263] Watson, G., & Robinson, I.S. (1990) A study of internal wave propagation in the strait of Gibraltar using shore-based marine radar images. *Journal of Physical Oceanography*, **20**, 374–395.
[264] Whitham, G.B. (1974) *Linear and Nonlinear Waves*. New York: J. Wiley and Sons.
[265] Weigand, J.G., Farmer, H.G., Prinsenberg, S.I., & Rattray, M., Jr. (1969) Effects of friction and surface tide angle of incidence on the coastal generation of internal tides. *Journal of Marine Research*, **2**, 241–259.
[266] Wessels, F., & Hutter, K. (1996) Interaction of internal waves with a topographic sill in a two-layered fluid. *Journal of Physical Oceanography*, **26**, 5–20.
[267] Wiegand, R.C., & Carmack, E. (1986) The climatology of internal waves in a deep temperature lake. *Journal of Geophysical Research*, **91**, 3951–3958.
[268] Willmot, A.J., & Edwards, B.D. (1987) A numerical model for the generation of tidally forced nonlinear internal waves over topography. *Continental Shelf Research*, **7** (5), 457–485.
[269] Wunsch, C. (1968) On the propagation of internal waves up a slope. *Deep-Sea Research*, **15** (3), 251–259.
[270] Xing, J., & Davies, A.M. (1998) A three-dimensional model of internal tides on the Malin-Hebrides shelf and shelf edge. *Journal of Geophysical Research*, **103** (C12), 27 821–27 847.
[271] Zeilon, N. (1912) On tidal boundary waves and related hydrophysical problems. *Kungl. Svenska Vetenskapsakademiens Handlingar*, **4** (47), 1–46.
[272] Zhou, X., & Grimshaw, R. (1989) The effect of variable currents on internal solitary waves. *Dynamics of Atmospheres and Oceans*, **14**, 17–39.
[273] Zubow, N.N. (1932). Hydrological investigations in the south-western part of the Barents Sea during the summer 1928. *Transactions of the Oceanographical Institute, Moscow*, **2** (4) (in Russian).

Index